D0768896

Springer Series on Environmental Management

Robert S. DeSanto, Series Editor

James W. Moore S. Ramamoorthy

Heavy Metals in Natural Waters

Applied Monitoring and Impact Assessment

With a Contribution by E.E. Ballantyne

With 48 Figures

Springer-Verlag
New York Berlin Heidelberg Tokyo

James W. Moore
Alberta Environmental Centre
Vegreville, Alberta
T0B 4L0 Canada

S. Ramamoorthy
Alberta Environmental Centre
Vegreville, Alberta
T0B 4L0 Canada

Library of Congress Cataloging in Publication Data
Moore, James W., 1947-
Heavy metals in natural waters.
Includes bibliographical references and index.
1. Water chemistry. 2. Heavy metals.
I. Ramamoorthy, S. II. Ballantyne, E. E. III. Title.
GB855.M65 1983 551.46′01 83-14842

Typeset by Progressive Typographers, Emigsville, Pennsylvania.
Printed and bound by Halliday Lithographers, West Hanover, Massachusetts.
Printed in the United States of America.

9 8 7 6 5 4 3 2 1

ISBN 0-387-90885-4 Springer-Verlag New York Berlin Heidelberg Tokyo
ISBN 3-540-90885-4 Springer-Verlag Berlin Heidelberg New York Tokyo

Series Preface

This series is dedicated to serving the growing community of scholars and practitioners concerned with the principles and applications of environmental management. Each volume is a thorough treatment of a specific topic of importance for proper management practices. A fundamental objective of these books is to help the reader discern and implement man's stewardship of our environment and the world's renewable resources. For we must strive to understand the relationship between man and nature, act to bring harmony to it, and nurture an environment that is both stable and productive.

These objectives have often eluded us because the pursuit of other individual and societal goals has diverted us from a course of living in balance with the environment. At times, therefore, the environmental manager may have to exert restrictive control, which is usually best applied to man, not nature. Attempts to alter or harness nature have often failed or backfired, as exemplified by the results of imprudent use of herbicides, fertilizers, water, and other agents.

Each book in this series will shed light on the fundamental and applied aspects of environmental management. It is hoped that each will help solve a practical and serious environmental problem.

Robert S. DeSanto
East Lyme, Connecticut

Preface

A biological monitoring system, which had the potential to save lives, was developed over one hundred years ago. The capital investment in this program amounted to a few pounds (in 1880 pounds) and operating costs were negligible, even by 19th century standards. The early coal mines of Great Britain became a safer place to work because canaries could be used to detect dangerous levels of carbon monoxide. Although biological monitoring in this form had a number of drawbacks, such as poor dctcction limits at low CO conccntrations, it did produce reproducible results at toxic levels. Consequently, the bird-CO system survived with minimal modifications well into the 20th century. Parakeets for example saw action in World War II submarines. It therefore comes as a shock to realize that this cost-effective program has no aquatic counterpart. Today, capital and operating costs of biological monitoring programs may be very high. Yet, economic and political factors together with intrinsic weaknesses in study design and methods may significantly reduce the effectiveness of the recommendations generated by such investigations. Hence the simplicity and cost-effectiveness of the bird-CO system is perhaps a utopian example which aquatic toxicologists cannot duplicate.

Biological monitoring and impact assessments have been carried out in lakes, rivers, and estuaries since the early part of this century and are now common features of most environmental programs. In most instances, the methods involved in both types of study are similar. In general terms, biological monitoring should occur before, during, and after environmental disturbances in order to avoid significant environmental impacts. Biological assessments on the other hand measure the extent of potentially major impacts which have already occurred.

Those knowledgeable in the field might well wonder why we need another review on monitoring and impact assessment. The literature on chemical and physical disturbances is already bulging with site specific studies and several major review articles have been published in recent years. Our response to the question is that we have emphasized an interdisciplinary approach to monitoring and impact assessments in this book. There are lengthy reviews on environmental chemistry, the pollution-ecology of algae, invertebrates and fish, and on aquatic toxicology, genetic toxicology, and the pathology of fishes and invertebrates in relation to heavy metals. Such an approach has the potential of overcoming intrinsic weaknesses in conventional programs which emphasize species diversity and community structure analysis. In the final chapters, we review biological, chemical, and political criteria which should be considered in the development of monitoring and impact assessment programs.

Because the book is written for managers and scientists with broad environmental interests, some topics such as pharmaco-kinetics and toxic mechanisms are not covered in great depth. Such information falls within the domain of the specialist and cannot be put to immediate use by agencies involved in monitoring and impact assessment. In addition, although every effort has been made to give a balanced review of the literature, we have possibly omitted relevant papers from our citations. This was done to improve the flow of the text and shorten the reference lists.

We would like to acknowledge the assistance of staff from the Alberta Environmental Centre in the preparation of this volume. We relied heavily on Sita Ramamoorthy and Joan O'Brien for the compilation and indexing of the literature. Sita Ramamoorthy and Jim Bradley also proofread the various drafts. The library staff, Mrs. Diana Lee and Harriet Judge, handled all of our literature requests, Mrs. Arhlene Hrynyk coordinated typing of the drafts, and Mr. Terry Zenith was responsible for figure preparation. Ann Wheatley collated data on the genetic toxicology of metals. Finally we would like to acknowledge Dr. R.S. Weaver, Executive Director, Alberta Environmental Centre, and Dr. L.E. Lillie, Head, Animal Sciences Wing, for their support and encouragement during this project.

James W. Moore
S. Ramamoorthy
Vegreville, Alberta

Contents

1
Introduction

Interdisciplinary Environmental Studies

In an earlier era, terminating in the 1970's in North America, uncontrolled industrial wastes were either deposited into or stored at a variety of land and water-based disposal sites. There was a general feeling, particularly in the Western world, that Nature could effectively treat or otherwise handle hazardous substances. Although there were several early indications, particularly from Japan, that this Naive Era of Waste Control was drawing to a close, the need for economic development apparently either blinded or distorted our view of what constituted a good life. Furthermore, Nature did perform quite well in treating some substances, which further contributed to our disregard for the hazards posed by wastes. One main consequence of such thinking was that comprehensive monitoring programs were either poorly supported or ineffective in detecting and preventing significant environmental disruptions. For example, the contamination of fish in the lower Great Lakes with mercury and PCB's came as a surprise to many people, including regulatory agencies which eventually pressed for compensation for local fisherman.

The end of the Naive Era seems to have occurred at roughly the same time throughout Western society. In the United States, the end was firmly established when the Office of Technology Assessment counted 243 cases of chemical contamination involving people since 1968, the most significant being the Michigan and Love Canal incidents (Smith, 1980). By the end of the 1970's, most industrialized nations were attempting to develop comprehensive rules and regulations regarding the use, storage and disposal of chemical wastes. This was the start of the Adjustment Era. Since then, other incidents involving the chemical contamination and physical disturbance of food, water and other resources have appeared with alarming frequency. However, these products of the Naive Era may eventually be controlled and we shall then have to contend with the subtleties of cost–benefit factors involved in controlling novel chemicals and other environmental disturbances (Gori, 1980).

No one will doubt that there has been a rapid change in the relative importance of environmental issues over the last few decades. Depending on place and time, environmentalists have been primarily concerned with eutrophication, acid-rain, heavy metals, synthetic chemicals, radionuclides, sedimentation, and hot water. It therefore comes as a surprise to find that many methods and concepts used during the Naive Era to investigate environmental problems in lakes and rivers still persist into the 1980's. Such relicts will in all likelihood slow our adjustment to novel and ongoing but serious environmental problems. One of the most pressing tasks scientists will have to face in future years is the re-evaluation of the framework for the assessment of environmental hazards. In an era of limited financial support, aquatic biologists will, for example, have to ask themselves whether the determination of the diversity of insect communities in a polluted stream is really worthwhile. Should there instead be investigations into the effects of pollutants at the cellular or molecular level? Should there be an emphasis on mutagen testing? Surely a chemical which disrupts the diversity of insect communities could induce acute, chronic and/or genetic toxicity to higher animals. Despite its effectiveness, the bird–CO system was eventually replaced by better systems.

It is therefore almost redundant to say that environmental studies need to be comprehensive and multidisciplinary in nature. Apart from the occupational and accidental aspects of hazardous chemical exposure to humans, contaminants also obviously influence water, soil, and air quality and the survival of a variety of plants and animals. The investigation of the effects of these chemicals can be conducted at any level of organization, from molecules to populations. Yet one simply has to scan the most recent issues of prominent environmental journals to see that the majority of investigations are restricted (largely out of financial necessity) to the effects of one disturbance on one particular organizational level of an individual species. As mentioned earlier, scientists have to critically re-evaluate the basis of the methods used in the evaluation of the hazards posed by modern environmental problems.

The purpose of this book is threefold. Firstly, there is an attempt to provide the reader with a critical review of a large amount of environmental data on heavy metals from several different disciplines. Such information should prove useful in the design and implementation of research projects, and the preparation of scientific papers and reports. We have chosen for discussion the eight most common heavy metals outlined in the EPA's priority list of pollutants: arsenic, cadmium, chromium, copper, lead, mercury, nickel, and zinc. Environmental scientists and managers will almost certainly have to deal with this group during the course of routine monitoring and impact assessment. In addition, since a substantial body of information is available on the chemistry, uptake and toxicity of the eight metals, it is possible to assess their impact on aquatic systems. Some highly specific fields, such as pharmaco-kinetics, are not considered in detail in this volume. Such material, although relevant to several fields of study, cannot be directly used in monitoring and impact assessment.

The second purpose of this book is to provide a review on the status and likely value of current methods used in monitoring and impact assessment. There is a critical evaluation of techniques which have been in use for many years and may no longer be relevant to current environmental problems. Our final purpose is to show that the comprehensive, multi-disciplinary approach is an effective means of detecting and evaluating potential hazards to aquatic resources and the users of the resources.

References

Gori, G. B. 1980. The regulation of carcinogenic hazards. *Science* **208**:256–261.
Smith, R. J. 1980. Swifter action sought on food contamination. *Science* **207**:163.

2
Arsenic

Chemistry

Arsenic, a member of Group Vb of the periodic classification, undergoes multiple electron transfer reactions. The variable chemical behaviour of arsenic is exemplified by its hard acid behaviour in As(+3) and soft base behaviour in compounds such as R_3As, where R = alkyl or aryl group (Ahrland, 1966). Arsenic is widely distributed in the environment, including plant and animal tissues. It forms a variety of inorganic and organic compounds of different toxicity to aquatic organisms. This is due to the varying physico-chemical properties of the arsenicals in different valency states. It appears that the stable, soluble inorganic arsenites and arsenates are readily absorbed by the intestinal tract and muscle tissue. Arsenate is excreted faster than arsenite, mostly through urine, because of its poor affinity for thiol groups. Thus arsenate is less toxic than arsenite and does not inhibit any enzyme system. However, arsenate inhibits ATP synthesis by uncoupling oxidative phosphorylation and replacing the stable phosphoryl group, whereas arsenite inhibits thiol-dependent enzymes, retained in the body tissue proteins, such as keratin disulfides in hair, nails and skin. Arsenic compounds in the environment are vulnerable to chemical and biological transformations. In spite of its known toxicity, arsenic has been used for its medicinal virtues in the form of organic arsenicals.

Arsenic hydride (arsine, AsH_3) is extremely poisonous. Arsine readily undergoes alkylation and arylation in the environment and interacts with inert carbonyls of Pd, Pt, etc., replacing carbon monoxide. These carbonyls are used as catalysts in industrial application. Arsenic forms oxyacids, the salts of which are known as arsenites and arsenates. The oxyanions comprise most of the aqueous chemistry of arsenic.

Production, Uses, and Discharges

Production

In the early part of the century, production of white arsenic (arsenious oxide) was around 12,000 metric tons per year. Germany was the leading producer, with the United States and Canada becoming increasingly important. Due to the heavy demand for use in pesticides, world production rose to 25,000 metric tons in 1920 and subsequently to 45,000 tons in 1930. Tremendous war demands for non-ferrous metals increased by-product uses. In 1941, US production of white arsenic alone was 29,466 metric tons, an increase of 86% from 1930 figures. Although world levels of white arsenic were similarly high ($>$ 80,000 tons) (US Minerals Yearbooks), production dropped sharply after the war to an estimated 43,000 metric tons in 1950 (Table 2-1). US levels were down to 12,041 tons, reflecting progressive replacement of lead and calcium arsenical insecticides by organic insecticides such as DDT, hexachlorobenzene, 2,4-D, 2,4,5-T, chlordane, toxaphane, and parathion.

World production (excluding USA) of white arsenic has gradually decreased since 1970. Figures from the US are not available and the withholding of such data continues (US Minerals Yearbooks). The public hearings in early 1975, held by the Occupational Safety and Health Administration, on the proposed standards for exposure to inorganic arsenic has had enormous impact on the consumption of white arsenic since 1975. The drop in US consumption decreased total production of white arsenic sharply to an estimated 33,000 tons in 1979.

Annual world production of total arsenic was approximately 12,600 metric tons during the period 1911–20, increasing to 39,400 metric tons by 1940 (Table 2-2). Production remained relatively constant until 1960, the last date for which US figures are available.

Table 2-1. World production of white arsenic (10^3 metric tons).

1940	1950	1970	1975	1976	1979
54[e]	43	49.6*	42.2*	34.3*	32.9*[e]

Source: US Minerals Yearbooks.
[e] estimated.
* Data do not include withheld US production figures.

Table 2-2. World production of total arsenic — 10 year annual average.

Period	Metric tons $yr^{-1} \times 10^3$
1911–1920	12.6
1921–1930	21.7
1931–1940	39.4
1941–1950	44.0
1951–1960	34.4
1961–1970	42.7*
1971–1980	41.1*

Source: US Minerals Yearbooks.
* Excluding US Production.

Uses

Uses of arsenic include: (i) medicinal preparations, (ii) lead alloys for bullets and shot, (iii) pyrotechnic and boiler compositions, (iv) depilatory agent, (v) paint pigments, (vi) opal glass and enamels, (vii) textile dyeing and calico printing, (viii) bronzing agent or decolorizing agent for glass, (ix) insecticides and rodenticides.

Arsenic is used extensively in insecticides as calcium arsenate to control boll-weevil in cotton fields, and in herbicides and plant dessicants as lead arsenate to control codling moth, plum curculio, cabbage worm, potato bug, tobacco hornworm, and other pests that attack fruits and vegetables. Arsenic as sodium arsenite continues to be employed as a herbicide (mainly for railroads and telephone posts), fungicide, and wood preservative. Total US consumption of calcium and lead arsenate in 1942 was approximately 41,000 and 55,000 tons, respectively.

The trend in US consumption of arsenic for end-use categories from 1940–1979 is given in Table 2-3. Consumption of arsenic compounds such as lead arsenate and calcium arsenate was not published and thus is not included in the table. However, the trend can be estimated from the consumption figures for white arsenic which is essentially used to produce calcium and lead arsenates. The sharp reduction in US consumption since 1975 is reflected in world production of white arsenic (Table 2-1). US demand for white arsenic exceeded supply in 1978 and 1979 and came from the cotton-growing and wood-preserving industries (US Minerals Yearbook, 1978–79). Domestic and foreign producers met the demand from their available supplies. Major demand for metallic arsenic was from the automobile battery industries.

Major uses of arsenic compounds derived from arsenic trioxide are in the manufacture of agricultural chemicals (herbicides and plant dessicants)—70%; industrial chemicals (wood preservatives and mineral flotation re-

Table 2-3. US imports of arsenicals for consumption by class (in metric tons).

Class	1940	1950	1970	1975	1976	1979
White arsenic (As_2O_3)	11976	13431	17057	10921	3875	11205
Metallic arsenic	8	62	414	439	261	368
Sulfide	203	69	8	0.2	250	35
Sheep dip	81	35	—	—	—	—
Lead arsenate	—	—	—	—	—	—
Arsenic acid	—	1	—	<0.2	18	160
Calcium arsenate	—	104	—	—	—	—
Sodium arsenate	—	50	85	0.5	18	1
Paris green	—	40	—	—	—	—
Arsenic compounds, n.c.c.	—	—	19	69	37	1

Source: US Minerals Yearbooks.

agents)—20%; glass and glassware—5%; and other uses (feed additives and pharmaceuticals)—2%. The major uses of arsenic metal are as an alloying agent in non-ferrous (lead- and copper-based) alloys and for electronic applications. These uses account for 3% of total consumption. Arsenical wood preservatives include chromated copper arsenate (CCA) and fluorchrome arsenate phenol (FCAP). US consumption of CCA for 1977 and 1978 was 11,263 and 11,358 metric tons, respectively. Use of FCAP was withheld in 1977 and was 102 metric tons in 1978.

Discharges

Anthropogenic sources of arsenic include (i) pesticides, (ii) plant dessicants, (iii) poultry feed additives and pharmaceuticals, (iv) coal and petroleum, (v) mine tailings and smelting, and (vi) detergents. Several common presoaks and household detergents were shown to contain 10–70 mg kg^{-1} of arsenic. Most anthropogenic inputs reach rivers, lakes and oceans. The total arsenic content of the world's oceans is approximately 2.8×10^9 metric tons. Other releases of arsenic include smelting and roasting of mineral ores, combustion of fossil fuels, leaching of mine wastes, and land erosion. Coal burning has contributed ~290,000 metric tons of arsenic in the last 70 years (Ferguson and Gavis, 1972). The sum of all anthropogenic contributions is estimated to be 110,000 tons yr^{-1} for this century. This is approximately 2.5 times the contribution due to weathering.

Arsenic transport from the continents to the oceans arises from natural processes, such as weathering and vulcanism. Although vulcanism contributed to the arsenic cycle over geological times this input is small compared to continental weathering. About 45,000 metric tons of arsenic is weathered per year, 73% of which is in the dissolved form (Ferguson and Gavis, 1972).

In the last five years, environmental restrictions and regulations have limited the amount of arsenic produced and used. Natural and anthropogenic emissions have resulted in metal-containing air-borne particulates. The residence time of the particulates vary and eventually they are deposited on the lithosphere through rain or snowfall. Urban air particulates in the United States are estimated to be enriched in arsenic by a factor of 10 relative to the earth's crust. Most of the input is anthropogenic but the volatility of some of the inorganic and organic forms of arsenic has to be taken into account in budgeting the atmospheric burden of arsenic. An additional source of atmospheric metal enrichment of special importance is fossil-fuel burning. Coal and oil burning contribute 5000 and 10 metric tons of arsenic per year (Bertine and Goldberg, 1971). High temperature processes involved in cement production add another 3,200 metric tons of arsenic per year.

Arsenic in Aquatic Systems

Speciation in Natural Waters

Arsenic can undergo eight electron reductions from the +5 to the −3 state and occurs in +5, +3, 0 and −3 states in aquatic systems. The metal is extremely rare whereas As(−3) is found only at extremely low Eh values. Arsenate species are stable at the high Eh values encountered in oxygenated waters. Under mildly reducing conditions, arsenites predominate. Oxidation of arsenite to arsenate is slow at natural pH values, but is faster in strongly alkaline or acidic solutions. Copper salts and carbon catalyze the reaction.

Arsenic combines strongly with sulfur and carbon in organic compounds. Many thousands of As(+3) and As(+5) compounds with carbon-arsenic bonds have been synthesized and tested for their effectiveness against various agricultural pests. Phenyl arsenic acid, substituted phenyl and diphenyl-diarseno compounds as chemotherapeutic agents, cacodylic acid [$(CH_3)_2AsOOH$], methane arsenic acid [$CH_3AsO(OH)_2$], lead and calcium arsenates as pesticides, and Lewisite [$CH_3 - CH = CHAsCl_2$] as a chemical warfare agent, are some of the well-known and widely used arsenicals.

Transport in Natural Waters

Arsenic species, As(+5), As(+3) and methylated As, occur in natural waters. Arsenic is a good tracer for geochemical and biological processes and for rare-element transportational pathways. In a study of ten major rivers in the southeastern USA, arsenate-As was the only dissolved species found (Waslenchuk, 1979). Concentrations ranged from 0.15–0.45 $\mu g\ L^{-1}$ and were similar for other world rivers (Ferguson and Gavis, 1972). Absence of metabolically produced arsenite and dimethylarsenic reflects the relative inactivity of biologically mediated redox and alkylation reactions. The rivers contain relatively high amounts of organic matter and Fe. Some are pristine and others support industry and adjacent cultivated lands. It was observed that mechanical weathering of moist piedmont rocks does not release arsenic to the extent of chemical weathering of wet coastal plain rocks. Higher dissolved levels (average 4.3 $\mu g\ L^{-1}$) were reported in some silty western US rivers which flow over rapidly weathering volcanic-sedimentary terrain. Similarly a good correlation between arsenic content and river discharge (flushout from rainfall) was shown (Waslenchuk, 1979).

Arsenic complexes with low molecular weight dissolved organic matter (Waslenchuk and Windom, 1978). The conservative behaviour of arsenic in surface waters is altered by organic complexation which reduces surface interactions with solid phase components of the water column. Arsenic also binds to particulate matter and the association is reported to be non-labile, with no significant exchange to the dissolved phase (Waslenchuk and Windom, 1978). Moreover, the particulate load could deposit entirely in estuaries and coastal zones delivering only the dissolved fraction to the ocean.

Absence of free cationic arsenic rules out direct complexation with humic or fulvic acid (Wagemann, 1978). In the presence of significant concentrations of mediating organic ligands, arsenic may be essentially in the dissolved form. In this condition, common to many freshwaters, total dissolved arsenic will be controlled by precipitation of arsenates in the order:barium > chromium > iron.

Concentrations of arsenic in river-borne particulates vary greatly (3–74 $mg\ kg^{-1}$ dry weight) with river discharge, grain size and organic content of the particulates (Crecelius *et al.*, 1975). Levels are higher during periods of low discharge when grain size is at a minimum and organic content is at a maximum. The percent of particulate-bound arsenic and subsequent release as the river mixes with sea water depends on the nature of inorganic and organic load. The heavily contaminated Rhine River contains about 67% of the total As in particulates, whereas relatively non-contaminated Puget Sound rivers contain only 33% in particulates.

Arsenic in Sediments

Relatively high levels of arsenic are found in freshwater sediments. Sediments from a warm, shallow lake contained <0.5 to 59 mg As kg^{-1} with an average of 22 $mg\ kg^{-1}$ (Ruppert *et al.*, 1974). Concentrations were positively correlated with decreasing particle size of the sediment. The increase in arsenic content compared to the surrounding soils and bedrock was related to the spraying of sodium arsenate (an aquatic biocide) in the 1950's and 1960's.

Total arsenic content in sediments of ten selected freshwater lakes in Canada ranged from 2.7 to 13.2 $mg\ kg^{-1}$ (Huang and Liaw, 1978). Concentrations were lower in lakes situated on the Precambrian shield than in agricultural areas. Hence, land erosion and run-off from farmlands probably contributed to arsenic burdens. The bulk of extractable arsenic was associated with the colloidal ($<0.2\ \mu m$) and non-colloidal ($0.2-<50\ \mu m$) fractions of sesquioxide components. Apatite in sediments held non-extractable arsenic, implying desorption from lake sediments dependent upon the stability of the As-bearing components of the sediments.

Organic matter of lake sediments is less important in the sorption of arsenic than other metals (Huang and Liaw, 1979). The porous sesquioxides and silico-sesquioxidic complexes present in a series of particle size fractions (clay, silt and sand) are the primary components in sorbing arsenic. The sorption-desorption of arsenic from these components will determine the level of arsenite in freshwater and the movement to the food chain. The bottom sediments of five lakes from Saskatchewan (Canada) were shown to abiotically oxidize the most toxic As(+3) state to relatively non-toxic As(+5) (Oscarson *et al.*, 1980). Thus the potential toxicity due to lesser sorption of arsenite could be alleviated by its abiotic oxidation to As(+5) in aquatic environments.

Desorption from contaminated sediments was highly dependent on the reduction of Fe(+3) to Fe(+2) in the system (Clement and Faust, 1981). Anaerobic conditions produced ten times higher concentrations of As(+3) compared to aerobic conditions. The ratio of anaerobic As/aerobic As decreased with decreasing temperatures. Arsenic in the aerobic reservoirs was about 70% arsenate and 20% organic-arsenic (Clement and Faust, 1981). Little pH effect was observed in the range of 6.0–8.5.

Transformations

Biological transformation was first observed in the early nineteenth century in the use of wall papers containing arsenic pigments. Methylation of arsenic by fungi, sewage fungi, soil microorganisms and bacteria (McBride *et al.*, 1978; Chau and Wong, 1978) to produce dimethyl- and trimethylarsine has been reported. Natural sediments can methylate arsenic, producing either non-volatile or/and volatile methylated arsenic compounds. Factors controlling arsenic methylation are not completely understood. Pure bacterial cultures of *Aeromonas* and *Flavobacterium* sp., isolated from lake water, and *Escherichia coli* methylate arsenic under aerobic conditions. Dimethyl arsenic acid was formed in the medium on the addition of arsenic compounds, yielding arsine and trimethylarsine in the head space of the culture flasks (Chau and Wong, 1978).

Methylated forms of arsenic have been detected in natural waters, bird eggshells, sea shells, and human urine (Braman and Foreback, 1973). Studies with aerobic and anaerobic organisms have shown that the former produce trimethylarsine and the latter dimethylarsine. Figure 2-1 postulates an arsenic cycle in nature based on both biotic and abiotic reactions. This

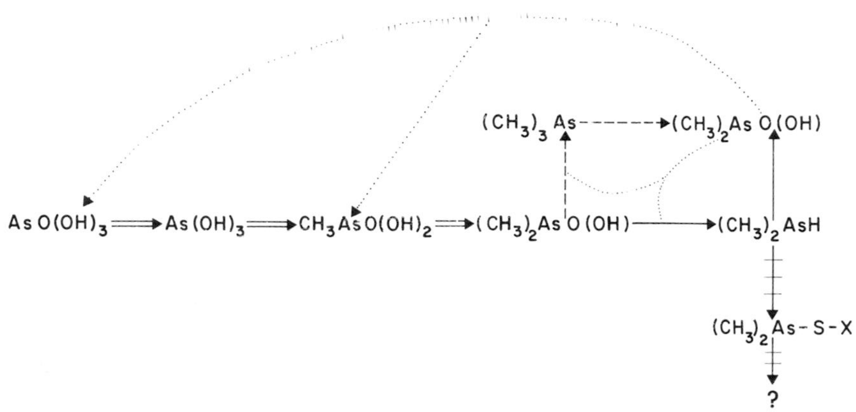

Figure 2-1. Biological arsenic cycle: (=) aerobic or anaerobic; (...) aerobic biotic or abiotic; (—) anaerobic; (---) aerobic; (++) probably abiotic. (From McBride *et al.*, 1978.)

scheme is not restricted to any single compartment but rather to ecosystems which vary in aerobicity or anaerobicity due to the availability of oxygen in the system. This will determine the nature of microbial flora and thus influence the fate and movement of arsine.

Arsenate, arsenite and methyl arsenate behave similarly under aerobic and anaerobic conditions up to the production of cacodylate [$(CH_3)_2AsO(OH)$] (Figure 2-1). Aerobes reduce and methylate this compound to trimethylarsine but anaerobes reduce it to dimethylarsine. It is evident that microflora in the environment have the ability to transform arsenic and its derivatives to gaseous arsines. Arsine is chemically and biologically different from the parent compounds.

Residues

Water, Precipitation and Sediments

Dissolved arsenic concentrations in water are often high in the vicinity of metal mines which process arsenic-bearing ores. Residues of over 5000 $\mu g\ L^{-1}$ were reported for a small subarctic lake receiving mine wastes (Wagemann *et al.*, 1978), whereas Aston *et al.* (1975) reported concentrations of $>500\ \mu g\ L^{-1}$ for a mining area in the UK. Relatively high levels (10–33 $\mu g\ L^{-1}$) have also been reported for streams receiving runoff from agricultural areas treated with the herbicide sodium arsenite and the dessicant arsenic acid (Richardson *et al.*, 1978). Rivers flowing through mixed industrial areas generally carry residues of 1–20 $\mu g\ L^{-1}$ whereas concentrations in most unpolluted freshwaters fall below 1 μg L^{-1} (van der Veen and Huizenga, 1980; Waslenchuk, 1979). Open ocean waters contain slightly higher average residues (2–3 μg L^{-1}) than freshwater reflecting the desorption of arsenic from particulate matter (Bryan, 1976; Andreae, 1979).

Arsenic is not a significant contaminant of precipitation, except at site-specific emission points. Total concentrations in rural areas of Europe, Canada and the USA ranged from 0.007 to 0.1 mg L^{-1} (Vallee, 1973; NAS, 1977; Johnson and Bramam, 1975). Higher values are often reported for air in the vicinity of some metal smelters, resulting in the contamination of nearby lakes. For example, there was a linear decrease in maximum arsenic residues (17–30 μg L^{-1}) in surface waters of three lakes located 5–20 km from a gold mine smelter in northern Canada (Moore *et al.*, 1978). Much higher concentrations in precipitation (63–812 μg L^{-1}) have been found in the vicinity of active volcanos (Boyle and Jonasson, 1973).

Metal mining is the principal source of arsenic-bearing solid wastes in both freshwater and marine sediments (Table 2-4). Concentrations in excess of 3000 mg kg^{-1} have been reported for several areas throughout the world. Another significant source of contamination in sediments is sodium arsenite. This compound, when used as a herbicide, has resulted in sediment residues of >500 mg kg^{-1} (Table 2-4). By contrast, concentrations in uncontaminated bottom material generally range from 5 to 15 mg kg^{-1}.

Table 2-4. Total arsenic levels (mg kg^{-1} dry weight) in freshwater and marine sediments.

Location	Average (range)	Polluting source
Carnon River (UK)[1]	1540 (1000–2300)	Pb–Zn mine
Streams (UK)[2]	–(<50–>5000)	metal mines
Great Slave Lake (Canada)[3]	2600 (50–3000)	gold mine
Great Bear Lake (Canada)[4]	1010 (400–3700)	silver mine
Kam Lake (Canada)[5]	1300 (40–3500)	gold mine
Chautauqua Lake (USA)[6]	22.1 (<0.5–58.8)	$NaAsO_2$ herbicide
Big Cedar Lake (USA)[7]	373 (150–659)	$NaAsO_2$ herbicide
Lake Michigan, nearshore (USA)[8]	20 (10.9–42.5)	municipal wastes, detergents
Lake Michigan, offshore (USA)[8]	6.6 (5.2–9.2)	natural
10 Saskatchewan lakes (Canada)[9]	–(2.7–13.2)	natural
Hayle Estuary (UK)[10]	1000 (12–4080)	metal mines
Restronguet Estuary (UK)[1]	1080 (90–>5000)	Pb–Zn mine

Sources: [1] Thornton *et al.* (1975); [2] Aston *et al.* (1975); [3] Moore (1979); [4] Moore (1981); [5] Wagemann *et al.* (1978); [6] Ruppert *et al.* (1974); [7] Kobayashi and Lee (1978); [8] Christensen and Chien (1979); [9] Huang and Liaw (1978); [10] Yim (1976).

Aquatic Plants

Arsenic is not a significant contaminant of plant tissues, except at local point source discharges. Consequently, residues are low (<50 mg kg^{-1} dry weight) in most industrial-zone waters. However concentrations of 150–3700 mg kg^{-1} dry weight were reported for macrophytes inhabiting a series of lakes (Canada) containing mine wastes (Wagemann *et al.,* 1978). Somewhat lower values (30–650 mg kg^{-1}) occurred in macrophytes from the Waikato river, New Zealand (Reay, 1972). The river was contaminated through geothermal activity, resulting in maximum water and sediment levels of 0.08 mg L^{-1} and 550 mg kg^{-1} dry weight, respectively. Hence concentration factors (CF) ranged from 100 to 20000, depending on species and site. Aerial emissions from a smelter in northern Canada produced average residues of 263 mg kg^{-1} dry weight in remote populations of *Myriophyllum exalbescens* (Franzin and McFarlane, 1980) while phytoplankton in Lake Michigan and Lake Superior had burdens of only 4.2–9.6 and 3.2–4.3 mg kg^{-1} dry weight (Seydel, 1972). CF's in the last two lakes were 1,750–12,800.

Estuarine waters along the SW coast of England produced five attached species with residues ranging from 59 to 189 mg kg^{-1} dry weight (Klumpp and Peterson, 1979). Substantially lower values (11–47 mg kg^{-1}) were recorded for five species of Rhodophyceae and Phaeophyceae from the Solent (UK) whereas residues in *Fucus serratus* inhabiting the highly polluted waters of the Severn Estuary reached 54 mg kg^{-1} (Leatherland and Burton, 1974). Similarly, several species of microscopic algae collected from coastal waters (USA) had burdens of 0.2–32 mg kg^{-1} (Sanders, 1979). Laboratory exposure of seven species of marine and freshwater phytoplankton to 0.001–0.003 mg As L^{-1} gave residues in the lipid phase of 0.4–4.8 mg kg^{-1} (Lunde, 1972). Hence CF's ranged from 200–5000.

Dimethyl arsenate accounted for 86–97% of total arsenic in five species of marine algae (Klumpp and Peterson, 1979). The remaining fraction consisted of arsenate and methylarsonate. By contrast the most common arsenic metabolites of *Dunaliella tertiolecta* in the non-lipid extract were monomethylarsonic acid, dimethylarsinic acid, arsenite and arsenate (Wrench and Addison, 1981). Edmonds and Francesconi (1981) suggested that the brown alga *Ecklonia radiata* was capable of processing arsenate into sugar derivatives.

The partitioning of arsenic in aquatic plants is not well-known. Maximum residues in *Ascophyllum nodosum* from SW England occurred in the oldest basal part of the plant (Klumpp and Peterson, 1979). This implied that the duration of exposure significantly influenced arsenic burdens. Although the same general trend was recorded for two species of *Fucus*, high levels were also found in the sporangia. Conway (1978) suggested that about 36% of arsenic in *Asterionella formosa* was bound to cellular material, whereas the remainder was associated with the cell wall. Seven species of phytoplankton concentrated arsenic in the lipids (Lunde, 1972) in a form which could be transferred to higher trophic levels (Wrench *et al.*, 1979).

This in part reflected the association of arsenic with the sugar moiety of algal glycolipids (Edmonds and Francesconi, 1981; Wrench and Addison, 1981).

Rate of uptake depends on concentration of arsenic in the media and duration of exposure. Conway (1978) found that sorption by the diatom *Asterionella formosa* was initially rapid, followed by a period of equilibrium 4–22 days after treatment. This change in rate of sorption was probably due to saturation of arsenic uptake sites. Hence *A. formosa,* and presumably other species, have a mechanism to maintain the cellular arsenic content at a non-toxic level.

Invertebrates

Although arsenic may be transferred to invertebrates from food and water, there is no evidence for extreme bioaccumulation in the majority of species. Hence arsenic is not a significant contaminant of tissues in most industrial areas. Wagemann *et al.* (1978) reported residues of 700–2400 and $<1-1300$ mg kg^{-1} dry weight in zooplankton and benthos from several lakes in northern Canada. The lakes were contaminated with arsenic-bearing mine wastes, resulting in high concentrations in water (0.7–5.5 mg L^{-1}) and sediments (6–3500 mg kg^{-1}). Thus, CF's for the plankton and benthos ranged up to 3400 and 1900, respectively, using water as the source of contamination, and 400/220 using sediment as the contaminating source. Cherry and Guthrie (1977) showed that the deposition of coal ash in the vicinity of the Savannah River (USA) produced average arsenic concentrations >20 mg kg^{-1} in various invertebrate species. Subsequent dredging of bottom sediments resulted in an increase to 60 mg kg^{-1}, reflecting the increased bioavailability of arsenic compounds. In unpolluted freshwaters, residues generally ranged from <0.5 to 20 mg kg^{-1}.

Comparable levels are generally reported for marine and estuarine invertebrates. Several different mollusc and crustacean species collected from the contaminated coastal waters of SW England had burdens of up to 65 and 35 mg kg^{-1} dry weight, respectively (Klumpp and Peterson, 1979). Leatherland and Burton (1974) reported that concentrations in gastropods and lamellibranchs from the Solent (UK) ranged up to 19 mg kg^{-1}, whereas collections made in other UK estuaries yielded residues of 3–48 mg kg^{-1}. Mussels *Mytilus edulis* sampled from Germany and Australian coastal waters had only 1–14 mg As kg^{-1} in soft tissues (Karbe *et al.,* 1977; Edmonds and Francesconi, 1977). Comparably low values (6–8 mg kg^{-1}) were recorded for copepods, amphipods and chaetognaths collected from a high arctic estuary (Bohn and McElroy, 1976).

Distribution of arsenic in tissues is species dependent. For example, approximately 87% of ^{74}As in the marine gastropod *Littorina littoralis* was associated with the digestive gland and gonad, 9% with the shell and 4% with the foot (Klumpp, 1980). However, the corresponding values for *Nucella lapillus* were 3, 85 and 7%, respectively. While residues in other groups of

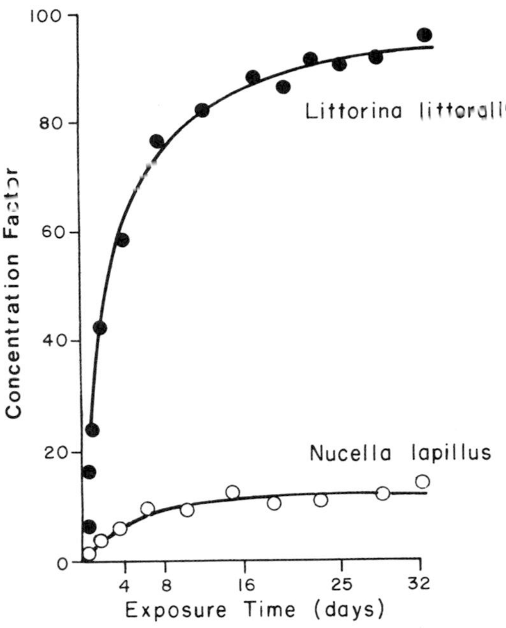

Figure 2-2. Accumulation of ^{74}As in the molluscs *Littorina littoralis* and *Nucella lapillus*. External arsenate 3 μg L^{-1}. (From Klumpp, 1980.)

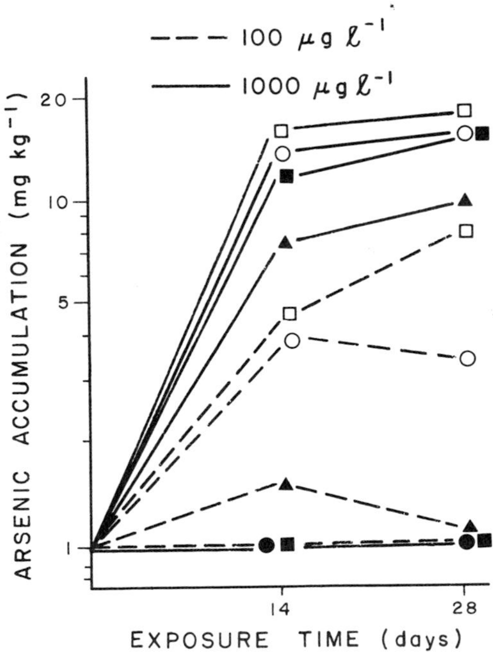

Figure 2-3. Accumulation of total arsenic in the snail *Stagnicola emarginata* exposed to two concentrations of As(II) (○), dimethyl sodium methyl arsenate (■), arsenicV (□), and sodium dimethyl arsenate (▲), and control value (●) for 28 days. (From Spehar *et al.*, 1980.)

invertebrates are probably greatest in tissues with high metabolic activity, there is little conclusive evidence to confirm this point.

Food is generally the main source of contamination, but most species also derive considerable amounts of arsenic directly from water. During initial exposure, uptake from water increases linearly with increasing external arsenic concentration (Figures 2-2, 2-3). Thereafter, rate of sorption decreases followed by a period of equilibrium. Uptake is also directly dependent on metabolic rate of the invertebrate (Klumpp, 1980). Hence, low temperatures and the administration of metabolic inhibitors such as cyanide generally cause a reduction in tissue residues. The same process may occur in marine invertebrates at suboptimal salinity levels.

Rate of uptake of different arsenic forms decreases in the order: As(+5), As(+3), disodium methyl arsenate = sodium dimethyl arsenate (Spehar *et al.,* 1980; Klumpp, 1980). Although the reverse order can probably be applied to rates of elimination, there are few specific data to confirm this point. Klumpp (1980) reported that the half-life of labeled arsenic (^{74}As) in two species of marine mollusc was only 4–13 days. Such rapid clearance partially accounts for the decrease of consistency in residues as invertebrates grow (Leatherland and Burton, 1974). Tissue burdens may also fall with increasing trophic position, again reflecting rapid clearance.

Fish

Arsenic does not usually accumulate in either freshwater or marine species. Hence, arsenic is not a threat to fisheries resources, except in cases of extreme ambient pollution. Some of the highest residues on record for freshwater were observed in the Canadian subarctic in a small lake contaminated with mine wastes (Wagemann *et al.,* 1978). Concentrations of up to 220 mg kg^{-1} dry weight were found in slimy sculpins, yielding CF's of 2–35. High levels (0.5–2.0 mg kg^{-1} wet weight) were also found in coregonids inhabiting the upper Great Lakes (Walsh *et al.,* 1977). Since there was no large industrial source of arsenic in this area, such residues probably reflected high binding efficiency in coregonid tissues. In unpolluted or mildly contaminated waters, levels generally range from <0.1–0.4 mg kg^{-1} wet weight.

Several marine species maintain relatively high burdens of arsenic. Of 95 fish (comprising 9 species) collected from Australian waters, 20 had residues exceeding 1.1 mg kg^{-1} (Bebbington *et al.,* 1977). Similarly, concentrations in black marlin from the same area ranged from 0.1 to 2.75 mg kg^{-1}, while several species inhabiting the Middle Atlantic Bight had maximum residues of 21 mg kg^{-1} (Greig *et al.,* 1976; Mackay *et al.,* 1975). Comparably high values were recorded for Arctic cod and shorthorn sculpins collected from uncontaminated coastal waters in the high arctic (Bohn and Fallis, 1978; Bohn and McElroy, 1976). These levels exceed various health standards (usually 0.5 mg kg^{-1}) and possibly imply that some marine species are unfit for human consumption. It has been noted however that toxic inorganic arsenicals are rapidly transformed to organic arsenic in fish (Oladimeji *et al.*, 1979). Since these latter forms are readily excreted by humans and are of low toxicity, the standard of 0.5 mg kg^{-1} may not be consistent with possible hazards posed to people. Additional work is required to clarify these points.

Arsenic levels in liver are not consistently higher than those in muscle. The ratio of arsenic in the liver : muscle of black marlin was 1.7 : 1 (Mackay *et al.*, 1975), whereas the corresponding values for shorthorn sculpins, and Arctic char were 2.0 : 1 and 1.4 : 1, respectively (Bohn and Fallis, 1978). By contrast, the liver : muscle ratio in Arctic cod was 0.10 : 1 (Bohn and McElroy, 1976) and 0.2–0.5 : 1 in various deep sea species (Greig *et al.*, 1976). Hence monitoring programs for arsenic should involve the analysis of a variety of tissues in fish.

The rapid change of inorganic arsenic to organic forms occurs in several species (Spehar *et al.*, 1980; Oladimeji *et al.*, 1979). Ingestion of ^{74}As by rainbow trout resulted in high levels of inorganic arsenic in various tissues within 6 h (Table 2-5). This was followed by a gradual increase in the proportion of organic arsenicals and a corresponding decrease in the amount of inorganic forms. As indicated earlier, it would be useful to know if such rapid conversion occurs in commercial species which have a high total arsenic level in their tissues.

Arsenic is sorbed primarily through food rather than water (Isensee *et al.*,

Table 2-5. Percentage of total radioactivity present as various forms of arsenic in tissues of rainbow trout following an oral dose of 20 μCi ^{74}As.

Time after dosing (h)	Inorganic arsenic	Mono-methylated arsenic	Unidentified arsenic	Unidentified organic arsenic 2
Muscle				
6	49.1	16.9	0.0	34.0
12	19.9	5.2	2.9	71.9
24	6.8	1.0	0.0	92.2
96	0.0	0.0	0.0	98.5
Liver				
6	17.2	7.2	1.1	74.5
12	4.5	5.3	0.6	89.6
24	13.2	7.4	3.7	74.6
96	3.1	5.2	0.0	91.7
Kidney				
6	30.1	7.7	5.0	57.3
12	29.7	11.8	5.6	52.9
24	6.9	15.2	0.0	75.7
96	6.1	6.9	2.9	84.0

Source: Oladimeji *et al.* (1979).

1973; Woolson, 1975). Since rate of uptake is greatest in young fish, residues in both liver and muscle decline with size, when expressed on a unit weight basis, in almost all species. Depuration of most arsenicals from fish tissues is rapid. Barrows *et al.* (1980) reported that the half-life of As_2O_3 in bluegill muscle was only 1 day. Similarly, there was an 80% reduction in arsenic residues in whole rainbow trout within 96 h of administration of an oral dose of ^{74}As (Oladimeji *et al.,* 1979); the corresponding values for whole blood, plasma and red blood cells were 67%, 66% and 69%, respectively. Although data are limited, excretion through gills appears to be the main route of arsenic elimination in rainbow trout and presumably other species.

Toxicity

Aquatic Plants and Invertebrates

Toxicity of arsenicals varies with valency state. Under comparable conditions, arsenite is more toxic to aquatic organisms than arsenate. However, differences in experimental format among studies have led to highly variable toxicity data for most arsenic compounds. For example, inhibition of the alga *Chlorella vulgaris* by Na_2HAsO_4 occurred at 0.06 mg L^{-1} (de Jong, 1965) whereas the corresponding values for As_2O_3 and $NaAsO_3$ were 0.5–1.0 mg L^{-1} (Brown and Rattigan, 1979; Nasu and Kugimoto, 1981). Although dibasic sodium arsenate inhibited *Chlamydomonas* sp. at concentrations of 1 μM (Christensen and Zielski, 1980), growth limitation in *Chlamydomonas reinhardtii* occurred at arsenate levels of 1–100 μM (Planas and Healey, 1978). Similarly, "no effect" concentrations for As(+3), As(+5), and total arsenic ranged from 0.16 to 1000 mg L^{-1} in several algal species (Luh *et al.,* 1973; Conway, 1978). This inconsistency in data reflects differences in test conditions and emphasizes the need for quantification of As(+3)/As(+5) levels in water. Furthermore, the effectiveness of management techniques involving arsenic-bearing wastes can be enhanced with a description of the chemical species in effluents.

Inhibitory effects of $NaAsO_2$, As_2O_3 and AsS_3 have been noted for several invertebrate species at concentrations of 0.4–40 mg As L^{-1}. These include reports on aquatic insects (Gilderhus, 1966; Sohacki, 1968), molluscs (Calabrese *et al.,* 1973; MacInnes and Thurberg, 1973), rotifers (Luh *et al.,* 1973) and zooplankton (Gilderhus, 1966). Exposure of amphipods *Gammarus pseudolimnaeus* to 0.96 mg As(+3) L^{-1} resulted in 100% mortality within 14 days (Spehar *et al.,* 1980). However, there was no mortality in two species of gastropod and one immature insect species following similar treatment. Additional exposures of the same organisms with 0.97 mg As(+5) L^{-1} did not reduce survival. The high toxicity of As(+3) to aquatic animals is due to its reaction with sulfhydryl groups of proteins which results in enzyme inhibition. As(+5) does not appear to react with sulfhydryl groups.

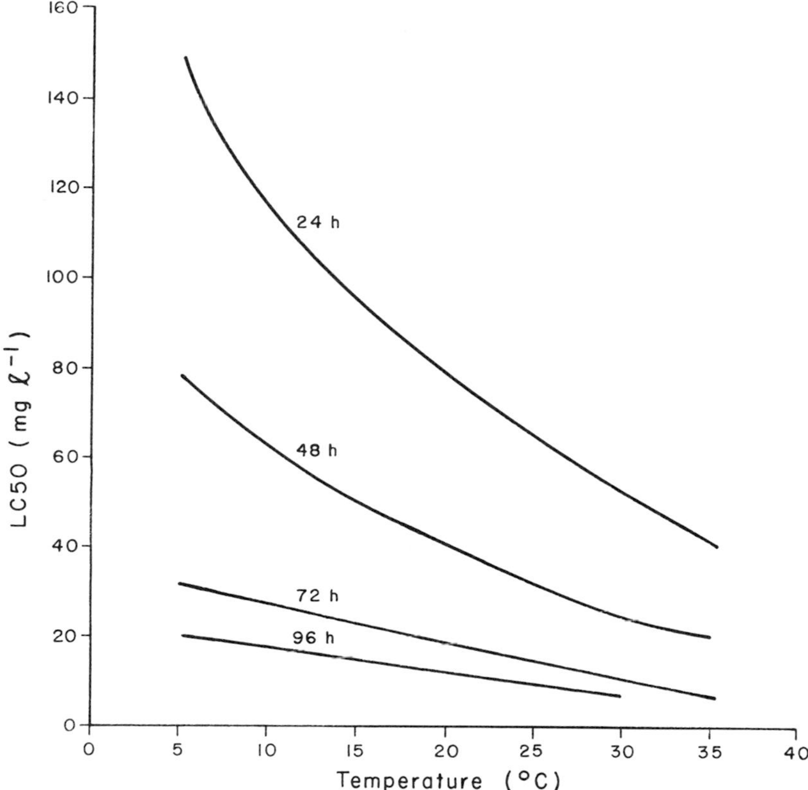

Figure 2-4. Effect of exposure time on temperature-dependent toxicity of arsenate to the rotifer *Philodina roseola*. (From National Research Council of Canada, 1978.)

Toxicity to aquatic plants and invertebrates generally decreases with an increase in the pH of the medium, reflecting concomitant formation of higher arsenic oxidation states. Similarly, the addition of phosphate to culture water also produces an antagonistic effect on toxicity to aquatic plants. This is due to the fact that phosphate inhibits biotransformations of arsenicals to more toxic states. In addition, because arsenate enters phosphorylation reactions, high phosphorus loadings decrease the relative level of competition for reaction sites. Although cations and organic chelators also probably act as antagonists to arsenicals, there are few confirmatory data for either plants or animals. As with most metals, the effect of temperature on arsenic toxicity is species dependent. Schaefer and Pipes (1973) showed that the 24 h LC_{50} of As(+5) to the rotifer *Philodina roseola* varied from 40 to 150 mg L^{-1} at a temperature range of 5–35°C (Figure 2-4). A similar but less exaggerated response was noted following 48–96 h of exposure.

Table 2-6. Acute toxicity of arsenic compounds to fish.

Species	Compound	Toxicity† (mg L^{-1})	Temperature (°C)	pH	Total hardness (mg L^{-1})	Dissolved oxygen (mg L^{-1})
Muskellunge fry[1]						
2 weeks old	sodium arsenite	1.1	15–17	7.2–7.9	149–191*	8.0–10.5
5 weeks old	sodium arsenite	2.6	15–17	7.2–7.9	149–191*	8.0–10.5
12 weeks old	sodium arsenite	16.0	15–17	7.2–7.9	149–191*	8.0–10.5
Muskellunge fry[2]	sodium arsenite	0.05(120h LC_{50})	15	7.2–7.7	149–190*	8.0–10.5
Mosquito fish[3]	sodium arsenite	59	21	NQ	NQ	NQ
Mosquito fish[3]	monosodium methyl arsonate	182–1300	21	NQ	NQ	NQ
Deep water ciscoes[4]	arsenic trioxide	17	6–8	NQ	40–48	NQ
Fathead minnow[5]	arsenic trisulfide	135	22–25	7.2–7.9	40–48	NQ

Sources: [1] Spotila and Paladino (1979); [2] Paladino and Spotila (1978); [3] Johnson (1978); [4] Passino and Kramer (1980); [5] Curtis *et al.*, (1979).
† Toxicity expressed as 96h LC_{50}, unless otherwise indicated.
* Total alkalinity. NQ—not quoted.

Fish

The LC_{50}'s of sodium arsenite and arsenic trioxide range from 0.05 to 59 mg L^{-1}, depending on age, species and test conditions (Table 2-6). While the LC_{50} of arsenate falls within a similar range (5 – 15 mg L^{-1}), arsenic trisulfide and methylated arsenicals appear to be less hazardous to fish (Spehar *et al.*, 1980; Luh *et al.*, 1973; Gilderhus, 1966). In general, total arsenic is acutely toxic at 1 – 50 mg L^{-1} though some species can tolerate 1000 mg L^{-1} (Sorensen, 1976; Luh *et al.*, 1973; Hale, 1977). Because the toxicity of arsenicals varies significantly with oxidation state, such general data are of limited value to managers who must assess potential impacts in receiving waters.

Toxicity of most organic arsenicals to aquatic organisms has also not been evaluated in detail. One of the few available studies (Kenaga and Moolenaar, 1979) indicated that organoarsenical concentrations of 0.5 mg L^{-1} caused 100% mortality in several algal species; the corresponding values for the cladoceran *Daphnia,* fathead minnows and emerald shiners ranged from 0.1 to 1.0 mg L^{-1}. Since these concentrations are comparable to those reported for inorganic As(+ 3), further toxicity testing should be performed.

Green sunfish collected from a contaminated lake (1 – 20 mg As L^{-1}) showed substantial damage to hepatic tissue (Sorensen *et al.*, 1980). There was extensive displacement of nuclei and formation of abnormally shaped parenchymal hepatocytes. The majority of specimens also developed necrotic foci, autophagic vacuoles, dense granulocytes and swollen hepatocytes. These changes imply that chronic exposures lead to a significant deviation in liver function in green sunfish.

Humans

Arsenates do not bind to thiol and hydroxy groups and thus do not inhibit enzyme systems. However, arsenates inhibit ATP synthesis by oxidative uncoupling reactions. In contrast, arsenites bind strongly to thiol groups and tissue proteins, such as carotin in skin, nail and hair. This is why hair is used to measure arsenic levels in humans. Arsenites have a longer half-life in mammalian tissues than other arsenicals. Symptoms of chronic intoxication in mammals include decreased motor coordination, nervous disorders, respiratory distress, and damage to kidneys and respiratory tract.

Although arsenic is a suspected carcinogen, the inability to confirm epidemiological evidence with experimental data has led some authors to suggest that it is a co-carcinogen. Tseng (1977) reported a positive dose-response relationship between arsenic concentrations (maximum 1.82 mg As L^{-1}) in well water in Taiwan and the incidence of skin cancer. A similar positive dose response occurred between the rate of mortality from lung cancer and exposure to arsenic among lead and calcium arsenate production workers (Ott *et al.*, 1974). Such trends have also been noted in copper smelters (Axelson *et al.*, 1978; Pinto *et al.*, 1977). Administration of arsenic trioxide, copper ore, and flue dust via the respiratory tract to rats resulted in one adenocarcinoma (Ishinishi *et al.*, 1977).

Petres *et al.* (1977) reported that inorganic arsenic interfered with DNA repair. Sodium arsenate reduced the incorporation of radioactively-labelled nucleotides in dermal cells and lymphocytes. It was suggested that arsenic may replace phosphorus in the DNA chain. By contrast, Löfroth and Ames (1978) found no mutagenic response for inorganic trivalent or pentavalent arsenicals in *Salmonella typhimurium;* Beaudoin (1974) reported that the embryotoxicity and teratogenic effects of sodium arsenate in rats were dose-dependent. Teratogenic effects included eye defects (anophthalmia and microphthalmia), exencephaly, and renal and gonadal agenesis. Atlas bones were rudimentary or missing in 63% of the fetuses examined for skeletal defects. Ferm (1974) observed the following malformations in hamsters subjected to injections of sodium arsenate late in embryogenesis: cleft palate and lip, micro-ophthalmia, ear deformities and genito-urinary abnormalities.

The Occupational Safety and Health Administration (OSHA) of the USA promulgated a final standard on the occupational exposure to inorganic arsenic, effective August 1, 1978, based on the results of studies of the carcinogenicity of arsenic. The maximum exposure to arsenic was lowered from 500 $\mu g\ m^{-3}$ of air to 10 $\mu g\ m^{-3}$ over an 8-hour time period. The new arsenic standard will have significant impact on Cu, Zn, Au, and Pb smelters where arsenic is a by-product and on consumers of arsenic trioxide. Although few smelter plants have initiated measures to comply with the new standard, firm regulations are now in place in Canada.

References

Ahrland, S. 1966. Factors contributing to (b)-behaviour in acceptors. *Structure and Bonding* **1**:207–220.

Andreae, M.O. 1979. Arsenic speciation in seawater and interstitial waters: the influence of biological-chemical interactions on the chemistry of a trace element. *Limnology and Oceanography* **24**:440–452.

Aston, S.R., I. Thornton, J.S. Webb, B.L. Milford, and J.B. Purves. 1975. Arsenic in stream sediments and waters of south west England. *The Science of the Total Environment* **4**:347–358.

Axelson, O., E. Dahlgren, C.D. Jansson, and S.O. Rehnlund. 1978. Arsenic exposure and mortality: a case-referent study from a Swedish copper smelter. *British Journal of Industrial Medicine* **35**:8–15.

Barrows, M.E., S.R. Petrocelli, K.J. Macek, and J.J. Carroll. 1980. Bioconcentration and elimination of selected water pollutants by Bluegill Sunfish *(Lepomis macrochirus). In:* R. Haque (Ed.), *Dynamic exposure and hazard assessment of toxic chemicals.* Ann Arbor Science, Ann Arbor, pp. 379–392.

Beaudoin, A.R. 1974. Teratogenicity of sodium arsenate in rats. *Teratology* **10**:153–158.

Bebbington, G.N., N.J. Mackay, R. Chvojka, R.J. Williams, A. Dunn, and E.H. Auty. 1977. Heavy metals, selenium and arsenic in nine species of Australian commercial fish. *Australian Journal of Marine and Freshwater Research* **28**:277–286.

Bertine, K.K., and E.D. Goldberg. 1971. Fossil fuel combustion and the major sedimentary cycle. *Science* **173**:233–235.

Bohn, A., and B.W. Fallis. 1978. Metal concentrations (As, Cd, Cu, Pb, and Zn) in shorthorn sculpins, *Myoxocephalus scorpius* (Linnaeus) and Arctic char, *Salvelinus alpinus* (Linnaeus), from the vicinity of Strathcona Sound, Northwest Territories. *Water Research* **12**:659–663.

Bohn, A., and R.O. McElroy. 1976. Trace metals (As, Cd, Cu, Fe, and Zn) in Arctic cod, *Boreogadus saida,* and selected zooplankton from Strathcona Sound, northern Baffin Island. *Journal of the Fisheries Research Board of Canada* **33**:2836–2840.

Boyle, R.W., and I.R. Jonasson. 1973. The geochemistry of arsenic and its use as an indicator element in geochemical prospecting. *Journal of Geochemical Exploration* **2**:251–296.

Braman, R.S., and C.C. Foreback. 1973. Methylated forms of arsenic in the environment. *Science* **182**:1247–1249.

Brown, B.T., and B.M. Rattigan. 1979. Toxicity of soluble copper and other metal ions to *Elodea canadensis. Environmental Pollution* **20**:303–314.

Bryan, G.W. 1976. Some aspects of heavy metal tolerance in aquatic organisms. *In*: A.P.M. Lockwood (Ed.), *Effects of pollutants on aquatic organisms.* Cambridge University Press, Cambridge, pp. 7–34.

Calabrese, A., R.S. Collier, D.A. Nelson, and J.R. MacInnes. 1973. The toxicity of heavy metals to embryos of the American oyster, *Crassostrea virginica. Marine Biology* **18**:162–166.

Chau, Y.K., and P.T.S. Wong. 1978. Occurrence of biological methylation of elements in the environment. *In*: F.E. Brinckman and J.M. Bellama (Eds.), *Organometals and organometalloids.* American Chemical Society Symposium Series No. 82, American Chemical Society, Washington, D.C., pp. 39–53.

Cherry, D.S., and R.K. Guthrie. 1977. Toxic metals in surface waters from coal ash. *Water Resources Bulletin* **13**:1227–1236.

Christensen, E.R., and N.K. Chien. 1979. Arsenic, mercury and other elements in dated Green Bay sediments. *Proceedings of the International Conference on Heavy Metals in the Environment,* London, pp. 373–376.

Christensen, E.R., and P.A. Zielski. 1980. Toxicity of arsenic and PCB to a green alga *(Chlamydomonas). Bulletin of Environmental Contamination and Toxicology* **25**:43–48.

Clement, W.H., and S.D. Faust. 1981. The release of arsenic from contaminated sediments and muds. *Journal of Environmental Science and Health. Part A.* **16**:87–122.

Conway, H.L. 1978. Sorption of arsenic and cadmium and their effect on growth, micronutrient utilization, and photosynthetic pigment composition of *Asterionella formosa. Journal of the Fisheries Research Board of Canada* **35**:286–294.

Crecelius, E.A., M.H. Bothner, and R. Carpenter. 1975. Geochemistries of arsenic, antimony, mercury, and related elements in sediments of Puget Sound. *Environmental Science and Technology* **9**:325–333.

Curtis, M.W., T.L. Copeland, and C.H. Ward. 1979. Acute toxicity of 12 industrial chemicals to freshwater and saltwater organisms. *Water Research* **13**: 137–141.

de Jong, L.E. 1965. Tolerance of *Chlorella vulgaris* for metallic and non-metallic ions. *Journal of Microbiology and Seriology* **31**:301–313.

Edmonds, J.S., and K.A. Francesconi. 1977. Methylated arsenic from marine fauna. *Nature* **265**:436.

Edmonds, J.S., and K.A. Francesconi. 1981. Arseno-sugars from brown kelp *(Ecklonia radiata)* as intermediates in cycling of arsenic in a marine ecosystem. *Nature* **289**:602–604.

Ferguson, J.F., and J. Gavis. 1972. A review of the arsenic cycle in natural waters. *Water Research* **6**:1259–1274.

Ferm, V.H., 1974. Effects of metal pollutants upon embryonic development. *Reviews in Environmental Health* **1**:237–259.

Franzin, W.G., and G.A. McFarlane. 1980. An analysis of the aquatic macrophyte, *Myriophyllum exalbescens,* as an indicator of metal contamination of aquatic ecosystems near a base metal smelter. *Bulletin of Environmental Contamination and Toxicology* **24**:597–605.

Gilderhus, P.A. 1966. Some effects of sublethal concentrations of sodium arsenite on bluegills and the aquatic environment. *Transactions of the American Fisheries Society* **95**:289–296.

Greig, R.A., D.R. Wenzloff, and J.B. Pearce. 1976. Distribution and abundance of heavy metals in finfish, invertebrates and sediments collected at a deepwater disposal site. *Marine Pollution Bulletin* **7**:185–187.

Hale, J.G. 1977. Toxicity of metal mining wastes. *Bulletin of Environmental Contamination and Toxicology* **17**:66–73.

Huang, P.M., and W.K. Liaw. 1978. Distribution and fractionation of arsenic in selected fresh water lake sediments. *Internationale Revue der Gesamten Hydrobiologie* **63**:533–543.

Huang, P.M., and W.K. Liaw. 1979. Adsorption of arsenite by lake sediments. *Internationale Revue der Gesamten Hydrobiologie* **64**:263–271.

Isensee, A.R., P.C. Kearney, E.A. Woolson, G.E. Jones, and V.P. Williams. 1973. Distribution of alkyl arsenicals in model ecosystems. *Environmental Science and Technology* **7**:841–845.

Ishinishi, N., Y. Kodama, K. Nobutomo, and A. Hisanaga. 1977. Preliminary experimental study on carcinogenicity of arsenic trioxide in rat lung. *Environmental Health Perspectives* **19**:191–196.

Johnson, C.R. 1978. Herbicide toxicities in the mosquito fish, *Gambusia affinis. Proceedings of the Royal Society of Queensland (Australia)* **89**:25–27.

Johnson, D.L., and R.J. Braman. 1975. Alkyl and inorganic arsenic in air samples. *Chemosphere* **4**:333–338.

Karbe, L., Ch. Schnier, and H.O. Siewers. 1977. Trace elements in mussels *(Mytilus edulis)* from coastal areas of the North Sea and the Baltic. Multielement analyses using instrumental neutron activation analysis. *Journal of Radioanalytical Chemistry* **37**:927–943.

Kenaga, E.E., and R.J. Moolenaar. 1979. Fish and *Daphnia* toxicity as surrogates for aquatic vascular plants and algae. *Environmental Science and Technology* **13**:1479–1480.

Klumpp, D.W. 1980. Accumulation of arsenic from water and food by *Littorina littoralis* and *Nucella lapillus*. *Marine Biology* **58**:265–274.

Klumpp, D.W., and P.J. Peterson. 1979. Arsenic and other trace elements in the waters and organisms of an estuary in SW England. *Environmental Pollution* **19**:11–20.

Kobayashi, S., and G.F. Lee. 1978. Accumulation of arsenic in sediments of lakes treated with sodium arsenite. *Environmental Science and Technology* **12**:1195–1200.

Leatherland, T.M., and J.D. Burton. 1974. The occurrence of some trace metals in coastal organisms with particular reference to the Solent region. *Journal of the Marine Biological Association of the United Kingdom* **54**:457–468.

Löfroth, G. and B.N. Ames. 1978. Mutagenicity of inorganic compounds in *Salmonella typhimurium:* arsenic, chromium and selenium. *Mutation Research* **53**:65–66.

Luh, M.D., R.A. Baker, and D.E. Henley. 1973. Arsenic analysis and toxicity—a review. *The Science of the Total Environment* **2**:1–12.

Lunde, G. 1972. The analysis of arsenic in the lipid phase from marine and limnetic algae. *Acta Chemica Scandinavica* **2**:2642–2644.

MacInnes, J.R., and F.P. Thurberg. 1973. Effects of metals on the behavior and oxygen consumption of the mud snail. *Marine Pollution Bulletin* **4**:185–186.

Mackay, N.J., N.M. Kazacos, R.J. Williams, and M.I. Leedow. 1975. Selenium and heavy metals in black marlin. *Marine Pollution Bulletin* **6**:57–60.

McBride, B.C., H. Merilees, W.R. Cullen, and W. Pickett. 1978. Anaerobic and aerobic alkylation of arsenic. *In:* F.E. Brinckman and J.M. Bellama (Eds.), *Organometals and organometalloids.* American Chemical Society Symposium Series No. 82, American Chemical Society, Washington, D.C., pp. 94–115.

Moore, J.W. 1979. Diversity and indicator species as measures of water pollution in a subarctic lake. *Hydrobiologia* **66**:73–80.

Moore, J.W. 1981. Epipelic algal communities in a eutrophic northern lake contaminated with mine wastes. *Water Research* **15**:97–105.

Moore, J.W., D. Sutherland, and V. Beaubien. 1978. A biological and water quality survey of Prosperous lake, Walsh lake, and the Yellowknife river. Manuscript Report. Environmental Protection Service, Environment Canada, Yellowknife, 40 pp.

Nasu, Y., and M. Kugimoto, 1981. *Lemna* (duckweed) as an indicator of water pollution. I. The sensitivity of *Lemna paucicostata* to heavy metals. *Archives of Environmental Contamination and Toxicology* **10**:159–169.

National Academy of Sciences. 1977. *Arsenic.* Committee on Medical and Biologic Effects of Environmental Pollutants. National Academy of Sciences, Washington, D.C., 332 pp.

National Research Council of Canada. 1978. *Effects of arsenic in the Canadian environment.* Publication No. NRCC 15391 of the Environmental Secretariat. Ottawa, Canada, 349 pp.

Oladimeji, A.A., S.U. Qadri, G.K.H. Tam, and A.S.W. DeFreitas. 1979. Metabolism of inorganic arsenic to organoarsenicals in rainbow trout *(Salmo gairdneri). Ecotoxicology and Environmental Safety* **3**:394–400.

Oscarson, D.W., P.M. Huang, and W.K. Liaw. 1980. The oxidation of arsenite by aquatic sediments. *Journal of Environmental Quality* **9**:700–703.

Ott, M.G., B.B. Holder, and H.L. Gordon. 1974. Respiratory cancer and occupational exposure to arsenicals. *Archives of Environmental Health* **29**:250–255.

Paladino, F.V., and J.R. Spotila. 1978. The effect of arsenic on the thermal tolerance of newly hatched muskellunge fry *(Esox masquinongy). Journal of Thermal Biology* **3**:223–227.

Passino, D.R.M., and J.M. Kramer. 1980. Toxicity of arsenic and PCB's to fry of deepwater ciscoes *(Coregonus). Bulletin of Environmental Contamination and Toxicology* **24**:527–534.

Petres, J., D. Baron, and M. Hagedorn. 1977. Effects of arsenic cell metabolism and cell proliferation: cytogenetic and biochemical studies. *Environmental Health Perspectives* **19**:223–227.

Pinto, S.S., P.E. Enterline, V. Henderson, and M.O. Varner. 1977. Mortality experience in relation to a measured arsenic trioxide exposure. *Environmental Health Perspectives* **19**:127–130.

Planas, D., and F.P. Healey. 1978. Effects of arsenate on growth and phosphorous metabolism of phytoplankton. *Journal of Phycology* **14**:337–341.

Reay, P.F. 1972. The accumulation of arsenic from arsenic-rich natural waters by aquatic plants. *Journal of Applied Ecology* **9**:557–565.

Richardson, C.W., J.D. Price, and E. Burnett. 1978. Arsenic concentrations in surface runoff from small watersheds in Texas. *Journal of Environmental Quality* **7**:189–192.

Ruppert, D.F., P.K. Hopke, P. Clute, W. Metzger, and D. Crowley. 1974. Arsenic concentrations and distribution in Chutauqua Lake sediments. *Journal of Radioanalytical Chemistry.* **23**:159–169.

Sanders, J.G. 1979. The concentration and speciation of arsenic in marine macroalgae. *Estuarine and Coastal Marine Science* **9**:95–99.

Schaefer, E.D., and W.O. Pipes. 1973. Temperature and the toxicity of chromate and arsenate to the rotifer, *Philodina roseola. Water Research* **7**:1781–1790.

Seydel, I.S. 1972. Distribution and circulation of arsenic through water, organisms and sediments of Lake Michigan. *Archiv fuer Hydrobiologie* **71**:17–30.

Sohacki, L.D. 1968. Dynamics of arsenic in the aquatic environment. Ph.D. Thesis, Michigan State University, Lansing, Michigan, 101 pp.

Sorensen, E.M.B. 1976. Toxicity and accumulation of arsenic in green sunfish, *Lepomis cyanellus,* exposed to arsenate in water. *Bulletin of Environmental Contamination and Toxicology* **15**:756–761.

Sorensen, E.M.B., R.R. Mitchell, C.W. Harlan, and J.S. Bell. 1980. Cytological changes in the fish liver following chronic, environmental arsenic exposure. *Bulletin of Environmental Contamination and Toxicology* **25**:93–99.

Spehar, R.L., J.T. Fiandt, R.L. Anderson, and D.L. DeFoe. 1980. Comparative toxicity of arsenic compounds and their accumulation in invertebrates and fish. *Archives of Environmental Contamination and Toxicology* **9**:53–63.

Spotila, J.R., and V.F. Paladino. 1979. Toxicity of arsenic to developing muskellunge fry *(Esox masquinongy). Comparative Biochemistry and Physiology* **62C**:67–69.

Thornton, I., H. Watling, and A. Darracott. 1975. Geochemical studies in several rivers and estuaries used for oyster rearing. *The Science of the Total Environment* **4**:325–345.

Tseng, W.P. 1977. Effects and dose-response relationships of skin cancer and blackfoot disease with arsenic. *Environmental Health Perspectives* **19**:109–119.

United States Minerals Yearbooks. 1911–1979. Bureau of Mines, US Department of the Interior, Washington, D.C.

Vallee, B.L. 1973. *Arsenic.* Air Quality Monograph No. 73–80, American Petroleum Instiute, Washington, D.C., 36 pp.

Van der Veen, C., and J. Huizenga. 1980. Combating river pollution taking the Rhine as an example. *Progress in Water Technology* **12**:1035–1059.

Wagemann, R. 1978. Some theoretical aspects of stability and solubility of inorganic arsenic in the freshwater environment. *Water Research* **12**:139–145.

Wagemann, R., N.B. Snow, D.M. Rosenberg, and A. Lutz. 1978. Arsenic in sediments, water and aquatic biota from lakes in the vicinity of Yellowknife, Northwest Territories, Canada. *Archives of Environmental Contamination and Toxicology* **7**:169–191.

Walsh, D.F., B.L. Berger, and J.R. Bean. 1977. Mercury, arsenic, lead, cadmium, and selenium residues in fish, 1971–73-National pesticide monitoring program. *Pesticides Monitoring Journal* **11**:5–34.

Waslenchuk, D.G. 1979. The geochemical controls on arsenic concentrations in southeastern United States rivers. *Chemical Geology* **24**:315–325.

Waslenchuk, D.G., and H.L. Windom. 1978. Factors controlling the estuarine chemistry of arsenic. *Estuarine and Coastal Marine Science* **7**:455–464.

Woolson, E.A. 1975. Bioaccumulation of arsenicals. *In:* E.A. Woolson (Ed.), *Arsenical pesticides.* American Chemical Society Symposium Series No. 7, American Chemical Society, Washington, D.C., pp. 97–107.

Wrench, J.J., and R.F. Addison. 1981. Reduction, methylation, and incorporation of arsenic into lipids by the marine phytoplankton *Dunaliella tertiolecta. Canadian Journal of Fisheries and Aquatic Science* **38**:518–523.

Wrench, J.J., S.W. Fowler, and M.Y. Unlu. 1979. Arsenic metabolism in a marine food chain. *Marine Pollution Bulletin.* **10**:18–20.

Yim, W.W.S. 1976. Heavy metal accumulation in estuarine sediments in a historical mining of Cornwall. *Marine Pollution Bulletin.* **7**:147–150.

3
Cadmium

Chemistry

Cadmium is the second member of the Group IIb triad (Zn, Cd, Hg) in the periodic classification of elements. The stable state of cadmium in the natural environment is Cd(+2). Cadmium is silvery white and ductile with a faint blue tinge. It has a medium class b character compared to zinc and mercury. This imparts moderate covalency in bonds and high affinity for sulfhydryl groups, leading to increased lipid solubility, bioaccumulation, and toxicity. Cadmium accumulates in liver and kidney through its strong binding with cysteine residues of metallothionein. Since the metabolism of cadmium is closely related to zinc metabolism, metallothionein binds and transports both cadmium and zinc. Cadmium seems to displace zinc in many vital enzymatic reactions, causing disruption or cessation of activity.

Although they are metals, zinc and cadmium are softer, lower melting, and more electropositive than their neighbouring transition group metals. The chemistry of cadmium is homologous to that of zinc and different from mercury in both the properties of the element and its compounds. $Cd(OH)_2$ is more basic than $Zn(OH)_2$ whereas $Hg(OH)_2$ is an extremely weak base. The halides of zinc and cadmium are essentially ionic whereas $HgCl_2$ is covalent and almost undissociated in aqueous solution. Complexes of Hg(+2) are generally several orders of magnitude more stable than those of Zn(+2) and Cd(+2). The properties of mercury may be considered as a manifestation of the "inert pair" effect, yielding an unusually high ionization potential and a high negative electrode potential in its complexes. Due to its superior tendency to form covalent bonds, mercury forms a large number of organometallic compounds, R_2Hg and RHgX, which are stable in air and water. The corresponding zinc and cadmium compounds are unstable in air and water. The unusual stability of mercury compounds must be attributed also to the very low affinity of mercury for oxygen. The possibility of biological methylation of elements has been discussed in terms of the relative ease of formation of metal-carbon bonds (Wood, 1974) and the reduction potential of elements (Ridley *et al.,* 1977). The metals, mercury, tin, palladium, platinum, gold, and thallium, and the metalloids, arsenic, selenium, tellurium, and sulfur, have been postulated to accept methyl groups from methyl donors such as methyl cobalamine in biological methylation whereas lead, cadmium and zinc have been predicted to be incapable of environmental methylation due to the extreme instability of their monoalkyl derivatives.

Production, Uses, and Discharges

Production

Cadmium is most commonly found associated with zinc in carbonate and sulfide ores. Cadmium is also obtained as a by-product in the refining of other metals. Thus humans, through their production of metals like copper, lead, and zinc for several centuries, were unknowingly polluting the environment with cadmium.

Cadmium and its compounds have found increasing applications in a variety of industrial products and operations, causing a marked increase in direct production. Hence total production was $<0.9 \times 10^5$ metric tons from 1911 to 1950, but increased to 1.5×10^5 tons during 1971–80 (Table 3-1).

Table 3-1. Global cadmium production.

Period	Quantity (metric tons $\times 10^5$)
1911–1920	0.01
1921–1930	0.07
1931–1940	0.26
1941–1950	0.48
1951–1960	0.84
1961–1970	1.40
1971–1980	1.50

Sources: US Minerals Yearbooks; Nriagu (1979).

Uses

Electroplating. Cadmium is deposited either electrolytically or mechanically on objects to provide bright appearance and resistance to corrosion. The end-products include parts and finishes in automobile and aircraft industries, industrial and builders' hardward, marine hardware, radio and television parts, and household appliances. Cadmium is also used in the packing industries except food packing.

Pigments. Cadmium sulfides give yellow-to-orange colours and cadmium sulfoselenides give pink-to-red and maroon. These pigments are used in the plastic industry, ceramics, paints, and coatings. Cadmium pigments are used in traffic paints, high quality industrial finishes, and in the glass enamel red label on 'Coca-Cola' bottles.

Plastic Stabilizers. Cadmium stearates are used as stabilizers in the production of polyvinyl chloride plastics (PVC). They stabilize the double bonds in the polymer by displacing the labile allyl chlorine atoms. Addition of barium (or zinc) salts, epoxy compounds, and organic phosphite esters protects the polymer from excess chlorine or chloride formed. The flexible PVC is used extensively in the making of calenders and plastisols. However, cadmium-based stabilizers are not used in the manufacture of flexible PVC as food containers. Japan shows a marked decline since 1970 in the use of cadmium in PVC industries.

Batteries. Due to its perfectly reversible electrochemical reactions at a wide range of temperature, low rate of self discharge, and easy recovery from dead batteries, cadmium is employed extensively in batteries. Such batteries are used in consumer items such as battery-operated toothbrushes, shavers, drills and handsaws, medical appliances, communication devices and emergency lighting supplies, airplanes, satellites and missiles, and ground equipment for polar regions.

Other Uses. Other uses of cadmium are: (a) cadmium phosphors as tubes in television sets, fluorescent lamps, x-ray screens, cathode-ray tubes, and phosphorescent tapes, (b) cadmium alloys in Cd-Ag solders, automatic sprinkler systems, fire-detection apparatus, valve seals for high-pressure gas containers, trolley and telephone wires, and in automobile radiator finstock, (c) electrical and electronic applications such as heavy duty relays, switches, automobile distributor contacts, and solar and photocells.

After dipping in 1977, consumption in the USA rose by 18% in 1978 and continues to increase (Table 3-2). The transportation category which includes cadmium from each of the remaining end-use categories accounted for 17% of total consumption (US Minerals Yearbook, 1978–79). The breakdown of the other categories is: (i) electrically or mechanically plated

Table 3-2. Cadmium consumption by industrial categories in US.

Industry	Consumption (10^3 kg)						
	1965	1970	1972	1974	1976	1978	1979
Electroplating	2447	2052	2869	2718	2690	1533	1638
Pigments	1160	587		997		586	626
			{1917}		{1631}		
Stabilizers	385	1081		906		496	530
Batteries	279	136		544		992	1060
			{953}		{1060}		
Alloys and others	408	251		441		135	145
Transportation						767	819

Source: US Minerals Yearbooks.

hardward—34%, (ii) nickel-cadmium, silver-cadmium, and mercury-cadmium batteries—22%, (iii) pigments—13%, (iv) plastics and synthetic products—11%, and (v) alloys and other uses—3%.

Cadmium in Aquatic Systems

Speciation in Natural Waters

Binding to Inorganic Ligands. Cadmium is an oxyphilic and sulfophilic element. It undergoes multiple hydrolysis at pH values encountered in the environment. Furthermore, within the triad Zn, Cd and Hg, marked differences in binding to ligands exist. Mercuric ion hydrolyses in the pH range 2–6, yielding $Hg(OH)_2$ as the final species at pH 6. Cadmium (+2) is present totally as the divalent species up to pH 8, in the absence of any precipitating anions such as phosphate and sulfide. Cadmium begins to hydrolyse at pH 9, forming $Cd(OH)^+$ species. Higher hydroxy species of cadmium are not relevant at the pH values commonly found in the environment.

The speciation of cadmium at different chloride concentrations at pH 8.5 is given in Figure 3-1. It should be noted that extremely low or high values of pH and chloride concentrations are never encountered in natural waters. In the absence of any precipitating anions, Cd(+2) will be available for sorption onto suspended solids and complexation with organic matter and will be transported in those forms.

Chlorides are more selective than many organic complexing agents in their interaction with cadmium. Among heavy metals, the degree of covalency of the metal-chloride bond varies markedly. Thus, chloride complexes compete with sparingly soluble precipitates and organic complexes of heavy metals in the order Hg > Cd > Pb > Zn. Moreover, chloride complexes of heavy metals are highly mobile and persistent unlike NTA (nitrilotriacetic

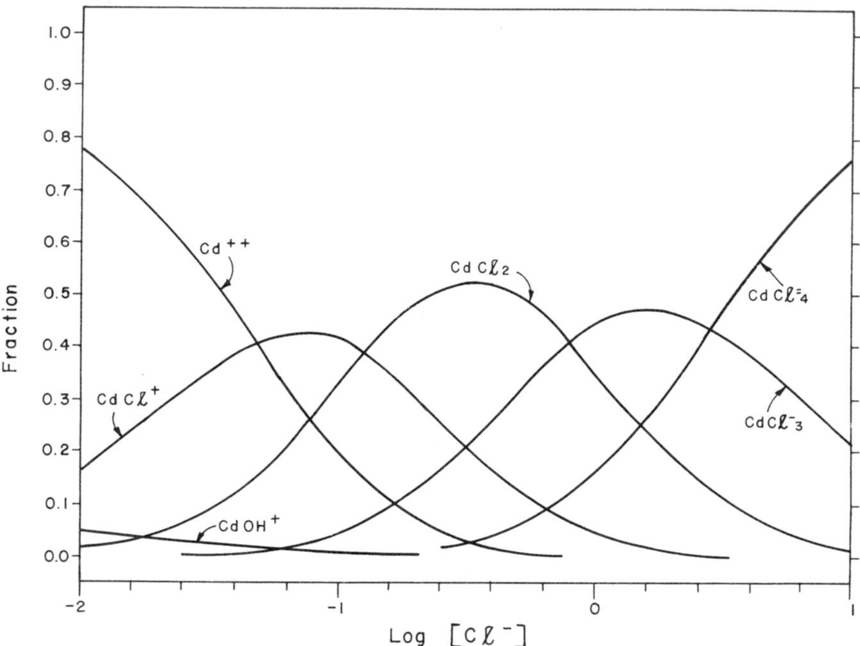

Figure 3-1. Simultaneous distributions of hydroxy and chloride complexes of Cd(+2) at pH 8.5 and different chloride concentrations. (From Hahne and Kroontje, 1973.)

acid) which is biodegradable. Therefore, chlorides could be a more important factor in the distribution of heavy metals than has been considered to date.

Binding to Organic Ligands. Cadmium forms moderately stable complexes with a variety of organic compounds. Being a soft acid acceptor in co-ordination reactions, it prefers soft donor atoms such as sulfur, selenium and nitrogen. Cadmium interacts strongly with sulfydryl groups such as in cysteine. The organic matter found in natural waters, including amino acids, aminosugars, polysaccharides, hydroxy and carboxylic acids of aliphatic and aromatic nature, also contains suitable donor atoms for complex formation with cadmium.

Binding of cadmium to humic substances from sea, river and lake waters was studied by Mantoura *et al.* (1978). In general, stabilities of the humic complexes of the various metals followed the Irving-Williams order of stabilities of chelates:

$$\text{Mg} < \text{Ca} < \text{Cd} \sim \text{Mn} < \text{Co} < \text{Zn} \sim \text{Ni} < \text{Cu} < \text{Hg}.$$

In typical lake water, humic acid complexation of cadmium accounted for only 2.7% of total Cd, dropping to < 1% in simulated estuarine water. It was concluded that humic substances vary in their interaction with metals, depending on their source. Hence studies on cadmium speciation in any particular water should use data for humic substances from that water.

Synthetic chelating agents, which show little or no degradation, are used in several industrial applications. Phosphates promote algal growth and are among those agents responsible for accelerated eutrophication of lakes. In some countries, nitrilotriacetic acid (NTA) is used as a surfactant, substituting wholly or partially the polyphosphates in commercial detergents. Since detectable levels of NTA occur in inland, estuarine and coastal waters, relatively stable Cd-NTA chelates may also be formed in significant quantities. However, Chau and Shiomi (1972) reported that bacterial degradation of Cd-NTA chelates required only 60 days. Hence, the use of NTA may not constitute an environmental problem in many countries. This is balanced by the fact that the biodegradability of NTA in activated sludge processes and presumably natural waters decreases at low temperatures and increasing concentrations of NTA. It has also been suggested that the formation of strong mixed ligand complexes with phosphates and synthetic agents may interfere in phosphate removal, depending on the origin and composition of waste waters.

Physico-Chemical Speciation in Surface Waters. Fresh ground water generally contains a greater proportion (up to 90%) of free Cd(+2) ions than sewage effluents (62–71%), whereas humic acid complexes may account for only 37–39% of total cadmium (Gardiner, 1974). Although Ramamoorthy and Kushner (1975a,b) similarly showed that water contaminated with sewage strongly bound cadmium, canal water contaminated with de-icing salt bound only ~50% of added cadmium. The binding components in the two waters were of different molecular size fractions (< 1400 and 1400–16,000 molecular weight, respectively).

An analytical scheme of speciating cadmium was applied to fresh waters and estuarine regions of the Yarra River, Australia (Hart and Davies, 1981). Total cadmium concentrations in the water ranged from 0.29 to 0.55 $\mu g\ L^{-1}$ with a mean of 0.42 $\mu g\ L^{-1}$. Filterable cadmium for the same study area was 0.21–0.47 $\mu g\ L^{-1}$ (mean, 0.33 $\mu g\ L^{-1}$). Approximately 75% of the filterable cadmium was ion-exchangeable (Figure 3-2). This finding was consistent with the theoretical calculations of cadmium speciation using stability constants of known complexes (CO_3, HCO_3, OH and Cl). More than 80% was free Cd(+2) species with small amounts of $CdCl^+$ species (Figure 3-3). Organic ligands competed in complexing cadmium at concentrations in excess of 10^{-6} M. It was also shown that cadmium speciated differently between fresh water and estuarine sections, with a considerably higher proportion of non-dialysable metal (MW $\geq$ 1000) present in fresh water. The average Cd(total) and Cd(filterable) concentrations in the estuary were 50% lower than the corresponding fractions in the fresh water section of the

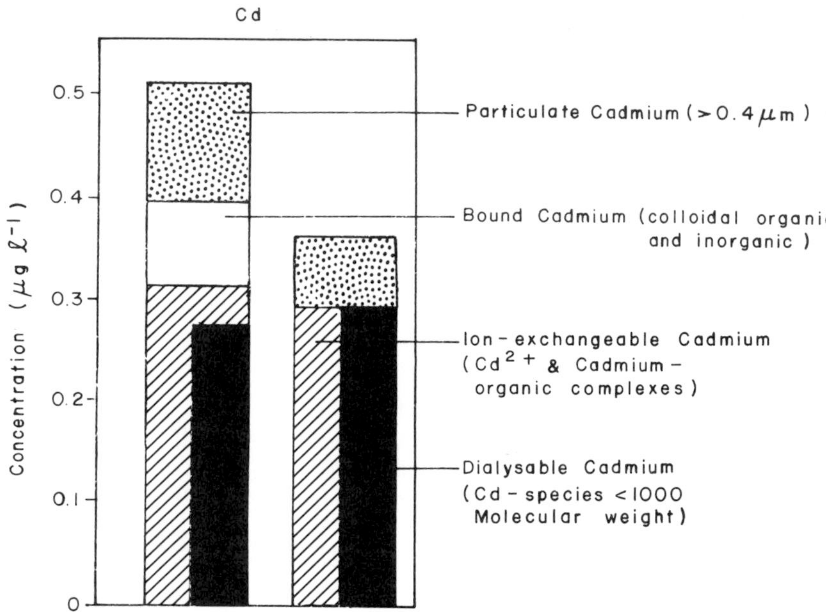

Figure 3-2. Average cadmium fractions taken from freshwater (F) and estuarine (E) sections of the Yarra River, Australia. (With permission, Hart and Davies, 1981. Copyright: Academic Press, Inc. (London) Ltd.)

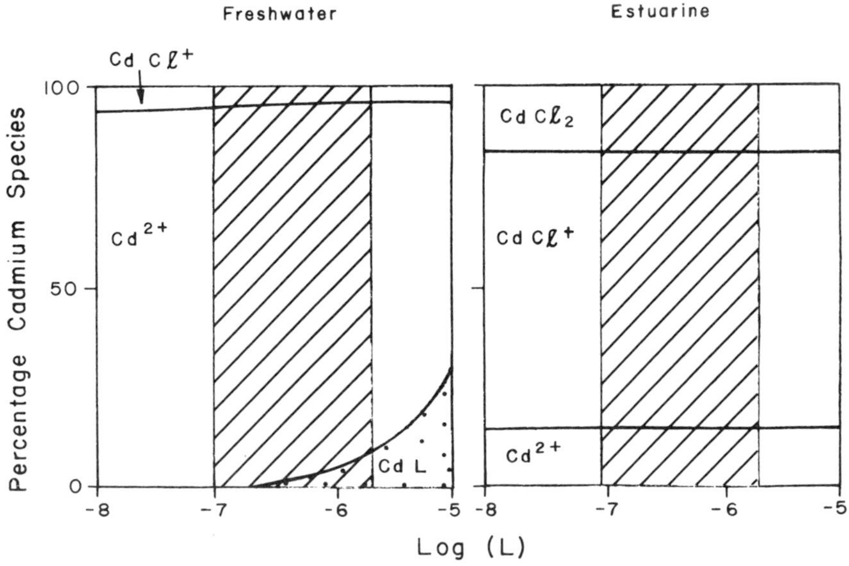

Figure 3-3. Calculated equilibrium distribution of ion-exchangeable Cd in the freshwater and estuarine sections of the Yarra River, as a function of organic ligand (L) concentration. (With permission, Hart and Davies, 1981. Copyright: Academic Press Inc. (London) Ltd.)

Table 3-3. Speciation of cadmium in contaminated lakes (US).

Location	% of total dissolved cadmium						
	Cd(+2)	$CdSO_4$	$CdCO_3$	$CdOH^+$	$CdCl^+$	Cd–organic	% Cd accounted for
Palestine Lake, Indiana	27.6–61.5	5.9–17.6	8.1–17.1	2.0–11.3	1.0–2.2	1.8–53.0	95.9–100.6
Grand Calumet River	70.0	7.0	4.9	6.8	2.1	4.5	95.3
Little Center Lake	49.5	15.4	8.6	7.7	9.7	15.2	105.1
Lake Michigan, Indiana Dunes	81.3	9.0	2.1	2.8	1.6	3.2	100.0
Lake Michigan, Michigan City	86.0	8.8	3.2	4.2	1.5	3.8	10[illegible].5

Source: Shephard *et al.* (1980)

river. This was explained by increased coagulation and sedimentation of the bound fraction. Ion-exchangeable and dialysable cadmium concentrations did not vary between the two types of water. Under low flow conditions, ~30% of cadmium in solution (bound to $\leq 0.4\ \mu m$) was sedimented in the upper part of the estuary.

Shephard *et al.* (1980) reported dominance of Cd(+2) species in a contaminated lake, excepting one site where the organic binding of cadmium accounted for 53% of the total Cd (Table 3-3). The relatively high percentage of organic complexing was related to the production of autochthonous organic matter by aquatic macrophytes and phytoplankton. It was concluded that the presence of high concentrations of Cd-binding ligands by itself may not be sufficient to reduce the free Cd concentration below levels possibly harmful to the biota of the system.

Cadmium in Sediments

Sorption. Adsorption to sediments increases with pH and beyond a threshold point (pH $\geq$ 7 for Cd), virtually all the metal ion is sorbed. In the presence of sorbents, the pH values required for precipitation of Cd species are greatly reduced. Addition of ligands which yield cationic complexes (e.g., glycine, α, α–dipyridyl) suppresses precipitation of Cd-hydroxide, yet the sorption increases with increasing pH. This suggests that formation of Cd-monohydroxide is not a prerequisite for sorption by clays. Formation of an anionic complex (e.g., with EDTA) results in zero uptake of cadmium over a wide pH range. This observation is consistent with cadmium sorption sites on clays being negative in sign.

Correlation of Cd sorption to organic content occurs in a variety of soils and sediments. Although Bower *et al.* (1978) found high concentrations (max. 908 mg kg^{-1}) in surface sediments near the discharge site of a Ni-Cd battery factory, levels fell by fourfold moving from the outfall to the westernmost part of the cove (~1 km). Levels decreased close to background (2 mg kg^{-1}) in deeper sediments (~20 cm). Similarly cadmium in the cores of pre-industrial sediments from Wisconsin lakes (Iskandar and Keeney, 1974) and shales from the Great Plains region (Tourtelot *et al.*, 1964) were 1.7 mg kg^{-1} and 2.0 mg kg^{-1}, respectively. Bower *et al.* (1978) concluded that in a shallow, fresh water cove, cadmium was transported by association with suspended solids, assisted by tidal currents. In the area of immediate alkaline waste discharge, cadmium was less mobile than nickel, apparently due to the formation of an insoluble $CdCO_3$ phase. Budgeting showed that 20–50 metric tons of Cd were present in the sediments of the cove.

High percentages of the total particulate cadmium were found in the exchangeable (26 and 28%) and carbonate (49 and 31%) fractions of suspended sediments in two Canadian rivers (Tessier *et al.*, 1980). Lower values occurred in the residual (12 and 27%) and Fe-Mn oxide (12 and 13%)

fractions. Cadmium bound to organic matter was undetectable. Gardiner (1974) concluded that sorption-desorption of cadmium from river muds was rapid and humic acid was responsible for the sorption. CF's of 5000 to 50,000 were reported for river mud-Cd interaction whereas the studies of Tessier *et al.* (1980) reported no interaction with organic matter. Large amounts of cadmium found in exchangeable and carbonate fractions suggest that an appreciable percentage of particulate cadmium would be available through solubilization with increase in salinity or lower pHs.

Ramamoorthy and Rust (1978) similarly studied the sorption–desorption of cadmium in sand, silt and organic-rich sediments from a river. The sorption for all sediment samples were fitted to the linear form of the Langmuir equation;

$$\frac{C}{x/m} = \frac{1}{kb} + \frac{C}{b}$$

where C is the equilibrium concentration of the sorbate (Cd(+2)ion), x/m is the amount of Cd(+2) per unit mass of sorbent, k is the bonding energy constant and b is the sorption maximum. There was a good correlation of sorption parameters with organic content and grain size (Tables 3-4, 3-5).

Cadmium sorbed on suspended solids (at 10–100 mg L^{-1} levels) in tributaries may well desorb when river plumes dissipate in the clear waters of a lake where suspended solids are quite low (<1 mg L^{-1}) or in an estuary with increased salinity. A cadmium budget prepared by Muhlbaier and Tisue (1981) for southern Lake Michigan estimated that (i) the residence time for Cd in water was 8.5 years, (ii) input rate was 2.5 times greater than the combined rates of sedimentation and outflow, and (iii) Cd levels in the area could rise within a few decades to a level deleterious to zooplankton communities (Table 3-6). Flinn and Reiners (1977) projected the rate of increase of cadmium emissions from several sources to be 4.6% per annum.

Table 3-4. Cadmium sorption by sediments in the Ottawa River.

	Sediment properties		Sorption parameters			
Number	Organic matter (%)	Grain size (Ø) mean	Std. dev.	b	k	Correlation coefficient
1	0.6	1.18	0.58	1.90	2.50	0.99
2	3.2	7.09	1.18	19.49	2.42	0.95
3	5.2	3.00	0.98	3.29	0.26	0.99
4	35.7	2.63	0.69	3.48	2.46	0.99
5	2.4	1.89	0.35	1.18	1.38	0.99
6	1.3	6.55	1.81	9.37	8.08	0.88
7	9.9	2.44	0.93	4.17	0.23	0.93

Source: Ramamoorthy and Rust (1978).
Ø = $-\log_e^{mm}$; b = Sorption maximum; k = Bonding energy constant.

Table 3-5. Correlation of sorption parameters with organic content and grain size of sediments.

Factors correlated	Sorption maxima (*b*)	Bonding energy constant (*k*)
Organic matter (%)	0.996	0.810
Mean grain size (Ø)	0.780	0.820

Source: Ramamoorthy and Rust (1978).

Table 3-6. Mass balance of cadmium in the southern basin of Lake Michigan.

Sources (metric ton yr^{-1})		Sinks (metric ton yr^{-1})	
Rain	4.3 (40.1%)	Outflow	0.28 (6.9%)
Dry deposition	2.2 (20.8%)	Sedimentation	3.8 (93.1%)
Erosion	1.0 (9.4%)		4.08
Tributaries			
Soluble	1.3 (12.3%)		
Suspended	1.8 (17.0%)		
	10.6		

Source: Muhlbaier and Tisue (1981).

Desorption. Release of sorbed cadmium into the bulkwater is dependent on the partition coefficient, which is related to sediment characteristics and water quality parameters. Ramamoorthy and Rust (1978) studied the desorption of cadmium from river sediments in the presence of a synthetic chelating agent (NTA), other heavy metals such as Hg, Pb and Cu and increased salinity. They concluded that:

1. The ratio of desorption of Cl^- and NTA was 1:2. Chloride desorbed about 50% of Cd from sediments; hence the Cd-sediment binding constant must be $\leqslant 10^{1.8}$ (stability constant of Cd–Cl complex).
2. Partitioning of cadmium was not greatly perturbed by the presence of other heavy metal ions in concentrations about the same order of magnitude as the Cd concentration.
3. Sediments could act as heavy metal ion exchangers on a mass action basis.

Unlike mercury, alkylation and methylation of cadmium has not been reported in natural environments, either in water, sediments or biota. Huey *et al.* (1975) reported the transient formation of methylcadmium under laboratory conditions.

Residues

Precipitation, Water, and Sediments

Worldwide annual emissions of cadmium from natural sources are approximately 8.43×10^5 kg (Nriagu, 1979). Vegetation, airborne soil particles, volcanogenic aerosols, and forest fires contribute to natural emissions. The annual atmospheric input from industrial activities accounts for 7.19×10^6 kg. Of this, about 76% originates from non-ferrous metal industries and the rest from inadvertent emissions. Precipitation removes cadmium effectively from the atmosphere. Hence concentrations in rainwater may exceed $\geqslant 50\ \mu g\ L^{-1}$ (Table 3-7).

Dissolved cadmium levels in freshwater generally range from 10 to 500 ng L^{-1} (Table 3-8). However, in cases of extreme pollution, concentrations may exceed 17,000 ng L^{-1}. Oceanic waters show significant enrichment (up to 125 ng L^{-1}) at mid-depth (Table 3-8). It has been suggested that aquatic organisms incorporate Cd in their tissue and transport it to deeper waters through sinking of their detritus and debris.

Cadmium residues in freshwater sediments are highly variable, ranging from $\leqslant 0.1$ mg kg^{-1} dry weight to $\geqslant 3000$ mg kg^{-1} (Table 3-9). In the vicinity of Cd-Ni battery factories, however, levels may approach 50,000 mg kg^{-1}. Unpolluted marine sediments may contain as little as 0.01 mg Cd kg^{-1}, whereas in industrial areas levels may exceed 50 mg kg^{-1} (Table 3-9).

Aquatic Plants

Total residues in freshwater plants may vary from 0.15 to 342 mg kg^{-1} dry weight (Table 3-10). In general, there is a good correlation between total

Table 3-7. Cadmium levels (ng L^{-1}) in precipitation samples.

Location	Average (range)
Argonne, Illinois[1]	320 (70–1100)
North Dakota[2]	730
Minnesota[2]	180
Rural, UK[3]	500
Göttingen, FRG[4]	580 (100–1000)
Ontario, Canada[5]	800 (10–50000)
Indiana[6]	(630–1340)

Sources: [1]Muhlbaier and Tisue (1981); [2]Thornton *et al.* (1980); [3]Cawse (1977); [4]Ruppert (1975); [5]Kramer (1976); [6]Yost (1978).

Table 3-8. Dissolved cadmium levels (ng L^{-1}) in fresh and marine waters.

Location	Average	Polluting source
Freshwaters		
Rivers and harbours, Lake Michigan[1]	400	mixed industrial
Mid-lake, Lake Michigan[1,2]	20–70	mixed industrial
Lake Mendota, Wisconsin[1]	170	mixed industrial
Rivers and harbours, Lake Ontario[1]	450	mixed industrial
Mid-lake, Lake Ontario[1]	160	mixed industrial
Palestine Lake, Indiana[3]	17300	electroplating plant
Remote streams, California[4]	10–100	unpolluted
Oceans		
North Atlantic (50–5000m)[5]	20–150	unpolluted
Pacific, South of New Zealand (3–5271 m)[6]	1–12	unpolluted
Pacific, East of Japan (8–5446m)[6]	0.8–9.6	unpolluted
Pacific, North of Hawaii (surface)[7]	20	unpolluted
Indian Ocean (surface)[8]	70	unpolluted

Sources: [1]Elzerman and Armstrong (1979); [2]Muhlbaier and Tisue (1981); [3]Shephard *et al.* (1980); [4]Kennedy and Sebetich (1976); [5]Eaton (1976); [6]Boyle *et al.* (1976); [7]Knauer and Martin (1973); [8]Chester and Stoner (1974).

Table 3-9. Cadmium levels (mg kg^{-1} dry weight) in freshwater and marine sediments.

Location	Average (range)	Polluting source
Freshwater sediments		
Great Lakes[1,2]	2.7 (–)	mixed industrial
Palestine Lake, Indiana (USA)[3]	(3–2640)	electroplating
Sudbury lakes (Canada)[4]	(0.5–19.4)	industrial smoke stacks
Lake Tunis (North Africa)[5]	(0.1–13.3)	sewage effluent
Los Angeles River (USA)[6]	860	mixed industrial
Foundry Cove, New York (USA)[7]	(3000–50,000)	Ni-Cd battery plant effluent
Rivers, Tokyo (Japan)[8]	max. 368	Cd consuming industries
River Conway (UK)[9]	(3–95)	Pb-Zn mine
Marine Sediments		
Coast of Israel[10]	(0.3–2.2)	small, local industry
Estuaries (UK)[11,12]	(0.2–25)	mixed industrial
Tokyo Bay (Japan)[13]	(3.1–40.4)	mixed industrial
Baltic Sea[14]	(<0.01–8.1)	mixed industrial
Santa Monica Canyon (USA)[15]	(0.2–65)	mixed industrial

Sources: [1]Kemp *et al.* (1976); [2]Kemp and Thomas (1976); [3]Wentsel and Berry (1974); [4]Semkin and Kramer (1976); [5]Harbridge *et al.* (1976); [6]Chen *et al.* (1974); Kneip *et al.*, (1974); [8]Asami (1974); [9]Thornton *et al.* (1975); [10]Roth and Hornung (1977); [11]Steele *et al.* (1973); [12]Leatherland and Burton (1974); [13]Ishibashi *et al.* (1970); [14]Olausson *et al.* (1977); [15]Schafer and Bascom (1976).

Table 3-10. Total cadmium residues (mg kg^{-1} dry weight) in some marine and freshwater plants.

Species	Average (range)	Location	Polluting source
Macroscopic/vascular			
Fucus vesiculosus[1,2]	4.9 (1.9–13)	Baltic Sea, fjords (Scandinavia)	mixed industrial, smelting plant
Ascophyllum nodosum[2,3]	8.3 (0.7–16)	Hardangerfjord (Norway)	smelting plant
Phaeophyceae (3 species)[4]	0.31 (0.15–0.43)	The Solent (UK)	unpolluted
Rhodophyta, Chlorophyta[5]	1.35 (0.9–2.1)	Mediterranean Sea (Israel)	mixed industrial
Vascular plants (2 species)[6]	3.1 (0.65–6.08)	Nepisiguit River (Canada)	metal mines
Vascular plants (5 species)[7]	5.9 (0.6–14)	Thompson Lake (Canada)	base metal smelter
Nitella flexilis[8]	14 (6–21)	Derwent Reservoir (UK)	flourspar mine
Potamogeton crispus[9]	16.5 (0.29–89.6)	Palestine Lake (USA)	electroplating plant
Lemanea fluviatilis[10]	22.7 (2–342)	35 rivers (Europe)	metal mines, mixed industrial sources
Cladophora glomerata[11]	0.62 (0.29–0.94)	River Leine (FRG)	mixed industrial
Microscopic			
Benthic algae[12]	3.5 (1.7–9.5)	Neckar River (FRG)	mixed industrial
Oscillatoria sp.[13]	1.9 (–)	Savannah River (USA)	coal ash
Benthic diatoms[14]	(10.3–16.8)	Rhine River (FRG)	mixed industrial
Mixed phytoplankton[15]	0.6 (0.2–1.1)	Lake Baldegg (Switzerland)	experimental additions
Benthic algae[15]	0.6 (–)	Lake Baldegg (Switzerland)	experimental additions

Sources: [1]Phillips (1979); [2]Melhuus *et al.* (1978); [3]Haug *et al.* (1974); [4]Leatherland and Burton (1974); [5]Roth and Hornung (1977); [6]Ray and White (197[illegible]); [7]Franzin and McFarlane (1980); [8]Harding and Whitton (1978); [9]McIntosh *et al.* (1978); [10]Harding and Whitton (1981); [11]Abo-Rady (1980); [12]Bartelt and Förstner (1977); [13]Guthrie and Cherry (1979); [14]Vogt and Kittelberger 1977); [15]Gächter and Geiger (1979).

concentrations in water and plant tissues, as exemplified by data for European populations of *Lemanea fluviatilis* (Figure 3-4). Similarly, residues in benthic species are positively correlated with concentrations in sediments. McIntosh *et al.* (1978) showed that the population of *Potamogeton crispus* inhabiting an industrially contaminated lake had sorbed 1.5 kg of elemental cadmium and that release of cadmium from dead plants could raise water concentrations by a maximum of 1 μg L^{-1}. Although estuarine and coastal species have been used to monitor cadmium, there does not appear to be a strong relation between residues in sea water and plant tissue. Of the four species studied by Melhuus *et al.* (1978) in Sørfjorden (Norway), two showed maximum residues in areas of reduced contamination. In addition CF's (3500–13,000) were usually lowest in the zone of maximum pollution, possibly reflecting inhibition of cadmium uptake by other heavy metals.

Microscopic species may accumulate significant amounts of cadmium (Table 3-10). Although peak values are considerably lower than those reported for vascular macroscopic species, this may reflect an inadequate data base. For example, Guthrie and Cherry (1979) showed that average residues in vascular, microscopic and macroscopic, non-vascular species from the Savannah River (USA) were 0.4, 1.9, and 1.4 mg kg^{-1} dry weight, respectively. Residues in phytoplankton from Lake Baldegg (Switzerland) varied seasonally, reaching maximum and minimum values in the late fall

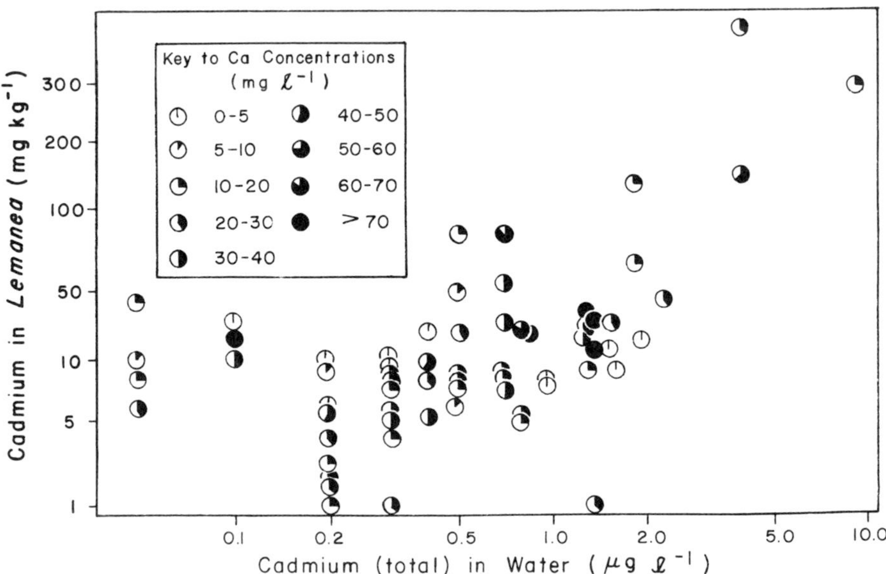

Figure 3-4. Correlation between mean cadmium concentrations in *Lemanea fluviatilis* and water of origin in various European rivers. (From Harding and Whitton, 1981.)

and spring, respectively (Gächter and Geiger, 1979). However, CF's ($<1-7 \times 10^5$) did not show seasonal trends.

Low temperatures reduce uptake, reflecting decreased binding efficiency and metabolism in tissues (Hart *et al.*, 1979). In addition, since differences in water temperature result in differential growth, fast-growing weeds may bear relatively low residues, particularly in spring and early summer. Sorption is generally slower in rapidly growing young algal cultures than older cultures approaching no-growth conditions (Phillips, 1979). Similarly, in some species, uptake is faster in hard water than in soft water and is dependent on cadmium levels in water.

The presence of manganese and iron in water inhibits cadmium uptake but zinc and cobalt may have no effect. The presence of other metals also probably results in competition for cadmium for uptake sites, though there are few data to confirm this point. Bruland *et al.* (1978) demonstrated that cadmium levels in marine phytoplankton were significantly correlated with phosphate and nitrogen in water. Thus the microplankton and their organic detritus were probably responsible for the oceanic distribution and cycling of cadmium. Such a relation may not be applicable to fresh waters because of extensive anthropogenic input of metals and organic material.

Invertebrates

Although cadmium is transferred to invertebrates through both food and water, CF's are low compared with those for organo-metals. Consequently, residues in invertebrates are often less than those reported for algae. Gächter and Geiger (1979) found that the CF's for zooplankton and phytoplankton in experimentally polluted Lake Baldegg ranged from $0.1-0.5 \times 10^5$ and <1 to 7×10^5, respectively. Similarly, the CF in several species of mollusc and Crustacea from the Severn Estuary (UK) was 0.9–3.5 times greater than those for attached algae (Butterworth *et al.*, 1972; Bryan, 1976). Despite relatively low CF's, high cadmium levels have been reported for invertebrates inhabiting heavily cadmium-polluted waters. In the Severn Estuary, littoral gastropods had residues ranging from 9 to 550 mg kg^{-1} dry weight, depending on location and species (Butterworth *et al.*, 1972). In the Bristol Channel (UK), levels in dog whelks (*Nucella lapillus*) varied from 500 to 1120 mg kg^{-1} (Stenner and Nickless, 1974).

Freshwater invertebrates usually have lower levels than those reported for marine species. Average concentrations for Trichoptera, Ephemeroptera, Diptera, Crustacea, molluscs and Hirudinea in an industrial-zone stream were 1.5, 5.9, 2.3, 2.2, 2.1, and 3.8 mg kg^{-1} dry weight, respectively (Anderson, 1977). Amphipods in the Rivers Werra and Weser (FRG) showed whole body residues of 0.2–1.1 mg kg^{-1} (Zauke, 1979) while, in a highly polluted section of the Savannah River (USA), cadmium levels in crayfish (*Procambarus* sp.) and carnivorous insects were 16.0 and

1.2 mg kg^{-1}, respectively (Guthrie and Cherry, 1979). Differences between residue levels in marine and freshwater biota are probably due to a shift in cadmium speciation as freshwater flows to the sea. Under oxygen-rich conditions, sediment suspensions released significantly greater amounts of cadmium as salinity increased. Similarly, De Groot and Allersma (1975) demonstrated that more than 90% of sediment-bound cadmium in the Rhine River (FRG) was released upon entering the North Sea, reflecting chloride and organic complexation. As outlined in Chapter 11, while managing cadmium waste discharge into freshwater, the concomitant change in chemical species in estuaries should be considered.

Maximum residues are found in the internal organs and specific cells of almost all species, reflecting the presence of cadmium-binding metallothioneins in major organs. Transport to the different tissues occurs mainly through the haemolymph. Depending on species, cadmium concentrations in specific organs may either increase or decrease with temperature and reproductive activity.

Uptake of cadmium may occur through water, food, or both. There is also significant sorption from sediments by benthic organisms. Uptake generally depends on the duration and intensity of exposure, and the presence of chelators in solution. The presence of zinc may lead to an increase in cadmium concentration in internal organs and depression in muscle tissue. Low temperatures at high salinities had no effect on sorption by the mussel *Mytilus edulis* but low temperatures at low salinities reduced uptake (Phillips, 1976). Wright (1980) suggested that calcium reduced the influx of cadmium into the amphipod *Gammarus pulex,* possibly through non-competitive inhibition.

Depuration is very slow in almost all species. Residues in experimentally contaminated shrimp *Pandalus montagui* showed no significant decline during a 75-day post-exposure period (Ray *et al.,* 1980). In contrast, cadmium residues in oysters *Crassostrea virginica* decreased by 71% in 44 days, reflecting the unusually high amount ($\sim$ 50%) of unbound metal in tissues, compared to a level of $<$10% in most other species (Mowdy, 1981).

Fish

Cadmium is accumulated primarily in major organ tissues of fish rather than in muscle. In experimentally contaminated perch, the distribution of cadmium in muscle, liver, kidney, gut, gills, bone, and skin was 1.2, 43.4, 1.6, 6.9, 11.3, 0.8 and 6.9%, respectively (remainder 27.8%) (Edgren and Notter, 1980). Although cadmium was not detected in the muscle of arctic char collected from an unpolluted high Arctic lake, concentrations of up to 2.3 mg kg^{-1} dry weight were found in livers (Bohn and Fallis, 1978). This latter study also showed that residues in the muscle and liver of shorthorn sculpins from a high arctic estuary averaged 1.4 and 4.1 mg kg^{-1} dry weight.

Because cadmium levels are normally low in edible muscle, accumulation is not a threat to most freshwater fishery resources. Walsh *et al.* (1977) found that only 0–4% of samples collected from throughout the USA contained concentrations of > 0.3 mg kg^{-1} and that residues had decreased from 1971 to 1973. There are a number of cadmium hot-spots in various industrialized regions of the world. Examples include the Severn Estuary (UK) where whiting contain up to 2.5 mg kg^{-1} in their muscle tissue (Badsha and Sainsbury, 1977) and NE coastal waters of the UK, where cod and flatfish have average burdens of 1.3 and 1.4 mg kg^{-1}, respectively (Wright, 1976). As discussed in the previous section, these relatively high values reflect high loading and the increased availability of cadmium in salt water.

In general, residues in fish muscle cannot be related to concentrations in water. In addition there is often no correlation between feeding habits and tissue levels in a wide range of teleost species. Although body burdens may increase with the size and age of fish, there are many reported exceptions to this relation. Water hardness did not influence the uptake rate of cadmium by nine species of freshwater fish (Wiener and Giesy, 1979), but the presence of the chelator EDTA significantly reduced sorption by carp (Muramoto, 1980). In laboratory experiments sorption in perch and presumably other species depended directly on temperature (Figure 3-5). By contrast, cadmium levels in whiting from the Severn Estuary were statistically similar regardless of time of year whereas those of five-bearded rockling increased in winter (Badsha and Sainsbury, 1977). Because of the highly variable and inconsistent nature of cadmium uptake, bio-monitoring programs involving fish have to be carefully designed to reflect ambient contamination levels.

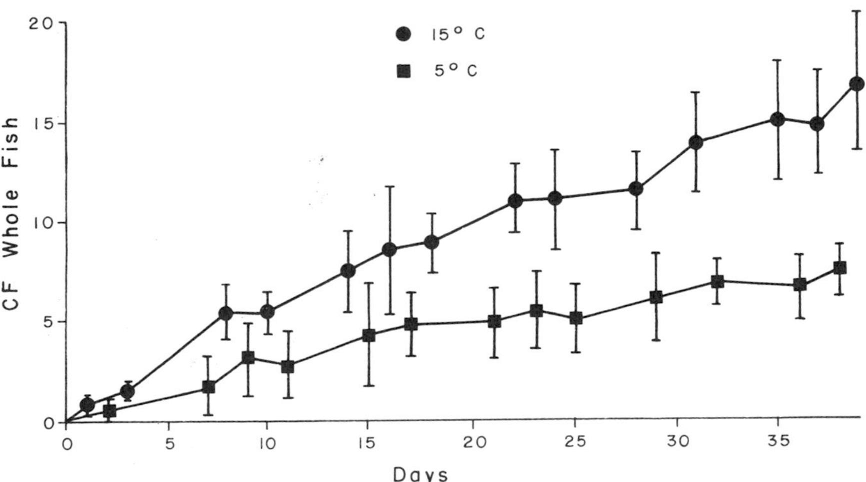

Figure 3-5. Accumulation of cadmium by fingerling yellow perch at different temperatures. (From Edgren and Notter, 1980.)

Toxicity

Aquatic Plants

Cadmium is usually less toxic to plants than methylmercury and copper, and similar in toxicity to lead, nickel and chromium(+3). Depending on species and test conditions, inhibition of growth and photosynthesis generally occurs at 0.02–1.0 mg L^{-1}. However, chronic effects have been reported for concentrations as low as 1 μg L^{-1}. The toxicity of cadmium may be attributed to the displacement of zinc from carrier proteins; consequently, the addition of zinc to media significantly reduces cell loss. Selenium may act as a synergist to cadmium, whereas cadmium may inhibit copper toxicity. Moshe *et al.* (1972) observed no synergistic effect in *Chlorella* when Cd, Cu, Cr, and Ni were added to culture media. Although low levels ($<$1.0 mg L^{-1}) of lead increase the toxicity of cadmium (0.1 mg L^{-1}) to phytoplankton, antagonism occurs when the concentration of lead exceeds that of cadmium. Pretreatment of algae with metals (Ni, Hg) reduces cadmium toxicity; this may reflect competition among metals for cellular binding sites.

Toxicity to duckweed (*Lemna*) was independent of temperature within the range 20–30° C (Nasu and Kugimoto, 1981). In many other species, toxicity probably decreases with temperature, though there are few data to confirm this point. Increasing pH usually increases mortality in a wide range of species.

Exposure of three species of green algae to $CdCl_2$ resulted in the formation of intramitochondrial granules containing cadmium (Silverberg, 1976). Swelling, vacuolization and degeneration of mitochondria were observed, implying significant cytotoxicity. By contrast, Li (1978) suggested that the algal membrane of *Thalassiorsira* was the main site of cadmium action. Although exposure of algae to heavy metals often results in the loss of cellular potassium, Overnell (1975) could find no evidence of K depletion in two species of marine algae, despite the fact that photosynthesis had been inhibited. Cadmium concentrations of 0.01–0.1 mg L^{-1} also reduce the concentration of ATP and chlorophyll in many species, and decrease oxygen production.

Invertebrates

LC_{50} values for freshwater invertebrates vary from 0.003 to $<$0.5 mg L^{-1} (Table 3-11). Although maximum toxicity is often associated with poorly buffered water of low pH, high LC_{50} values have been recorded for several common species when water hardness was $<$50 mg L^{-1} (Table 3-11). Wright and Frain (1981b) demonstrated that calcium had an antagonistic effect on cadmium toxicity to the amphipod *Gammarus pulex;* consequently, the post-molt animals were much more sensitive to cadmium than intermolt specimens. Similarly the 48h LC_{50} for the oligochaete *Tubifex tubifex*

Table 3-11. Acute toxicity of cadmium to some freshwater invertebrates.

Species	Toxicity* (mg L^{-1})	Temp (°C)	pH	Total hardness (mg L^{-1})	Dissolved oxygen (mg L^{-1})
Philodina acuticornis (rotifer)[1]	0.2–0.5	20	7.4–7.9	25	NQ
Nais sp. (oligochaete)[2]	1.7	17	7.6	50	6.2
Tubifex tubifex (oligochaete)[3]	0.003 (48h LC_{50})	20	6.3	0.1	NQ
	0.031 (48h LC_{50})	20	6.85	34.2	NQ
	0.045 (48h LC_{50})	20	7.2	34.2	NQ
	0.72 (48h LC_{50})	20	7.3	261	NQ
Physa integra (gastropod)[4]	0.010 (28-day LC_{50})	15	7.1–7.7	44–48	10–11
Physa gyrina (gastropod)[5]	0.43	20–22	6.7	200	10–14
Amnicola sp. (gastropod)[2]	8.4	17	7.6	50	6.2
Daphnia hyalina (cladoceran)[7]	0.055 (48h LC_{50})	10	7.2	NQ	NQ
Cyclops sp. (copepod)[8]	0.34	23	NQ	20	NQ
Cyclops abyssorum (copepod)[7]	3.8 (48h LC_{50})	10	7.2	NQ	NQ
Eudiaptomus padanus (copepod)[7]	0.55 (48h LC_{50})	10	7.2	NQ	NQ
Gammarus pulex (amphipod)[9]	0.12	10	NQ	NQ	NQ
Ephemerella sp. (mayfly)[4]	<0.003 (28-day LC_{50})	15	7.1–7.7	44–48	10–11
Tanytarsus dissimilis (dipteran)[6]	0.004	22	7.5	46.8	8.7
Chironomus sp. (dipteran)[2]	1.2	17	7.6	50	6.2

Sources: [1]Buikema *et al.* (1974); [2]Rehwoldt *et al.* (1973); [3]Brković-Popović and Popović (1977); [4]Spehar *et al.* (1978); [5]Wier and Walter (1976); [6]Anderson *et al.* (1980); [7]Baudouin and Scoppa (1974); [8]Fennikoh *et al.* (1978); [9]Wright and Frain (1981b).

* Toxicity expressed as 96h LC_{50}, unless otherwise indicated.

NQ—not quoted.

Table 3-12. Acute toxicity of cadmium to some marine invertebrates.

Species	Toxicity* (mg L^{-1})	Temp (°C)	pH	Salinity ($^{0}/oo$)	Dissolved oxygen (mg L^{-1})
Nereis diversicolor (polychaete)[1]	100.0 (192h LC_{50})	13	NQ	17.5	NQ
Nereis virens (polychaete)[2]	11.0	20	8.0	20	NQ
Mya arenaria (softshell clam)[3]	0.85	22	7.95	30	NQ
Crassostrea virginica (oyster)[4]	3.8(48h LC_{50})	26	7.0–8.5	25	NQ
Molluscs (4 species)[2]	2.2–25.0	20	8.0	20	NQ
Mysidopsis bahia (mysid)[5]	0.016	25–28	NQ	15–25	NQ
Marinogammarus obtusatus (amphipod)[6]	3.5–13.3	10	NQ	NQ	NQ
Crustaceans (4 species)[2]	0.32–4.1	20	8.0	20	NQ
Paragrapus quadridentatus (crab)[7]	1.17	17	NQ	35	NQ
Paragrapus gaimardii (crab)[8]	22.4	5	8.0	8.6	>95% sat.
	101.9	5	8.0	34.6	>95%
	15.7	19	8.0	8.6	>95%
	34.3	19	8.0	34.6	>95%
Asterias forbesi (starfish)[2]	0.82	20	8.0	20	NQ

Sources: [1]Bryan (1976); [2]Eisler (1971); [3]Eisler (1977); [4]Calabrese *et al.* (1973); [5]Nimmo *et al.* (1978); [6]Wright and Frain (1981a); [7]Ahsanullah and Arnott (1978); [8]Sullivan (1977).

* Toxicity expressed as 96h LC_{50}, unless otherwise indicated.

NQ—not quoted.

increased about 240 times when pH and total hardness were altered (Table 3-11). Susceptibility to intoxication also varies directly with metabolic activity; hence, sensitivity should vary seasonally with water temperature and dissolved oxygen levels. Although acute toxicity is also species dependent, it is difficult to determine which group of organisms are consistently sensitive to cadmium. This is partially related to the variable nature of test conditions on LC_{50} data.

Toxicity of cadmium to estuarine invertebrates varies by a factor of at least 5000 (Table 3-12). Crustaceans appear to be most sensitive, followed by molluscs and polychaetes (Table 3-12). As with freshwater invertebrates, calcium has a sparing action on cadmium toxicity. Increasing salinity also reduces toxicity to many estuarine species, whereas low temperatures enhance survival in most organisms. Cycling temperatures had a stimulating effect on the survival of crab larvae compared to constant temperatures (Rosenberg and Castlow, 1976). This implies that there is a degree of metabolic control over toxic effects.

Several species have the ability to adapt and survive in mildly contaminated media. This is probably related to the binding of cadmium to metallothioneins, which in turn act as a detoxification system. Exposure of invertebrates to cadmium may lead to the synthesis of low molecular weight cadmium-binding proteins. Some invertebrate species also have the ability to detect and avoid water with very low (0.001 mg L^{-1}) cadmium levels.

Fish

Acute 96h LC_{50} values vary from 0.09 to 105 mg L^{-1} for freshwater fish and 8 to 85 mg L^{-1} for estuarine species. In general, juveniles are more sensitive to cadmium than either the adult or egg stage. In chinook salmon and rainbow trout, newly-hatched alevins were relatively resistant to intoxication whereas the swim-up/parr stages were most sensitive (Table 3-13). It is

Table 3-13. Acute toxicity of cadmium to different stages of development in rainbow trout and chinook salmon[1].

Stage	Rainbow trout 96h LC_{50} ($\mu g\ L^{-1}$)	Chinook salmon 96h LC_{50} ($\mu g\ L^{-1}$)
Alevin	>27	>26
Swim-up	1.3	1.8
Parr	1.0	3.5
Smolt	2.9	2.9

Source: Chapman (1978).

[1] pH 7.1, total hardness 23 mg L^{-1}, dissolved oxygen 10.2 mg L^{-1}.

also known that reproductive activity in some species increases susceptibility. Such findings imply that the development of effluent criteria and the decision to discharge wastes should consider life cycle patterns of indigenous biota.

Water hardness plays a role in determining the toxicity of cadmium to adults and larvae. For example, the 48h LC_{50} for adult rainbow trout increased from 0.09 to 3.70 mg L^{-1} as hardness increased from 20 to 320 mg L^{-1} (Calamari *et al.*, 1980). It was also shown that trout acclimated to hard water retained some resistance to cadmium intoxication when tested in soft water. Carroll *et al.* (1979) concluded that the major characteristics of most hard waters were also the most important source of protection against cadmium toxicity. Magnesium ion, sulfate ion, sodium ion, and the carbonate system provided little or no protection. High salinity also usually reduces toxicity to estuarine species. In experiments with adult mummichog, the 192h LC_{50} was 15 mg L^{-1} at 5‰ and 28 mg L^{-1} at 35‰ (Eisler, 1971).

The presence of chelators and zinc in water reduces toxicity of cadmium to fish, but the addition of cadmium to a mixture of zinc and copper results in a significant increase in mortality. Although low dissolved oxygen levels probably increase toxicity to many species, Voyer *et al.* (1975) reported that survival of mummichog was similar regardless of oxygen concentration within the range of 3.5 to 9.4 mg L^{-1}.

Treatment of fish with cadmium generally reduces their ability to osmoregulate. While Kumada *et al.* (1980) reported no adverse effect on survival, growth, and tissue structure of rainbow trout exposed to concentrations of 4 μg Cd L^{-1} for 10 weeks, liver and kidney enzyme activity decreased following chronic exposures. In brook trout, hemorrhagic necrosis of testes may occur at 1 μg Cd L^{-1} (Sangalang and Freeman, 1974). This in turn retarded the rate of maturation and production of testosterone and 11-ketotestosterone. Mummichog exposed to 28 mg Cd L^{-1} showed necrosis and sloughing of mucosa of respiratory epithelium of gill filaments and lamellae (Voyer *et al.*, 1975). This accounts for the decrease in oxygen consumption and increase in ventilation frequency in chronically exposed fish.

Humans

There are few recorded instances of cadmium poisoning in humans following consumption of contaminated fish or water. The most significant example of intoxication, Itai-itai disease, was diagnosed in residents of Toyama Prefecture (Japan) from the 1940's to the 1960's. Untreated metal mine wastes had been discharged into local rivers (Friberg *et al.*, 1974). Potable water was heavily contaminated, and was also used to irrigate rice fields, which in turn became heavily contaminated. Patients suffering from Itai-itai disease showed signs of osteomalacia in bones and calcification and pyelonephritis in kidneys. This resulted in skeletal deformation and renal dysfunction. Cadmium-induced kidney damage was not reported in any other sector of the population (Yosumura *et al.*, 1980).

Because cadmium accumulates in organs and has a long half-life (10–30 years), the ingestion of small amounts of contaminated fish over long periods may lead to some form of cadmium intoxication. Consequently, regulatory standards restrict consumption of fish with residues of >0.5 mg kg^{-1} wet weight. This in turn implies that the liver and other organs of most fish are not fit for human consumption.

The teratogenic and embryotoxic effects of cadmium have been documented for a number of animal species. Gale and Layton (1980) administered a single dose (2 mg kg^{-1}) of cadmium sulfate to hamsters on the 8th day of gestation and observed the following abnormalities in the embryos: fetal resorption, exencephaly, cleft lip, fused ribs, hydrocephalus, and absence of digits. It has been suggested that cadmium may lead to cellular damage in fetal vascular endothelium, thereby decreasing utero-placental blood flow; fetal death may then be a result of anoxia or lack of essential nutrients. Machemer and Lorke (1981) found that cadmium chloride administered orally was toxic to female rats and significantly reduced fetal and placenta weight, and increased the occurrence of stunted forms and other malformations.

Several epidemiological studies have demonstrated a causal relationship between exposure to cadmium and cancer incidence. Kjellström *et al.* (1979) found the number of deaths due to prostatic cancers to be significantly higher than expected among workers in a cadmium alloy factory. It was also shown that the observed number of cases of nasopharynx cancers was significantly higher than expected in workers from a cadmium-nickel battery factory. Lemen *et al.* (1976) examined the histories of 292 male cadmium factory workers who had worked at least two years in the plant. The observed number of malignant neoplasms causing death was 27, as compared to the expected number of 17.6. Most of these were associated with the respiratory system. Cancer risk increased with the length of employment.

References

Abo-Rady, M.D.K. 1980. Aquatic macrophytes as indicator for heavy metal pollution in the River Leine (West Germany). *Archiv fuer Hydrobiologie* **89**:387–404.

Ahsanullah, M., and G.H. Arnott. 1978. Acute toxicity of copper, cadmium, and zinc to larvae of the crab *Paragrapsus quadridentatus* (H. Milne Edwards), and implications for water quality criteria. *Australian Journal of Marine and Freshwater Research* **29**:1–8.

Anderson, R.L., C.T. Walbridge, and J.T. Fiandt. 1980. Survival and growth of *Tanytarsus dissimilis* (Chironomidae) exposed to copper, cadmium, zinc, and lead. *Archives of Environmental Contamination and Toxicology* **9**:329–335.

Anderson, R.V. 1977. Concentration of cadmium, copper, lead and zinc in six species of freshwater clams. *Bulletin of Environmental Contamination and Toxicology* **18**:492–496.

Asami, T. 1974. Environmental pollution by cadmium and zinc discharged from a

Braun Tube factory. *Ibaraki Daigaku Nogakubu Gakujutsu Hokoku (Japan)* **22**:19–23.

Badsha, K.S., and M. Sainsbury. 1977. Uptake of zinc, lead and cadmium by young whiting in the Severn estuary. *Marine Pollution Bulletin* **8**:164–166.

Bartelt, R.D., and U. Förstner. 1977. Schwermetalle im staugeregleten Neckar. Untersuchungen an sedimenten, algen und wasserproben. *Jahresber. Mitt. Oberrheinischen Geo. Ver.* **59**:247–263.

Baudouin, M.F., and P. Scoppa. 1974. Acute toxicity of various metals to freshwater zooplankton. *Bulletin of Environmental Contamination and Toxicology* **12**:745–751.

Bohn, A., and B.W. Fallis. 1978. Metal concentrations (As, Cd, Cu, Pb, and Zn) in shorthorn sculpins, *Myoxocephalus scorpius* (Linnaeus) and Arctic char, *Salvelinus alpinus* (Linnaeus), from the vicinity of Strathcona Sound, Northwest Territories. *Water Research* **12**:659–663.

Bower, P.M., H.J. Simpson, S.C. Williams, and Y.H. Li. 1978. Heavy metals in the sediments of Foundry Cove, Cold Spring, New York. *Environmental Science and Technology* **12**:683–692.

Boyle, E. A., F. Sclater, and J.M. Edmond. 1976. On the marine geochemistry of cadmium. *Nature* **263**:42–44.

Brković-Popović, I., and M. Popović. 1977. Effects of heavy metals on survival and respiration rate of tubificid worms: Part I—Effects on survival. *Environmental Pollution* **13**:65–72.

Bruland, K.W., G.A. Knauer, and J.H. Martin. 1978. Cadmium in Northeast Pacific waters. *Limnology and Oceanography* **23**:618–625.

Bryan, G.W. 1976. Some aspects of heavy metal tolerance in aquatic organisms. *In*: A.P.M. Lockwood (Ed.), *Effects of pollutants on aquatic organisms.* Cambridge University Press, Cambridge, pp. 7–34.

Buikema, A.L., J. Cairns, and G.W. Sullivan. 1974. Evaluation of *Philodina acuticornis* (Rotifera) as a bioassay organism for heavy metals. *Water Resources Bulletin* **10**:648–661.

Butterworth, J., P. Lester, and G. Nickless. 1972. Distribution of heavy metals in the Severn estuary. *Marine Pollution Bulletin* **3**:72–74.

Calabrese, A., R.S. Collier, D.A. Nelson, and J.R. MacInnes. 1973. The toxicity of heavy metals to embryos of the American oyster *Crassostrea virginica. Marine Biology* **18**:162–166.

Calamari, D., R. Marchetti, and G. Vailati. 1980. Influence of water hardness on cadmium toxicity to *Salmo gairdneri* Rich. *Water Research* **14**:1421–1426.

Carroll, J.J., S.J. Ellis, and W.S. Oliver. 1979. Influences of hardness constituents on the acute toxicity of cadmium to brook trout *(Salvelinus fontinalis). Bulletin of Environmental Contamination and Toxicology* **22**:575–581.

Cawse, P.A. 1974–1977. *A survey of atmospheric trace elements in the U.K.* Results for 1972–73, 1974, 1975, 1976. AERE Publ. R-7669, R-8038, R-8393, R-8869. Her Majesty's Stationery Office, London.

Chapman, G.A. 1978. Toxicities of cadmium, copper and zinc to four juvenile stages of chinook salmon and steelhead. *Transactions American Fisheries Society* **107**:841–847.

Chau, Y.K., and Shiomi, M.T. 1972. Complexing properties of Nitrilotriacetic acid in the lake environment. *Water, Air, and Soil Pollution* **1**:149–164.

Chen, K.Y., C.S. Young, T.K. Jan, and N. Rohatgi. 1974. Trace metals in wastewater effluents. *Journal Water Pollution Control Federation* **46**:2663–2675.

Chester, R., and J.H. Stoner. 1974. Distribution of zinc, nickel, manganese, cadmium, copper, and iron in some surface waters from the world ocean. *Marine Chemistry* **2**:17–32.

De Groot, A.J., and E. Allersma. 1975. Field observations on the transport of heavy metals in sediments. *Progress in Water Technology* **7**:85–95.

Eaton, A. 1976. Marine geochemistry of cadmium. *Marine Chemistry* **4**:141–154.

Edgren, M., and M. Notter. 1980. Cadmium uptake by fingerlings of perch *(Perca fluviatilis)* studied by Cd-115m at two different temperatures. *Bulletin of Environmental Contamination and Toxicology* **24**:647–651.

Eisler, R. 1971. Cadmium poisoning in *Fundulus heteroclitus* (Pisces: Cyprinodontidae) and other marine organisms. *Journal of the Fisheries Research Board of Canada* **28**:1225–1234.

Eisler, R. 1977. Acute toxicities of selected heavy metals to the softshell clam *Mya arenaria. Bulletin of Environmental Contamination and Toxicology* **17**:137–145.

Elzerman, A.W., and D.E. Armstrong. 1979. Enrichment of Zn, Cd, Pb, and Cu in the surface microlayer of Lakes Michigan, Ontario, and Mendota. *Limnology and Oceanography* **24**:133–144.

Fennikoh, K.B., H.I. Hirshfield, and T.J. Kneip. 1978. Cadmium toxicity in planktonic organisms of a freshwater food web. *Environmental Research* **15**:357–367.

Flinn, J.E., and R.A. Reiners. 1977. U.S. Environmental Protection Agency Publication No. EPA-450/1-77-003.

Franzin, W.G., and G.A. McFarlane. 1980. An analysis of the aquatic macrophyte, *Myriophyllum exalbescens,* as an indicator of metal contamination of aquatic ecosystems near a base metal smelter. *Bulletin of Environmental Contamination and Toxicology* **24**:597–605.

Friberg, L., M. Piscator, G.F. Nordberg, T. Kjellström, and P. Boston. 1974. *Cadmium in the environment.* Second Edition. CRC Press, Cleveland, Ohio, 248 pp.

Gächter, R., and W. Geiger. 1979. Melimex, an experimental heavy metal pollution study: Behaviour of heavy metals in an aquatic chain. *Schweizerische Zeitschrift fuer Hydrologie* **41**:277–290.

Gale, T.F., and W.M. Layton. 1980. The susceptibility of inbred strains of hamsters to cadmium-induced embryotoxicity. *Teratology* **21**:181–186.

Gardiner, J. 1974. The chemistry of cadmium in natural water. I. A study of cadmium complex formation using the cadmium specific ion electrode. *Water Research* **8**:23–30.

Guthrie, R.K., and D.S. Cherry. 1979. Trophic level accumulation of heavy metals in a coal ash basin drainage system. *Water Resources Bulletin* **15**:244–248.

Hahne, H.C.H., and W. Kroontje. 1973. Significance of pH and chloride concentration on behavior of heavy metal pollutants: mercury(II), cadmium(II), zinc(II), and lead(II). *Journal of Environmental Quality* **2**:444–450.

Harbridge, W., O.H. Pilkey, P. Whaling, and P. Swetland. 1976. Sedimentation in the lake of Tunis: a lagoon strongly influenced by man. *Environmental Geology* **1**:215–225.

Harding, J.P.C., and B.A. Whitton. 1978. Zinc, cadmium and lead in water, sediments and submerged plants of the Derwent Reservoir, northern England. *Water Research* **12**:307–316.

Harding, J.P.C., and B.A. Whitton. 1981. Accumulation of zinc, cadmium, and lead by field populations of *Lemanea. Water Research* **15**:301–319.

Hart, B.T., and S.H.R. Davies. 1981. Trace metal speciation in the freshwater and estuarine regions of the Yarra River, Victoria. *Estuarine, Coastal and Shelf Science* **12**:353–374.

Hart, B.A., P.E. Bertram, and B.D. Scaife, 1979. Cadmium transport by *Chlorella pyrenoidosa. Environmental Research* **18**:327–335.

Haug, A., S. Melsom, and S. Omang. 1974. Estimation of heavy metal pollution in two Norwegian fjord areas by analysis of the brown alga *Ascophyllum nodosum. Environmental Pollution* **7**:179–192.

Huey, C.W., F.E. Brinckman, W.P. Iverson, and S.O. Grim. 1975. Bacterial volatilization of cadmium. *In*: T.C. Hutchinson (Ed.), *Proceedings of 1st International Conference on Heavy Metals in the Environment,* Abstracts, University of Toronto Institute for Environmental Studies, Toronto, Canada.

Ishibashi, M., S. Ueda, and Y. Yamamoto. 1970. The chemical composition and the cadmium, chromium and vanadium contents of shallow-water deposits in Tokyo Bay. *Journal of the Oceanographical Society of Japan* **26**:189–194.

Iskandar, I.K., and D.R. Keeney. 1974. Concentration of heavy metals in sediment cores from selected Wisconsin lakes. *Environmental Science and Technology* **8**:165–170.

Kemp, A.L.W., and R.L. Thomas. 1976. Impact of man's activities on the chemical composition in the sediments of Lakes Ontario, Erie and Huron. *Water, Air, and Soil Pollution* **5**:469–490.

Kemp, A.L.W., and R.L. Thomas, C.I. Dell, and J.M. Jaquet. 1976. Cultural impact on the geochemistry of sediments in Lake Erie. *Journal of the Fisheries Research Board of Canada* **33**:440–462.

Kennedy, V.C. and M.J. Sebetich. 1976. Trace elements in Northern California streams. *In*: Geological Survey Research 1976, Washington, D.C., pp. 208–209.

Kjellström, T., L. Friberg, and B. Rahnster. 1979. Mortality and cancer morbidity among cadmium-exposed workers. *Environmental Health Perspectives* **28**:199–204.

Knauer, G., and J. Martin. 1973. Seasonal variations of cadmium, copper, manganese, lead and zinc in water and phytoplankton in Monterey Bay, California. *Limnology and Oceanography* **18**:597–604.

Kneip, T.J., G. Re, and T. Hernandez. 1974. Cadmium in an aquatic ecosystem: Distribution and effects. *In*: D.D. Hemphill (Ed.), *Trace substances in environmental health,* Volume **8**, University of Missouri, Columbia, pp. 172–177.

Kramer, J.R. 1976. Fate of atmospheric sulfur dioxide and related substances as indicated by chemistry of precipitation. Unpublished Report, Department of Geology, McMaster University, Hamilton, Ontario, Canada.

Kumada, H., S. Kimura, and M. Yokote. 1980. Accumulation and biological effects of cadmium in rainbow trout. *Bulletin of the Japanese Society of Scientific Fisheries* **46**:97–103.

Leatherland, T.M., and J.D. Burton. 1974. The occurrence of some trace metals in coastal organisms with particular reference to the Solent region. *Journal of the Marine Biological Association of the United Kingdom* **54**:457–468.

Lemen, R.A., J.S. Lee, J.K. Wagoner, and H.P. Blejer. 1976. Cancer mortality among cadmium production workers. *Annals of the New York Academy of Sciences* **271**:273–279.

Li, W.K.W. 1978. Kinetic analysis of interactive effects of cadmium and nitrate on growth of *Thalassiosira fluviatilis* (Bacillariophyceae). *Journal of Phycology* **14**:454–460.

Machemer, L., and D. Lorke. 1981. Embryotoxic effect of cadmium on rats upon oral administration. *Toxicology and Applied Pharmacology* **58**:138–143.

Mantoura, R.F.C., A. Dickson, and J.P. Riley. 1978. The complexation of metals with humic materials in natural waters. *Estuarine and Coastal Marine Science* **6**:387–408.

McIntosh, A.W., B.K. Shephard, R.A. Mayes, G.J. Atchison, and D.W. Nelson. 1978. Some aspects of sediment distribution and macrophyte cycling of heavy metals in a contaminated lake. *Journal of Environmental Quality* **7**:301–305.

Melhuus, A., K.L. Seip, H.M. Seip, and S. Myklestad. 1978. A preliminary study of the use of benthic algae as biological indicators of heavy metal pollution in Sørfjorden, Norway. *Environmental Pollution* **15**:101–107.

Moshe, M., N. Betzer, and Y. Kott. 1972. Effect of industrial wastes on oxidation pond performance. *Water Research* **6**:1165–1171.

Mowdy, D.E. 1981. Elimination of laboratory-acquired cadmium by the oyster *Crassostrea virginica* in the natural environment. *Bulletin of Environmental Contamination and Toxicology* **26**:345–351.

Muhlbaier, J., and G.T. Tisue. 1981. Cadmium in the southern basin of Lake Michigan. *Water, Air, and Soil Pollution* **15**:45–59.

Muramoto, S. 1980. Effect of complexans (EDTA, NTA and DTPA) on the exposure to high concentrations of cadmium, copper, zinc and lead. *Bulletin of Environmental Contamination and Toxicology* **25**:941–946.

Nasu, Y., and M. Kugimoto. 1981. *Lemna* (duckweed) as an indicator of water pollution. I. The sensitivity of *Lemna paucicostata* to heavy metals. *Archives of Environmental Contamination and Toxicology* **10**:159–169.

Nimmo, D.R., R.A. Rigby, L.H. Bahner, and J.M. Sheppard. 1978. The acute and chronic effects of cadmium on the estuarine mysid, *Mysidopsis bahia*. *Bulletin of Environmental Contamination and Toxicology* **19**:80–85.

Nriagu, J.O. 1979. Global inventory of natural and anthropogenic emissions of trace metals to the atmosphere. *Nature* **279**:409–411.

Olausson, E., O. Gustafsson, T. Mellin, and R. Svensson. 1977. Current level of heavy metal pollution and eutrophication in the Baltic proper. *Medd Marinegeol. Labor, Göteborg* **9**:28 pp.

Overnell, J. 1975. The effect of some heavy metal ions on photosynthesis in a freshwater alga. *Pesticide Biochemistry and Physiology* **5**:19–26.

Phillips, D.J.H. 1976. The common mussel *Mytilus edulis* as an indicator of pollution by zinc, cadmium, lead and copper. II. Relationship of metals in the mussel to those discharged by industry. *Marine Biology* **38**:71–80.

Phillips, D.J.H. 1979. Trace metals in the common mussel, *Mytilus edulis* (L.), and in the alga *Fucus vesiculosus* (L.) from the region of the Sound (Öresund). *Environmental Pollution* **18**:31–43.

Ramamoorthy, S., and D.J. Kushner. 1975a. Heavy metal binding sites in river water. *Nature* **256**:399–401.

Ramamoorthy, S., and D.J. Kushner. 1975b. Heavy metal binding components of river water. *Journal of the Fisheries Research Board of Canada* **32**:1755–1766.

Ramamoorthy, S., and B.R. Rust. 1978. Heavy metal exchange processes in sediment-water systems. *Environmental Geology* **2**:165–172.

Ray, S., and W. White. 1976. Selected aquatic plants as indicator species for heavy metal pollution. *Journal of Environmental Science and Health. Part A* **11**:717–725.

Ray, S., D.W. McLeese, B.A. Waiwood, and D. Pezzack. 1980. The disposition of cadmium and zinc in *Pandalus montagui*. *Archives of Environmental Contamination and Toxicology* **9**:675–681.

Rehwoldt, R., L. Lasko, C. Shaw, and E. Wirhowski. 1973. The acute toxicity of some heavy metal ions toward benthic organisms. *Bulletin of Environmental Contamination and Toxicology* **10**:291–294.

Ridley, W.P., L.J. Dizikes, and J.M. Wood. 1977. Biomethylation of toxic elements in the environment. *Science* **197**:329–332.

Rosenberg, R., and J.D. Castlow, Jr. 1976. Synergistic effects of cadmium and salinity combined with constant and cycling temperatures on the larval development of two estuarine crab species. *Marine Biology* **38**:291–303.

Roth, I., and H. Hornung. 1977. Heavy metal concentrations in water, sediments and fish from Mediterranean coastal area, Israel. *Environmental Science and Technology* **11**:265–269.

Ruppert, H. 1975. Geochemical investigations on atmospheric precipitation in a medium-sized city (Göttingen, F.R.G.). *Water, Air, and Soil Pollution* **4**:447–460.

Sangalang, G.B., and H.C. Freeman. 1974. Effects of sublethal cadmium on maturation and testosterone and 11-ketotestosterone production *in vivo* in brook trout. *Biology of Reproduction* **11**:429–435.

Schafer, H.A., and W. Bascom. 1976. Sludge in Santa Monica Bay. *Annual Report, Southern California Coastal Water Research Project,* pp. 77–82.

Semkin, R.G., and J.R. Kramer. 1976. Sediment geochemistry of Sudbury-area lakes. *Canadian Mineralogist* **14**:73–90.

Shephard, B.K., A.W. McIntosh, G.J. Atchison, and D.W. Nelson. 1980. Aspects of the aquatic chemistry of cadmium and zinc in a heavy metal contaminated lake. *Water Research* **14**:1061–1066.

Silverberg, B.A. 1976. Cadmium-induced ultrastructural changes in mitochondria of freshwater green algae. *Phycologia* **15**:155–159.

Spehar, R.L., R.L. Anderson, and J.T. Fiandt. 1978. Toxicity and bioaccumulation of cadmium and lead in aquatic invertebrates. *Environmental Pollution* **15**:195–208.

Steele, J.H., A.D. McIntyre, R. Johnston, I.G. Baxter, G. Topping, and H.D. Dooley. 1973. Pollution studies in the Clyde Sea area. *Marine Pollution Bulletin* **4**:153–157.

Stenner, R.D., and G. Nickless. 1974. Absorption of cadmium, copper and zinc by dog whelks in the Bristol Channel. *Nature* **247**:198–199.

Sullivan, J.K. 1977. Effects of salinity and temperature on the acute toxicity of cadmium to the estuarine crab *Paragrapsus gaimardii* (Milne Edwards). *Australian Journal of Marine and Freshwater Research* **28**:739–743.

Tessier, A., P.G.C. Campbell, and M. Bisson. 1980. Trace metal speciation in the Yamaska and St. François Rivers (Quebec). *Canadian Journal of Earth Sciences* **17**:90–105.

Thornton, I., H. Watling, and A. Darracott. 1975. Geochemical Studies in several rivers and estuaries used for oyster rearing. *The Science of the Total Environment* **4**:325–345.

Thornton, J.D., S.J. Eisenreich, J.W. Munger, and G. Gorham. 1981. Trace metal and strong acid composition of rain and snow in northern Minnesota. *In*: S.J. Eisenreich (Ed.), *Atmospheric pollutants in natural waters.* Ann Arbor Science, Ann Arbor, Michigan, pp. 261–284.

Tourtelot, H., C. Huffman, and L. Rader. 1964. Cadmium in samples of the Pierre Shale and some equivalent stratigraphic units, Great Plains Region. U.S. Geological Survey Professional Paper 475-D.

United States Minerals Yearbooks. 1911 – 1979. Bureau of Mines, US Department of the Interior, Washington, D.C.

Vogt, G., and F. Kittelberger. 1977. Study about uptake and accumulation of heavy metals in typical algae associations of the Rhine between Germersheim and Gernsheim. *Fisch und Umwelt Heft* **3**:15 – 18.

Voyer, R.A., P.P. Yevich, and C.A. Barszcz. 1975. Histological and toxicological responses of the mummichog, *Fundulus heteroclitus* (L.) to combinations of levels of cadmium and dissolved oxygen in a freshwater. *Water Research* **9**:1069 – 1074.

Walsh, D.F., B.L. Berger, and J.R. Bean. 1977. Mercury, arsenic, lead, cadmium, and selenium residues in fish, 1971 – 73 – National Pesticide Monitoring Program. *Pesticides Monitoring Journal* **11**:5 – 34.

Wentsel, R.S., and J.W. Berry. 1974. Cadmium and lead levels in Palestine Lake, Palestine, Indiana. *Proceedings Indiana Academy of Science* **84**:481 – 490.

Wiener, J.G., and J.P. Giesy, Jr. 1979. Concentrations of Cd, Cu, Mn, Pb, and Zn in fishes in a highly organic softwater pond. *Journal of the Fisheries Research Board of Canada* **36**:270 – 279.

Wier, C.F., and W.M. Walter. 1976. Toxicity of cadmium in the freshwater snail, *Physa gyrina* Say. *Journal of Environmental Quality* **5**:359 – 362.

Wood, J.M. 1974. Biological cycles for toxic elements in the environment. *Science* **183**:1049 – 52.

Wright, D.A. 1976. Heavy metals in animals from the north east coast. *Marine Pollution Bulletin* **7**:36 – 38.

Wright, D.A. 1980. Cadmium and calcium interactions in the freshwater amphipod *Gammarus pulex. Freshwater Biology* **10**:123 – 133.

Wright, D.A., and J.W. Frain. 1981a. Cadmium toxicity in *Marinogammarus obtusatus:* effect of external calcium. *Environmental Research* **24**:338 – 344.

Wright, D.A., and J.W. Frain. 1981b. The effect of calcium on cadmium toxicity in the freshwater amphipod, *Gammarus pulex* (L.) *Archives of Environmental Contamination and Toxicology* **10**:321 – 328.

Yost, K.J. 1978. Some aspects of the environmental flow of cadmium in the United States. *Cadmium 77: Proceedings of the 1st Cadmium Conference.* Metal Bulletin Ltd., London, pp. 147 – 166.

Yosumura, S., D. Vartsky, K.J. Ellis, and S.H. Cohn. 1980. Cadmium in human beings. *In*: J.O. Nriagu (Ed.), *Cadmium in the environment. Part I. Ecological cycling.* Wiley, New York, pp. 12 – 34.

Zauke, G.P. 1979. Cadmium in Gammaridae (Amphipoda: Crustacea) of the rivers Werra and Weser. I. Evaluation of the sources of variance. *Chemosphere* **10**:769 – 775.

4
Chromium

Chemistry

Chromium is a white, hard and lustrous metal, melting at 1860°C. Many corrosive reagents render the metal passive and hence chemically inert. This accounts for its extensive use as an electroplated protective coating. However it is fairly active when not passivated and dissolves in non-oxidizing mineral acids. Chromium(+6) exists only as oxy species such as CrO_3, CrO_4^{2-} and $Cr_2O_7^{2-}$ and is strongly oxidizing. Chromium(+5) and (+4) exist in transient states only to disproportionate to Cr(+3) and Cr(+6). A fair number of chromous (+2) compounds are known, all of which are strong and rapid reducing agents. The most important oxidation state of Cr is +3 which forms large numbers of kinetically inert complexes.

Chromium is found in small quantities in RNA of a few organisms. Metal ions that biological systems can use must be both abundant in nature and relatively soluble in water. Abundance restricts the available metals to those of atomic number less than 40. The heavy metal ions with the exception of manganese remain in fixed stereochemical positions in biomolecules, while sodium, potassium, magnesium and calcium participate in biochemical reactions as mobile cations. The virtual absence of nickel and chromium (except in RNA and niacin) from living organisms is probably due to lower stability of their protein complexes resulting from the irregular geometry of the protein chelating sites compared to the octahedral sites provided by soil silicates. Chromium(+3) and (+6) are classified as hard acids. Thus chromium is one of the least toxic of the trace elements, considering its oversupply and essentiallity. Mammalian bodies can tolerate 100–200 times their total body content of Cr without adverse effects. The stomach acidity reduces Cr(+6) to the much less toxic (+3) form whose gastrointestinal absorption is less than 1%.

Production, Uses, and Discharges

Production

Chromite is the only commercially important ore mineral for the production of chromium. The composition is $FeOCr_2O_3$ with a chromic oxide (Cr_2O_3) content of 68%. If pure chromium is not required (as for use in ferrous alloys) chromite is simply pyrolytically reduced with carbon to yield the carbon-containing alloy ferrochrome. When pure chromium is required, chromite is first oxidized in molten alkali to convert Cr(+3) to (+6), which is precipitated as sodium dichromate. This is then reduced first with carbon and then with aluminium to metallic chromium.

US production of chromium was about 205 metric tons per year between 1900–1936. However production did rise to 82,500 metric tons during the First World War. US production of chromium started to increase again in 1936 to about 275,000 metric tons in 1941. Similarly imports of chromite rose from 38,500 metric tons in 1910 to about a million metric tons in 1941. In the first half of this century, the Philippine Islands, Cuba, Southern Rhodesia, and the USSR were the leading producers in the world. In recent times, South Africa has become the largest chromite producer with 3.2 million metric tons in 1979. It is followed by the USSR, Rhodesia, India, and the Philippines. Approximately 205 million metric tons of chromite were produced in the last sixty years of the century (Table 4-1). Production

Table 4-1. Global production of chromite.

Period	Quantity (metric tons $\times 10^3$)
1921–1930	4280
1931–1940	11630
1941–1950	17220
1951–1960	38690
1961–1970	46550
1971–1980*	86200
Total	204570

Sources: US Minerals Yearbooks; Canadian Minerals Yearbooks.

* Estimated figures for the year 1980.

has almost doubled in the last decade reflecting the increasing consumption of chromium in a wide range of industrial applications.

Uses

The earliest use of chromium was in the making of paint pigment around 1800 in France, Germany and England. During the next 25 years, chromium compounds were employed in textile colouring and leather tanning processes. The refractory use of chromium in furnace linings was introduced in the late nineteenth century. The metallurgical importance of chromium was recognized around 1910–1915 and since then consumption of chromium has shown sustained growth. The three major industrial uses are metallurgical, refractory and chemical; their respective percent of the total chromite consumption in the past decade are 58, 21, and 21.

Metallurgical. Metallurgical grade chromite is used in the production of ferroalloys. The principal alloys are high and low carbon ferrochrome and ferrochrome–silicon. Presence of chromium in iron castings improves the resistance to corrosion and oxidation and also increases the ability to withstand stress at elevated temperatures. The ferrochrome alloys are widely employed in the production of stainless steel and heat-resisting steels, which are then used in corrosive environments, in petrochemical processing, turbine and furnace parts, cutlery and decorative trim, mechanical tools, jet engines, etc. An energy efficient method, called argon–oxygen decarburization, is now widely used in the production of low-carbon ferrochrome for the manufacture of heat-resistant stainless steels. Argon and oxygen added to the melt preferentially oxidize carbon instead of chromium from high-carbon ferro alloys.

Refractory. Chromite under this category is mainly employed in the manufacture of refractory bricks, mortars, ramming gun mixes, and high-temperature furnace repair material. Refractories, consisting of chromite and magnesite, are used in ferrous and non-ferrous metal industries. In the ferrous industry, electric furnace production is phasing out the basic open-hearth furnace operations. Still, electrical furnaces require the use of chromite refractories and the consumption is expected to stabilize in the next few years. In the non-ferrous industry, chromite–magnesite brick is used in converters. Use of high-temperature oxygen-blowing converters will necessitate a change to higher magnesite content bricks, thereby reducing chromite usage.

Chromite refractory bricks are employed in the reheating chambers of the glass industry and in the recovery furnaces to resist chemical attack by spent liquors. Chromite mortars and gunning mixes are used in bonding and coating of chemically different bricks. Ramming mixes are used in open-hearth furnaces.

Chemical. Most chromium chemicals are manufactured from the pure sodium dichromate obtained from chemical-grade chromite. Chromium compounds are used as: pigments, mordants and dyes in textile industry, tanning agents in leather industry, and for chrome electroplating, anodizing, and dipping. Chromium compounds are employed as oxidants and catalysts in the manufacture of products such as saccharin, in bleaching and purification of oils, fats and chemicals and as agents to increase the anti-wetting by water insolubility of various products such as glues, inks and gels.

US consumption of chromite ore and concentrates increased sharply during 1940–50 and remained stable in 1950–1974. In 1970, the consumption was about 1.3 million metric tons of chromite or 0.39 million metric tons of chromium. In 1975, consumption dropped to 0.8 million metric tons of chromite and since then stabilized at a level of 0.9 million metric tons. The consumption by metallurgical industries dropped from 65% in 1970 to 58% in 1977. Fluctuations in the steel industry will reflect on the consumption of chromium. However, chromium is irreplaceable in many applications, especially in certain stainless steels used in energy-producing operations.

Discharges

Vast quantities of chromite ore are used for the production of stainless steel, chrome-plated metals, pigments and various chemicals. The major contributors to airborne chromium are in the order of decreasing total emissions: ferrochrome production and handling, production of refractory bricks, coal combustion and chromium-steel production. Due to its high boiling point, chromium vapor condenses rapidly as oxide on the surface of the airborne particles. Up to 3000 mg kg^{-1} of chromium are present in fine particulates and are thus likely to penetrate deeply into the respiratory tract (Jaworski, 1980). Airborne chromium levels around chrome factories may exceed 1 mg m^{-3} compared to the natural background level of 10^{-6} mg m^{-3}. Depending on climatic conditions, these particulates may become windblown over long distances or deposited back to the lithosphere through precipitation as rain or snowfall. The enrichment factor, accumulation in atmospheric particles relative to earth's crust for chromium, is 11. Compared to iron, the atmospheric chromium burden arises relatively more from anthropogenic sources. In contrast, metals such as mercury, selenium, and antimony are enriched in the atmosphere by four orders of magnitude. Emissions from fossil-fuel burning are estimated to contribute about 1450 metric tons of chromium annually to the atmospheric load (Bertine and Goldberg, 1971).

The principal chromium emissions into surface waters are from metal-finishing processes such as electroplating, pickling, and bright dipping. Uncontrolled emissions have great potential for contaminating the fresh waters with the relatively toxic form, Cr(+6). Other smaller discharges of

Cr(+6) are from the additive in circulating cooling waters, laundry chemicals, and animal glue manufacture. Sources of Cr(+3) contamination include liquid waste discharges from leather tanning and textile dyeing containing up to several thousand mg L^{-1} of chromium (Jaworski, 1980).

Soil contamination by chromium includes land disposal of slags (containing 2–6% of Cr) as by-products of ferrochrome and chromium steel production or deliberate use as mineral fertilizers. Certain phosphate fertilizers also contain high levels of chromium, 10^2–10^4 mg kg^{-1} (Jaworski, 1980).

Chromium in Aquatic Systems

Speciation in Natural Waters

Chromium(+3) is the most stable and important oxidation state. It forms a large number of relatively kinetically inert complexes which can be isolated as solids. The complexes have greater kinetic than thermodynamic stability. Cr(+3) is hexacoordinate in its reactions: the well-known complexes are amines, $[CrAm_{6-n-m}(H_2O)_nR_m]^{(3-m)+}$ where Am = NH_3 or polydentate amine such as ethylenediamine, R = acido ligand such as halide, nitro or sulfate ion. Chromium(+3) forms hydroxo and oxo-bridged polynuclear complexes. Being a hard acid, Cr(+3) forms relatively strong complexes with oxygen donor ligands. Neutral ligands such as ethylenediamine bind very strongly, $\beta_2 = 10^{30.5}$ (Table 4-2).

Transport in Natural Waters

The two important oxidation states of chromium in natural waters are +3 and +6. In well-oxygenated waters, Cr(+6) is the thermodynamically stable species. However Cr(+3), being kinetically stable, could persist bound to naturally occurring solids. Interconversions of Cr(+3) and Cr(+6) occur in conditions similar to natural waters.

Cr(+6) is easily reduced by Fe(+2), dissolved sulfides, and certain organic compounds with sulfydryl groups. By contrast Cr(+3) is oxidized rapidly by a large excess of MnO_2 and slowly by oxygen at conditions approximating to natural waters. Based on these results of interconversions, it is desirable that water quality standards should be based on total Cr rather than on Cr(+6).

Municipal waste waters release considerable amounts of chromium into the environment. For example 600–700 metric tons of chromium at concentrations of 40–800 $\mu g\ L^{-1}$, are discharged annually into coastal waters by the Southern California municipal treatment plants (Young *et al.*, 1973; Schafer, 1976). Speciation studies showed that chromium occurred principally (67–98%) in the particulate state in the municipal waters (Jan and

Table 4-2. Stability constants of Cr(+3)-organic complexes.

Ligand	Formula	Donor atom(s)	Log β_1	Log β_2	Log β_3
Oxalic acid	$C_2H_2O_4$	O	5.34	10.51	15.44
Glycine	$C_2H_5O_2N$	O,N	8.62	16.27	
α-Alanine	$C_3H_7O_2N$	O,N	8.53	15.97	
Serine	$C_3H_7O_3N$	O,N	8.0	14.2	19.4
Succinic acid	$C_4H_6O_4$	O	6.42	10.99	13.85
Phthalic acid	$C_8H_6O_4$	O,O	5.52	10.00	12.48
Aspargine	$C_4H_8O_3N_2$	O,N	7.7	13.6	18.5
Ethylenediamine	$C_2H_8N_2$	N,N	16.5	<30.5	
Sulfoxine	$C_9H_7O_4NS$	O,N,S	10.99	21.04	
Ethylenediamine tetra-acetic acid	$C_{10}H_{10}O_8N_2$	O,N	23.40		

Source: Sillén and Martell (1971).

Young, 1978). The major portion (97-99%) of the dissolved fraction was the Cr(+3) form. The relatively more toxic Cr(+6) constituted less than 1% of the total chromium in the waste waters. Chlorination did not increase the concentrations of Cr(+6). Clean coastal waters contained predominantly Cr(+6); the median concentrations of Cr(+3) and Cr(+6) were 0.045 and 0.14 $\mu g\ L^{-1}$, respectively. In contrast, sub-surface sea water samples characterized by high amounts of particulates and ammonia-nitrogen resulting from waste water plume, contained particulate-chromium up to 100 times greater than control levels with no enhancement of Cr(+6) concentrations. In summary, the relatively high dissolved Cr(+6) in coastal water is of natural origin and the municipal waste discharge does not increase the concentration of this toxic Cr(+6) form in the sea water.

In freshwaters, anthropogenically introduced soluble Cr(+6) is removed by reduction to Cr(+3) and subsequent sorption to particulates and sediment (Pfeiffer *et al.*, 1980). The domestic waste input into the rivers causes a sharp drop in the dissolved oxygen content with hydrogen sulfide production. This reduced the Cr(+6) from 87% to 34% of the total chromium in solution. About 3-13% retention of Cr(+6) by particulates for a mass of 0.084 to 2.4 mg L^{-1} of suspended load was also observed.

Chromium is transported in rivers primarily in the solid phase: e.g., 51-36% of total chromium in the Iowa River (Shuman and Dempsey, 1977) and 85% in the Amazon and Yukon rivers (Gibbs, 1977). The proportion of Cr(+3) in the dissolved Cr fraction may vary from 34 to 65% (Pankow *et al.*, 1977) and 44 to 95% (Shuman and Dempsey, 1977). Extensive studies have been conducted on the fate of ^{51}Cr in the Columbia River (Canada, USA) because of the nuclear operations discharging irradiated dissolved chromium in the cooling water to the river (Kopp and Kroner, 1968; Evans and

Cutshall, 1973). These studies have shown that the water contained 3.2 nM dissolved Cr (≥ 90% as Cr(+6)) and 1.8 nM particulate Cr whereas estuarine samples contained 2.4 nM dissolved Cr and 2.8 nM particulate Cr. Flocculation processes increased the particulate Cr and Cr(+3) concentration in the salinity range 8–10%. This explains the net increase in the particulate Cr in the mid-estuary (Cranston and Murray, 1980).

Recent work on the speciation of chromium in the Pacific Ocean and Japan Sea (Nakayama *et al.*, 1981) reported neither Cr(+6) nor organic species coprecipitate with ferric hydroxide in sea water. It was claimed that earlier studies did not analyse for chromium organic species. Chromium in both seas was 10–20% inorganic Cr(+3), 25–40% Cr(+6) and 45–65% organic Cr species. The concentrations of dissolved versus particulate chromium were 0.42:0.07 $\mu g\ L^{-1}$ (Pacific Ocean) and 0.42:0.08 $\mu g\ L^{-1}$ (Japan Sea). However the vertical distribution of Cr(+6) was different between the two seas. In the Pacific Ocean, chromium increased at a depth of ≥ 1000 m while in the Japan Sea it decreased. The ratio of Cr(+6)/Cr(+3) inorganic averaged 2.7 in the Pacific Ocean and 1.8 for the Japan Sea. The difference in Cr(+6) distribution was due to the abundance of highly oxidizing manganese dioxide at considerable depth in the Pacific Ocean and lack of it in the Japan Sea. The following model was proposed for the circulation of chromium in sea water (Figure 4-1).

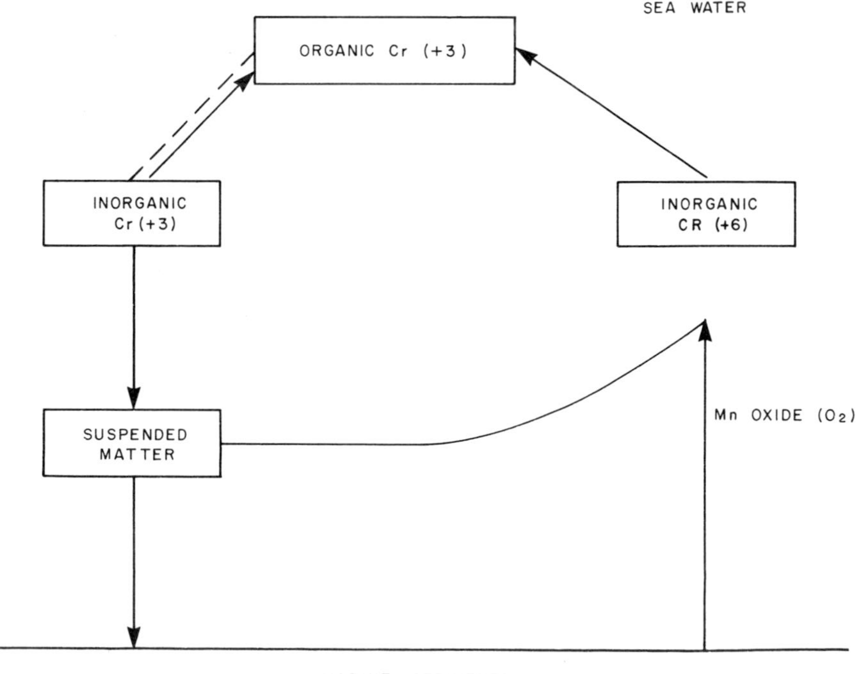

Figure 4-1. Circulation of chromium in sea water.

Residues

Water, Precipitation, and Sediments

There are few recorded instances of significant contamination of surface waters by chromium. Dissolved concentrations in unpolluted lakes and rivers generally vary from 1 – 2 μg L^{-1} compared with 0.05 – 0.5 μg L^{-1} for marine waters. Higher values (5 – 50 μg L^{-1}) have been reported for several major rivers, as they flow through industrial areas (Van der Veen and Huizenga, 1980; Wilbur and Hunter, 1977). This is usually due to the input of plating industry wastes into receiving waters. Concentrations in excess of 100 μg L^{-1} are seldom recorded, regardless of industry type.

Chromium residues in precipitation in Belgium ranged from 11 to 43 μg L^{-1} over a five year period (Kretzschmar *et al.,* 1980). Substantially lower values (1 – 2 μg L^{-1}) were reported for a town in New Jersey (USA), whereas meltwater from the Rhône glacier did not contain detectable amounts of chromium (Wilbur and Hunter, 1979; Lum-Shue-Chan, 1981). The relatively high rate of deposition in cities results in substantial contamination of soils. Andersen *et al.* (1978) for example reported average residues of 85 mg kg^{-1} for soils in Copenhagen, compared with 9.3 mg kg^{-1} for a town in New Jersey (Wilbur and Hunter, 1979). The high rate of loading in cities also contributes to the elevated concentrations of chromium (27 μg L^{-1}) in storm runoff and receiving waters (Wilbur and Hunter, 1979).

Elevated chromium levels from anthropogenic sources have been reported in sediments (Table 4-3). Sediment enrichment is correlated with fluxes of charcoal from different combustion processes: oil, coal, and wood burning. Installation of fly ash removal devices on stack emissions have led to slight decreases in Cr concentration in sediments since 1968. Sediments of the lower Saddle River (New York), receiving heavy urban discharge, showed a chromium enrichment of 510% (35 mg kg^{-1}) compared to control areas (Wilber and Hunter, 1979). Sequential extraction produced the following chromium fractionation: 0.5% in river water soluble fraction, 0.6% in ion-exchangeable fraction, 2.7% in easily reducible manganese oxide fraction, 28.3% in organic fraction bound to humic materials and 67.9% in easily reducible iron oxides. Only the first two fractions were determined to be bioavailable under aerobic conditions. For the total chromium of sediment levels of 2.90 mg kg^{-1}, this represented only 0.03 mg kg^{-1} of available chromium.

Total chromium levels of 8 – 241 mg kg^{-1} were reported for the sediments of the estuary and Gulf of St. Lawrence, Canada (Loring, 1979). About 2 – 11% of the total chromium was in the non-detrital fraction and potentially bioavailable. The remainder was held in the acetic acid insoluble detrital fraction, not readily bioavailable. Sedimentation of the detrital–chromium complex via chromite and magnetite has contributed to the enrichment of the sediments. Organic matter, mostly of terrestrial origin,

Table 4-3. Total chromium levels (mg kg^{-1} dry weight) in freshwater and marine sediments.

Location	Average (range)	Polluting source
Qishon River (Israel)[1]	191(3–650)	municipal, industrial
Illinois River (USA)[2]	17(2–87)	municipal, industrial
Saddle River (USA)[3]	832(530–1337)[a]	municipal
Thompson Lake (Canada)[4]	38(12–150)	gold mining
2 lakes (Netherlands)[5]	415(106–680)	mixed industrial
Lake Superior (Canada)[6]	48(26–60)	natural
Lake Huron (Canada)[6]	37(29–47)	natural
Baltimore Harbour (USA)[7]	76(–)	mixed industrial
Ems Estuary (Netherlands)[8]	96(84–107)	mixed industrial
Rhine Estuary (Netherlands)[8]	275(92–642)	mixed industrial
Bayou Chico Estuary (USA)[9]	–(17–174)	mixed industrial
Restronguet Estuary (UK)[10]	289(40–1060)	Pb–Zn mining

Sources: [1]Kronfeld and Navrot (1974); [2]Mathis and Cummings (1973); [3]Wilbur and Hunter (1979); [4]Moore (1980); [5]Salomons and Mook (1980); [6]Kemp *et al.* (1978); [7]Helz (1976); [8]Salomons and Mook (1977); [9]Pilotte *et al.* (1978); [10]Thornton *et al.* (1975).
[a] Particle size 0.01–0.15 μm.

and Fe–Mn oxide grain coatings transport the non-detrital-chromium to bottom sediments.

Chromium in Lake Ontario sediments was enriched in the humic and fulvic acids compared to the entrapping sediments (Nriagu and Coker, 1980). The enrichment factor was approximately 2. About 8–11% of total chromium in sediment was associated with the organic matter.

Aquatic Plants, Invertebrates and Fish

Chromium is not a significant contaminant of plant tissues, except at site-specific discharge points. Residues from freshwater industrial zone sources generally range up to 50 mg kg^{-1} dry weight and, in unpolluted areas, seldom exceed 5 mg kg^{-1}. Concentrations in marine plants are usually higher than those in freshwater, reflecting increased bioavailability of $CrCl_6$. Residues in 20 filamentous species from unpolluted Malayan waters ranged up to 58 mg kg^{-1} (Sivalingam, 1978). Although higher values (up to 140 mg kg^{-1}) were recorded for 19 species inhabiting Japanese coastal water (Gryzhankova *et al.,* 1973), levels of only 1–13 mg kg^{-1} occurred in *Ascophyllum nodosum* from Trondheimsfjorden, Norway (Lande, 1977).

Relatively little is known about factors influencing the uptake of chromium by aquatic plants. Sorption by marine grass occurred mainly through the roots, reflecting direct uptake from sediments (Montgomery and Price, 1979). The same situation probably exists in other benthic species, though there are few data to confirm this point. CF's based on water are generally

Table 4-4. Total chromium residues (mg kg^{-1} dry weight) in some marine invertebrates.

Species	Average (range)	Location	Poluting source
Polychaetes (whole body)			
Hermodice carunculata[1]	5.6 (–)	Mediterranean (Lebanon)	mixed, small-scale industry
Nereis diversicolor[2]	($\leqslant$ 0.1–2.4)	Cornwall (UK)	unpolluted
Molluscs (soft parts)			
Mytilus edulis[3]	19.9 (4–49)	Trondheimsfjorden (Norway)	metal mines/smelters
Mytilus edulis[4]	2.1 (0.4–21)	coastal waters (FRG)	mixed industrial
Crassostrea gigas[5]	– (0.1–4.5)	Tamar Estuary (Tasmania)	metal mine
Crustaceans (whole body)			
Carcinus maenas[3]	8 (–)	Trondheimsfjorden (Norway)	metal mines/smelter
brown shrimp[6]	2.1 (0.4–3.8)	Gulf of Mexico (USA)	unpolluted
rock shrimp[6]	2.8 (1.8–4.2)	Gulf of Mexico (USA)	unpolluted

Sources: [1]Shiber (1981); [2]Bryan and Hummerstone (1977); [3]Lande (1977); [4]Karbe *et al.* (1977); [5]Bertine and Goldberg (1972); [6]Horowitz and Presley (1977).

Table 4-5. Concentration (mg kg^{-1} dry weight) of chromium in tissues of different invertebrates.

Species	Location	Organ	Concentration average (range)
Scrobicularia plana[1] (bivalve)	SW England	digestive gland	3.6
		mantle and siphons	2.1
		foot and gonad	1.9
Mytilus edulis[2] (mussel)	Belgium	shell	(0.05–0.38)
		soft tissues	(0.8–1.9)
Ensis ensis[2] (shrimp)	Belgium	carapace	(ND–2.1)
		soft tissues	(ND–10.0)
Orconectes australis[3] (decapod)	Cave stream, USA	carapace	1.9
		muscle	2.7
		gill	3.4
		hepatopancreas	0.9
Cambarus tenebrosus[3] (decapod)	Cave stream, USA	carapace	0.5
		muscle	3.1
		gill	2.0
		hepatopancreas	0.5

Sources: [1]Bryan and Hummerstone (1978); [2]Bertine and Goldberg (1972); [3]Dickson *et al.* (1979).
ND - not detected.

much lower than those reported for mercury and cadmium. CF's for algae in the River Rhine (FRG) were only 200–300, whereas in a polluted part of the Savannah River (USA) they were $\leq$ 100 (Cherry and Guthrie, 1977; Vogt and Kittelberger, 1977).

Chromium is readily transferred through food to invertebrates. However, there is no indication of bioaccumulation in any species. Residues in invertebrates from polluted freshwaters generally range up to 25 mg kg^{-1} dry weight compared with $\leq$ 5 mg kg^{-1} for unpolluted waters. A similar range in concentrations has been reported in invertebrates from coastal and estuarine waters (Table 4-4). Unlike some metals, chromium does not concentrate strongly in specific tissues (Table 4-5).

In most species, food is probably a more significant source of chromium than water. Uptake is often temperature dependent, and thus there is a seasonal cycle in chromium levels in natural populations (Karbe *et al.*, 1977). Although rate of uptake is greatest in young individuals, residues decline with age, reflecting rapid elimination.

Chromium does not normally accumulate in fish; hence burdens are low in both marine and freshwater species collected from industrialized parts of the world. Concentrations in the muscle of freshwater fish generally fall below 0.25 mg kg^{-1} wet weight. Marine specimens collected from industrial areas often have slightly higher burdens than those found in freshwater species. Residues in wrass from the Mediterranean (Lebanon) reached 1.6 mg kg^{-1} wet weight whereas several species collected from the coast of Israel had up to 1 mg Cr kg^{-1} (Shiber, 1981; Roth and Hornung, 1977). There are numerous reports from other industrialized areas indicating residues in muscle do not exceed 0.5 mg kg^{-1}.

Levels in internal organs are not consistently higher than those in muscle. For example, average concentrations in the muscle, liver, kidney, heart, gonad, spleen, and gills of 8 marine species from New Zealand were 0.02, 0.1, 0.2, 0.3, 0.2, 1.2, and 0.5 mg kg^{-1} wet weight, respectively (Brooks and Rumsey, 1974). However residues in the muscle of seven deep-water species were comparable to those in the liver (Greig *et al.*, 1976). This implies that monitoring and impact assessment programs should include analysis of a variety of tissues.

Chromium levels in fish are usually not related to feeding habits. Thus five species of omnivorous fish from the Illinois River (USA) averaged 0.21 mg Cr kg^{-1} wet weight compared with 0.12 mg kg^{-1} for five carnivorous species (Mathis and Cummings, 1973). Similarly, omnivorous white suckers inhabiting the Kennebec River (USA) had burdens of $\leq 0.01-0.22$ mg kg^{-1} dry weight, whereas the corresponding values for carnivorous yellow perch and smallmouth bass were ≤ 0.01 and $0.15-1.67$ mg kg^{-1} (Friant, 1979).

Toxicity

Aquatic Plants, Invertebrates and Fish

Toxicity of chromium (Cr(+3), Cr(+6)) to aquatic organisms is generally low. Under most conditions, mercury, cadmium, copper, lead, nickel, and zinc are more toxic than chromium. Although growth inhibition in aquatic plants generally occurs at 0.5–5 mg Cr (+6) L^{-1}, $K_2Cr_2O_7$ may actually stimulate the growth of some species. Toxicity to plants depends on pH of the media and hence the availability of free and chelated ions. Other factors, such as the presence of organic chelators, cations, nutrients, and other heavy metals in solution likely influence toxicity to plants; however there is little specific information to confirm this point.

Acute toxicity of chromium to freshwater invertebrates is highly variable. Rehwoldt *et al.* (1973) found that the 96h LC_{50} of Cr(+3) for seven species ranged from 3 to 50 mg L^{-1} whereas the corresponding values for Cr(+6) generally fall within the range 0.1–20 mg L^{-1} (Buikema *et al.*, 1974; Sudo and Aiba, 1973). Although statistical analysis is lacking, there appears to be little difference in the toxicity of Cr(+3) and Cr(+6). Chromium is less toxic in salt water, due in part to competitive inhibition with cations. The 96h LC_{50} for blue crab (*Callinectes sapidus*) ranged from 31 to 106 mg L^{-1} (Frank and Robertson, 1979), compared with 5–35 mg L^{-1} for several species of mollusc and polychaete (Calabrese *et al.*, 1973; Reish *et al.*, 1976).

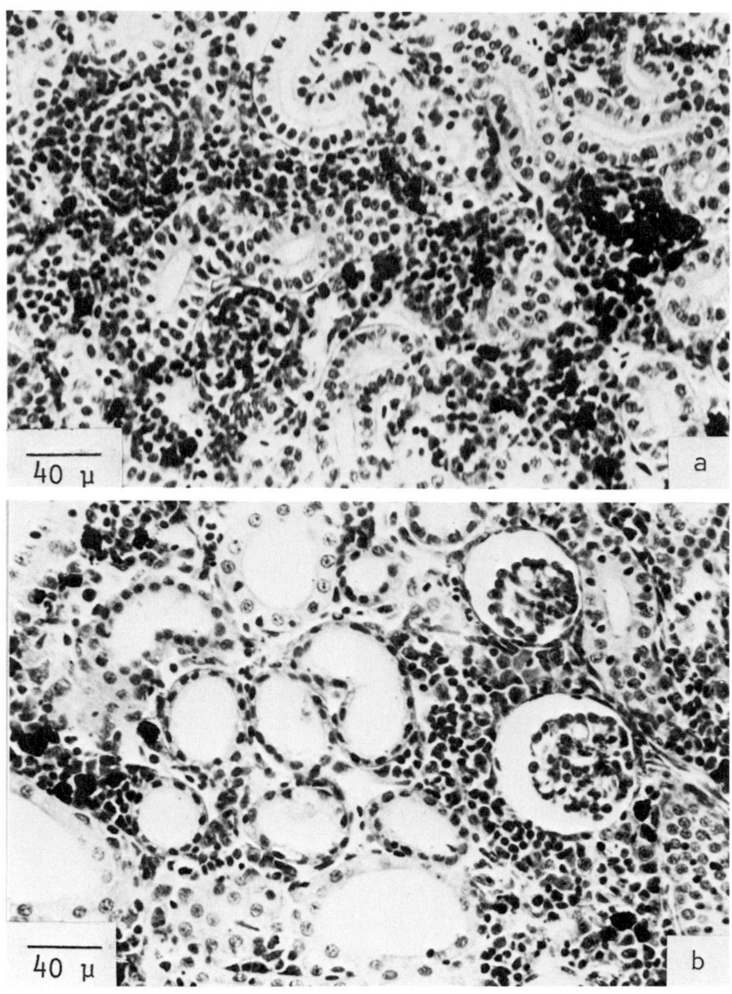

Figure 4-2. Kidney of rainbow trout. (a) Control fish. (b) Trout exposed to 44.8 mg L^{-1} Cr(+6) for 96 h at pH 7.8. Dilation of lumina of tubules and increase of nucleus-to-cytoplasm ratio of the epithelium. (From Van der Putte, 1981.)

Sublethal/chronic effects of chromium intoxication include decreased growth and body size. There may also be a significant reduction in the rate of reproduction and survival of progeny. Although sublethal exposure of Cr(+6) to the oligochaete *Tubifex tubifex* did not result in a change in respiration rate, such response is probably dependent on species and test regime (Brković-Popović and Popović, 1977b). Thus, the potential use of respiration rate as a monitoring tool must be prefaced by a careful evaluation of environmental conditions.

As with other metals, changing water hardness and salinity significantly influence toxic effects. The LC_{50} of the Cr(+6) to the oligochaete *Tubifex tubifex* decreased 750 times following an increase from 0.1 to 261 mg $CaCO_3$ L^{-1} (Brković-Popović and Popović, 1977a). Similarly the 96h LC_{50} for the blue crab *Callinectes sapidus* averaged 34 and 98 mg L^{-1} at salinities of 1 and 35 $^{o}/oo$, respectively (Frank and Robertson, 1979). As indicated earlier, competitive interaction, particularly from calcium, is partially responsible for reductions in toxicity. In addition, some marine species come under physiological stress when placed in water of low salinity, thereby increasing susceptibility to intoxication.

Fish are generally less susceptible to the toxic effects of chromium than invertebrates. The 96h LC_{50} of both Cr(+3) and Cr(+6) for freshwater species may vary from 3.5 to 118 mg L^{-1}. Much of this variability can be attributed to differential species response. The 96h LC_{50}'s for fathead minnows and bluegills were 17.6 and 118 mg L^{-1}, respectively, when tested under identical conditions (Pickering and Henderson, 1966). In addition, pH of the media and size of fish significantly influence the rate of intoxication. Van der Putte *et al.* (1981) found that the 96h LC_{50} of rainbow trout decreased from 53 to 16 mg L^{-1} as pH fell from 7.8 to 6.5. There was an additional fourfold increase in toxicity as fish weight decreased from 13 to 0.1 g.

Temperature changes significantly influence the susceptibility of fish to intoxication. However, the extent and nature of effect is species dependent. Thus, increasing temperature from 5 to 30° C caused a decline from 300 to 110 mg L^{-1} in the 24h LC_{50} for goldfish, whereas the corresponding values for rainbow trout increased from 20 to 90 mg L^{-1} (Smith and Heath, 1979). Toxic interactions of chromium with other pollutants have not been fully documented. In one study, Cr(+6) and cyanide interacted less than additively towards rainbow trout and fathead minnow (Broderius and Smith, 1979).

The site of toxic action in fish may depend on pH of the media. At pH $\geqslant$ 7.5, exposure to chromium results in histological damage to kidney and stomach (Figure 4-2). Although there appears to be little damage to gills, treatment with chromium at pH $\leqslant$ 6.5 results in significant hyperplasia (Figure 4-3). This in turn causes a decrease in plasma osmolarity and an increase in hematocrit.

Humans

Chromium is not acutely toxic to humans. This is due to the high stability of natural chromium complexes in abiotic matrices. In addition, the hard acid nature of chromium imparts strong affinity for oxygen donors rather than sulfur donors present in biomolecules. However, Cr(+6) is more toxic than Cr(+3) because of its high rate of adsorption through intestinal tracts. In the natural environment, Cr(+6) is likely to be reduced to Cr(+3), thereby reducing the toxic impact of chromium discharges.

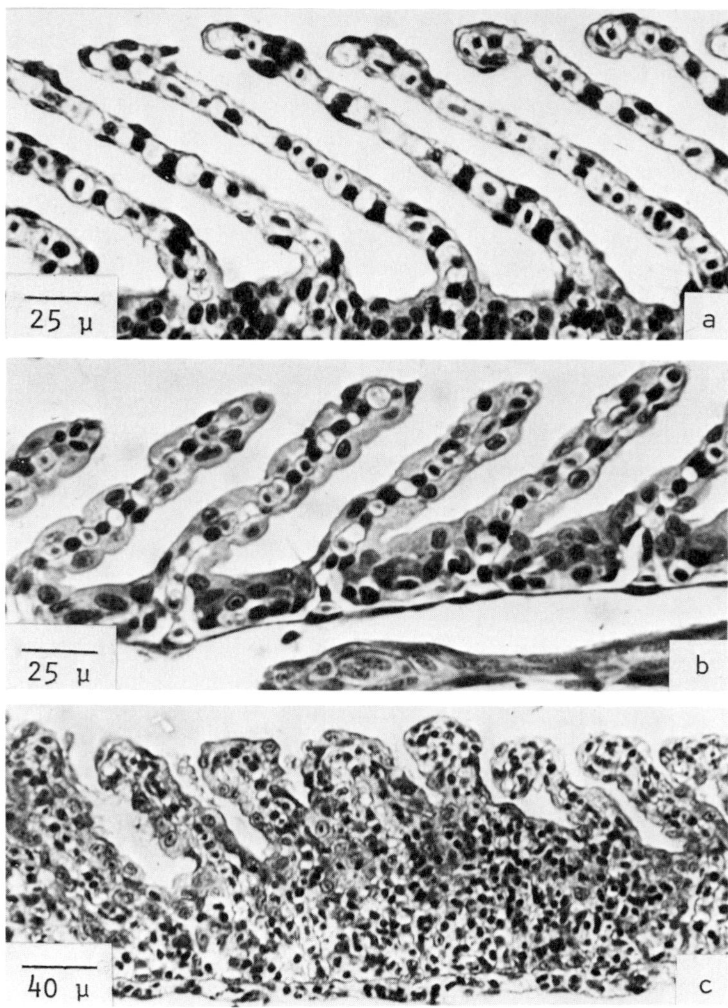

Figure 4-3. Gill lamellae of rainbow trout. (a) Control fish. (b) Trout exposed to 44.8 mg L^{-1} Cr(+6) for 96 h at pH 7.8. Gill lamellae show varying but minimal epithelial hypertrophy. (c) Trout exposed to 13.1 mg L^{-1} Cr(+6) for 96 h at pH 6.5. Extensive alteration of the normal architecture of the lamellae is evident with severe epithelial hyperplasia. (From Van der Putte, 1981.)

Chromate and dichromate were frameshift mutagens in the *Salmonella* plate incorporation test though trivalent chromium did not cause an increase in reversion frequency (Löfroth and Ames, 1978). It appears, based on experimental evidence, that hexavalent chromium also has mutagenic properties. Although the case for trivalent chromium is not as strong, its mutagenicity must be considered a possibility.

Epidemiological studies have shown a positive relationship between occupational exposure to chromates and cancer incidence (Sittig, 1980). Risk of lung cancer appears greatest among ferrochromium, chromate and chrome pigment workers. Slightly soluble hexavalent chromium salts, specifically calcium chromate, are the most potent carcinogens. These compounds have induced high levels of chromosomal aberrations in cultures of mammalian cells.

References

Andersen, A., M.F. Hovmand, and I. Johnsen. 1978. Atmospheric heavy metal deposition in the Copenhagen area. *Environmental Pollution* **17:**133–151.

Bertine, K.K., and E.D. Goldberg. 1971. Fossil fuel combustion and the major sedimentary cycle. *Science* **173:**233–235.

Bertine, K.K., and E.D. Goldberg. 1972. Trace elements in clams, mussels, and shrimp. *Limnology and Oceanography* **17:**877–884.

Brković-Popović, I., and M. Popović. 1977a. Effects of heavy metals on survival and respiration rate of tubificid worms: Part I. Effects on survival. *Environmental Pollution* **13:**65–72.

Brković-Popović, I., and M. Popović. 1977b. Effects of heavy metals on survival and respiration rate of tubificid worms: Part II. Effects on respiration rate. *Environmental Pollution* **13:**93–98.

Broderius, S.J., and L.L. Smith, Jr. 1979. Lethal and sublethal effects of binary mixtures of cyanide and hexavalent chromium, zinc, or ammonia to the fathead minnow (*Pimephales promelas*) and rainbow trout (*Salmo gairdneri*). *Journal of the Fisheries Research Board of Canada* **36:**164–172.

Brooks, R.R., and D. Rumsey. 1974. Heavy metals in some New Zealand commercial sea fishes. *N.Z. J. Mar. Freshwater Res.* **8:**155–166.

Bryan, G.W., and L.G. Hummerstone. 1977. Indicators of heavy-metal contamination in the Looe Estuary (Cornwall) with particular regard to silver and lead. *Journal of the Marine Biological Association of the United Kingdom* **57:** 75–92.

Bryan, G.W., and L.G. Hummerstone. 1978. Heavy metals in the burrowing bivalve *Scrobicularia plana* from contaminated and uncontaminated estuaries. *Journal of the Marine Biological Association of the United Kingdom* **58:**401–419.

Buikema, A.L. Jr., J. Cairns, Jr., and G.W. Sullivan. 1974. Evaluation of *Philodina acuticornis* (Rotifera) as a bioassay organism for heavy metals. *Water Resources Bulletin* **10:**648–661.

Calabrese, A., R.S. Collier, D.A. Nelson and J.R. MacInnes. 1973. The toxicity of heavy metals to embryos of the American oyster *Crassostrea virginica. Marine Biology* **18:**162–166.

Canadian Minerals Yearbooks. 1920–1979. Publishing Center, Department of Supply and Services, Ottawa, Ontario.

Cherry, D.S., and R.K. Guthrie. 1977. Toxic metals in surface waters from coal ash. *Water Resources Bulletin* **13:**1227–1236.

Cranston, R.E., and J.W. Murray. 1980. Chromium species in the Columbia River and estuary. *Limnology and Oceanography* **25:**1104–1112.

Dickson, G.W., L.A. Briese, and J.P. Giesy, Jr. 1979. Tissue metal concentrations in two crayfish species cohabiting a Tennessee cave stream. *Oecologia* **44**:8–12.

Evans, D.W., and N.H. Cutshall. 1973. Effects of ocean water on the soluble suspended distribution of Columbia radionuclides. *In: Radioactive contamination of the marine environment.* International Atomic Energy Agency, Vienna, pp. 125–140.

Frank, P.M., and P.B. Robertson. 1979. The influence of salinity on toxicity of cadmium and chromium to the blue crab, *Callinectes sapidus. Bulletin of Environmental Contamination and Toxicology* **21**:74–78.

Friant, S.L. 1979. Trace metal concentrations in selected biological, sediment, and water column samples in a northern New England river. *Water, Air and Soil Pollution* **11**:455–465.

Gibbs, R.J., 1977. Transport phases of transition metals in the Amazon and Yukon Rivers. *Geological Society of America Bulletin* **88**:829–843.

Greig, R.A., D.R. Wenzloff, and J.B. Pearce. 1976. Distribution and abundance of heavy metals in finfish, invertebrates and sediments collected at a deepwater disposal site. *Marine Pollution Bulletin* **7**:185–187.

Gryzhankova, L.N., G.N. Sayenko, A.V. Karyakin, and N.V. Laktionova. 1973. Concentration of some metals in the algae of the Sea of Japan. *Oceanology* **13**:206–210.

Helz, G.R. 1976. Trace element inventory for the northern Chesapeake Bay with emphasis on the influence of man. *Geochimica et Cosmochimica Acta* **40**:573–580.

Horowitz, A., and B.J. Presley. 1977. Trace metal concentrations and partitioning in zooplankton, neuston, and benthos from the south Texas outer continental shelf. *Archives of Environmental Contamination and Toxicology* **5**:241–255.

Jan, T.K., and D.R. Young. 1978. Chromium speciation in municipal wastewaters and seawater. *Journal of Water Pollution Control Federation* **50**:2327–2336.

Jaworski, J.F. 1980. *Effects of chromium, alkali halides, arsenic, asbestos, mercury, cadmium in the Canadian environment.* National Research Council of Canada Publication No. NRCC. 17585 of the Environmental Secretariat, Ottawa, Canada, 80pp.

Karbe, L., CH. Schnier and H.O. Siewers. 1977. Trace elements in mussels (*Mytilus edulis*) from coastal areas of the North Sea and the Baltic. Multielement analyses using instrumental neutron activation analysis. *Journal of Radioanalytical Chemistry* **37**:927–943.

Kemp, A.L.W., J.D.H. Williams, R.L. Thomas and M.L. Gregory. 1978. Impact of man's activities on the chemical composition of the sediments of Lakes Superior and Huron. *Water, Air and Soil Pollution* **10**:381–402.

Kopp, J.F., and R.C. Kroner. 1968. *Trace metals in waters of the United States.* Federal Water Pollution Control Administration, Division of Pollution Surveillance, Cincinnati, Ohio.

Kretzschmar, J.G., I. Delespaul, and T. de Rijck. 1980. Heavy metal levels in Belgium: a five year survey. *The Science of the Total Environment* **14**:85–97.

Kronfeld, J., and J. Navrot. 1974. Transition metal contamination in the Qishon River system, Israel. *Environmental Pollution* **6**:281–288.

Lande, E. 1977. Heavy metal pollution in Trondheimsfjorden, Norway, and the recorded effects on the fauna and flora. *Environmental Pollution* **12**:187–197.

Löfroth, G. and B.N. Ames. 1978. Mutagenicity of inorganic compounds in *Salmo-*

nella typhimurium: arsenic, chromium and selenium. *Mutation Research* 53:65–66.

Loring, D.H. 1979. Geochemistry of cobalt, nickel, chromium, and vanadium in the sediments of the estuary and open Gulf of St. Lawrence. *Canadian Journal of Earth Sciences* **16:**1196–1209.

Lum-Shue-Chan, K. 1981. Dissolved and particulate trace metals in meltwater from the Rhône glacier. *Verlag Hydrologie Art* **12:**(In Press).

Mathis, B.J., and T.F. Cummings. 1973. Selected metals in sediments, water and biota in the Illinois River. *Journal of Water Pollution Control Federation* **45:**1573–1583.

Montgomery, J.R., and M.T. Price. 1979. Release of trace metals by sewage sludge and the subsequent uptake by members of a turtle grass mangrove ecosystem. *Environmental Science and Technology* **13:**546–549.

Moore, J.W. 1980. Distribution and transport of heavy metals in the sediments of a small northern eutrophic lake. *Bulletin of Environmental Contamination and Toxicology* **24:**828–833.

Nakayama, E., H. Tokoro, T. Kuwamoto, and T. Fujinaga. 1981. Dissolved state of chromium in seawater. *Nature* **290:**768–770.

Nriagu, J.O., and R.D. Coker. 1980. Trace metals in humic and fulvic acids from Lake Ontario sediments. *Environmental Science and Technology* **4:**443–446.

Pankow, J.F., and others. 1977. Analysis of chromium traces in the aquatic ecosystem. 2. A study of Cr(III) and Cr(VI) in the Susquehanna River basin of New York and Pennsylvania. *The Science of the Total Environment* **7:**17–26.

Pfeiffer, W.C., M. Fiszman, and N. Carbonell. 1980. Fate of chromium in a tributary of the Irajá River, Rio de Janeiro. *Environmental Pollution (Series B)* **1:**117–126.

Pickering, Q.H., and C. Henderson. 1966. The acute toxicity of some heavy metals to different species of warmwater fishes. *Air and Water Pollution International Journal* **10:**453–463.

Pilotte, J.O., J. W. Winchester, and R.C. Glassen. 1978. Detection of heavy metal pollution in estuarine sediments. *Water, Air, and Soil Pollution* **9:**363–368.

Rehwoldt, R., L. Lasko, C. Shaw, and E. Wirhowski. 1973. The acute toxicity of some heavy metal ions toward benthic organisms. *Bulletin of Environmental Contamination and Toxicology* **10:**291–294.

Reish, D.J., J.M. Martin, F.M. Piltz, and J.Q. Word. 1976. The effect of heavy metals on laboratory populations of two polychaetes with comparisons to the water quality conditions and standards in southern California marine waters. *Water Research* **10:**299–302.

Roth, I., and H. Hornung. 1977. Heavy metal concentrations in water, sediments, and fish from Mediterranean coastal area, Israel. *Environmental Science and Technology* **11:**265–269.

Salomons, W., and W.G. Mook. 1977. Trace metal concentrations in estuarine sediments: mobilization, mixing or precipitation. *Netherlands Journal of Sea Research* **11:**119–129.

Salomons, W., and W.G. Mook. 1980. Biogeochemical processes affecting metal concentrations in lake sediments (Ijsselmeer, the Netherlands). *The Science of the Total Environment* **16:**217–229.

Schafer, H.A. 1976. Characteristics of municipal wastewater discharges, 1975. *Annual Report, Southern California Coastal Water Research Project,* El Segundo, pp. 57–60.

Shiber, J.G. 1981. Metal concentrations in certain coastal organisms from Beirut. *Hydrobiologia* **83**:181–195.

Shuman, M.S., and J. H. Dempsey. 1977. Column chromatography for field preconcentration of trace metals. *Journal Water Pollution Control Federation* **49**:2000–2006.

Sillén, L.G., and A.E. Martell. 1971. *Stability constants of metal-ion complexes, Supplement No.1.* Special Publication No.25, The Chemical Society, London, 865 pp.

Sittig, M. 1980. *Priority toxic pollutants. Health impacts and allowable limits.* Noyes Data Corporation, New Jersey, 370 pp.

Sivalingam, P.M. 1978. Biodeposited trace metals and mineral content studies of some tropical marine algae. *Botanica Marina* **21**:327–330.

Smith, M.J., and A.G. Heath. 1979. Acute toxicity of copper, chromate, zinc, and cyanide to freshwater fish: effect of different temperatures. *Bulletin of Environmental Contamination and Toxicology* **22**:113–119.

Sudo, R., and S. Aiba. 1973. Effect of copper and hexavalent chromium on the specific growth rate of ciliata isolated from activated sludge. *Water Research* **7**:1301–1307.

Thornton, I., H. Watling, and A. Darracott. 1975. Geochemical studies in several rivers and estuaries used for oyster rearing. *The Science of the Total Environment* **4**:325–345.

United States Minerals Yearbooks. 1920–1979. Bureau of Mines, US Department of the Interior, Washington, D.C.

Van Der Putte, I., M.A. Brinkhorst, and J.H. Koeman. 1981. Effect of pH on the acute toxicity of hexavalent chromium to rainbow trout (*Salmo gairdneri*). *Aquatic Toxicology* **1**:129–142.

Van der Veen, C., and J. Huizenga. 1980. Combating river pollution taking the Rhine as an example. *Progress in Water Technology* **12**:1035–1059.

Vogt, G. and F. Kittelberger. 1977. Study about uptake and accumulation of heavy metals in typical algae associations of the Rhine between Germersheim and Gernsheim. *Fisch und Umwelt Heft* **3**:15–18.

Wilber, W.G., and J.V. Hunter. 1977. Aquatic transport of heavy metals in the urban environment. *Water Resources Bulletin* **13**:721–734.

Wilber, W.G., and J.V. Hunter. 1979. The impact of urbanization on the distribution of heavy metals in bottom sediments of the Saddle River. *Water Resources Bulletin* **15**:790–800.

Young, D.R., J.N. Johnson, A. Soutar, and J.D. Isaacs. 1973. Mercury concentrations in dated varved marine sediment collected off southern California. *Nature* **244**:273–274.

5
Copper

Chemistry

Copper is widely distributed in nature in the free state and in sulfides, arsenides, chlorides, and carbonates. It is soft and ductile with a high thermal and electrical conductivity second only to silver. Copper is classified as intermediate between hard and soft acids in its chemical interactions with donor atoms. Hard acids prefer O donors forming essentially electrovalent bonds whereas soft acceptors prefer S or Se donor atoms forming covalent bonds. Copper belongs to the third transition metal series, exhibiting a wide variation in properties such as spectral, magnetic, complexing capacity, and oxidation states. These properties result from the partially filled d subshell. Copper complexes of oxidation states (+1), (+2), and (+3) are known, though Cu(+2) is most common. Cu(+1) is a typical soft acid.

Several copper-containing proteins have been identified in biological systems: oxygen-binding hemocyanin, cytochrome oxidase, tyrosinase, and laccase. Ceruloplasmin, a blue protein in mammalian serum, accounts for over 95% of the circulating copper in mammals. The level of ceruloplasmin is highly sensitive to a variety of pathological changes and reflects the disorders of copper metabolism; the reduction of both serum copper and ceruloplasmin copper are widely accepted as an important diagnostic tool. Copper and iron are involved in the natural selection of aerobic cells and the evolution of metalloproteins and metalloenzymes. The evolution resulted in the development of copper–zinc enzymes (superoxide dismutase), heme enzymes, iron–copper enzymes and oxygen-carrying proteins (Figure 5-1). This adaptation demanded elaborate storage and transport proteins, exclu-

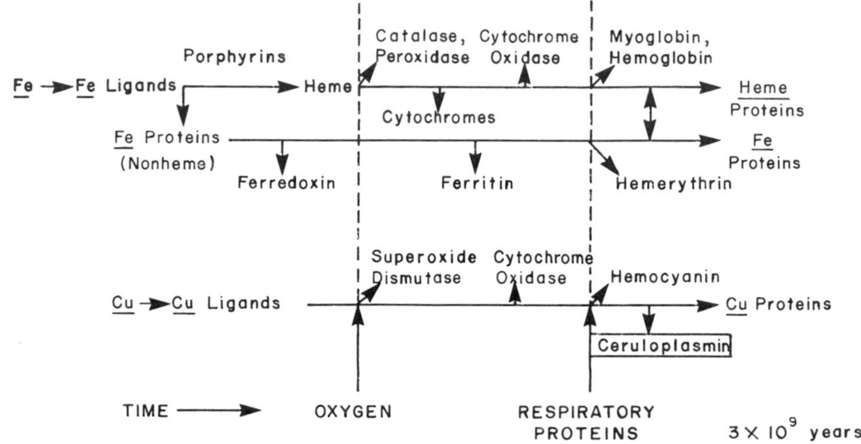

Figure 5-1. Evolution of copper and iron proteins. (From Frieden, 1979.)

sively for copper and iron. Ceruloplasmin represents the latest example of the link between copper and iron in the vertebrates. In all these proteins, copper is coordinated to nitrogen, oxygen and/or sulfur ligands. Hence the dominant role of copper in biological systems is its ability to stabilize sulfur radicals.

Production, Uses, and Discharges

Production and Uses

Magnetic hydrothermal deposits, containing two thirds of the world's copper reserves, are localized around the Pacific rim and the mountain belts in southeastern Europe and central Asia. Areas in Australia and Siberia also contain relatively large deposits of copper. The all-time mine production of

Table 5-1. All-time global copper production (in metric tons $\times 10^6$).

Period	Production	Period	Production
Pre-1850	45	1941-1950	23.8
1850-1900	13	1951-1960	32.4
1901-1910	7.5	1961-1970	61.4
1911-1920	11.3	1971-1980	82.5
1921-1930	13.5	Total (all-time)	306.7
1931-1940	16.3		

Sources: US Minerals Yearbooks; Nriagu (1979b).

copper is estimated to be 307 million metric tons (Table 5-1). About 81% of total production has been in the twentieth century and about 27% in the past decade (Table 5-1).

Properties such as malleability, ductility, conductivity, corrosion resistance, alloying qualities, and pleasing appearance make copper's use universal in the electrical, construction, plumbing, and automotive industries. Hence, total consumption has remained high, albeit relatively steady, throughout the last decade (Table 5-2). The largest single user of copper is the electrical industry accounting for ≥50% of total consumption (Table 5-3). Usage under this category includes power transmission, electronics and electrical equipment.

Generation and consumption of electrical energy requires vast quantities of copper for heat exchangers, bus bars, magnet wire and windings in motors, generators, and transformers. Construction and plumbing is the second largest user (Table 5-3). Copper is incorporated into hardware roofing and plumbing products, whereas alloys are used in bearings, fastenings, and fittings in marine hardware. The use of copper as a biocide has decreased considerably during this decade.

Table 5-2. US consumption of refined copper and global mine production during 1970–1979 (in metric tons × 10^6).

Year	US consumption	Global production	Year	US consumption	Global production
1970	1.86	6.03	1975	1.40	7.02
1971	1.84	6.03	1976	1.81	7.52
1972	2.03	6.66	1977	1.99	7.73
1973	2.21	7.17	1978	2.19	7.56
1974	1.99	7.32	1979	2.16	7.61

Source: US Minerals Yearbooks.

Table 5-3. Consumption of refined copper by class (in metric tons × 10^3).

Class of consumer	1970	1972	1974	1976	1978	1979
Wire mills	1217.0	1387.5	1340.3	1240.0	1517.4	1500.0
Brass mills	600.5	606.6	609.2	531.6	619.2	610.2
Chemical plants	2.0	0.8	0.6	0.5	0.4	0.4
Secondary smelters	6.4	9.0	8.3	6.8	7.5	6.3
Foundries	14.8	13.8	17.6	14.0	12.4	11.9
Miscellaneous*	16.8	17.6	18.6	17.9	32.3	30.1
Total	1857.5	2035.3	1994.6	1810.8	2189.2	2158.9

Source: US Minerals Yearbooks.

*Includes iron and steel plants, primary smelters producing alloys other than copper, consumers of copper powder and copper shot, and miscellaneous manufacturers.

Discharges

The total flux of copper to the atmosphere is aproximately 75,000 metric tons yr^{-1} of which 5000–13,000 tons are depostied into the ocean through both wet and dry depositions (Nriagu, 1979c). Approximately 75% of atmospheric emissions are from anthropogenic sources (Tables 5-4, 5-5). Production of nonferrous metals is the largest single source, followed by wood combustion and iron/steel production. The most important natural

Table 5-4. Global emissions of copper from natural sources.

Source	Global production (10^6 metric tons yr^{-1})	Global Cu emissions (10^3 metric tons)
Windblown dusts	500	12.0
Forest fires	36	0.3
Volcanic particles	10	3.6
Vegetation	75	2.5
Seasalt sprays	1000	0.08
Total		18.5

Source: Nriagu (1979a). Reprinted by permission from *Nature* **279**:409–411. Copyright © 1979, Macmillan Journals Limited.

Table 5-5. Global anthropogenic emissions of copper in 1975.

Source	Global production/ consumption (10^6 metric tons yr^{-1})	Cu emissions (10^3 metric tons)
Mining	16	0.8
Primary non-ferrous metal production		
Copper	7.9	19.7
Lead	4.0	0.29
Zinc	5.6	0.78
Secondary non-ferrous metal production	4.0	0.33
Iron and steel production	1300	5.9
Industrial applications	—	4.9
Coal combustion	3100	4.7
Oil (including gasoline) combustion	2800	0.74
Wood combustion	640	12.0
Waste incineration	1500	5.3
Phosphate fertilizer production	118	0.6
Total		56.04

Source: Nriagu (1979a). Reprinted by permission from *Nature* **279**:409–411. Copyright © 1979, Macmillan Journals Limited.

Table 5-6. All-time global anthropogenic emissions of copper (in metric tons × 10^3).

Period	Global emissions	% of total consumption
Pre-1850	319	0.709
1850–1900	92	0.708
1901–1910	53	0.707
1911–1920	80	0.708
1921–1930	96	0.711
1931–1940	116	0.712
1941–1950	169	0.710
1951–1960	230	0.710
1961–1970	435	0.708
1971–1980	585	0.709
Total (all-time) emissions	2175	

Source: Nriagu (1979a). Reprinted by permission from *Nature* **279**:409–411.

Table 5-7. Annual copper discharges on land by sources.*

Source	Global production/ discharge (10^3 metric tons)	Input into soils (10^3 metric tons)
Fertilizer production	94	9.4
Flyash and mine wastes	280	67
Municipal sewage	5	4.7
Sludge disposal		
Industrial sewage	5	4.7
Total		85.8

* Compiled from Nriagu (1979b,c).

discharge of copper to the atmosphere is wind-blown dust. Overall global emissions have more than tripled during the last 30 years (Table 5-6).

Discharge of mine tailings and flyash is the major source of solid copper waste (Table 5-7). Disposal in this category accounts for ≥ 75% of total discharge. Other sources of solid wastes include fertilizer production and municipal and industrial sewage. Approximately 17,000 metric tons of solid copper wastes are deposited annually into the oceans (Nriagu, 1979c). The residence time of copper in the oceans is reported to vary from 1500 to 78,000 years (Nriagu, 1979b). This variation arises from differences in copper levels in the oceans.

Copper in Aquatic Systems

Speciation in Natural Waters

In aquatic environments, copper can exist in three broad categories: particulate, colloidal, and soluble. The dissolved phase could contain both the free ion as well as copper complexed to organic and inorganic ligands. Speciation of copper in natural waters is determined by the physico-chemical, hydrodynamic characteristics and the biological state of the water.

Binding to Inorganic and Organic Ligands. Copper forms complexes with hard bases like carbonate, nitrate, sulfate, chloride, ammonia, and hydroxide (Table 5-8). Neutral ligands such as ethylenediamine, ammonia, and pyridine form strong complexes with copper which are typically 4-coordinate complexes. Copper interacts strongly with sulfur forming relatively stable insoluble sulfides ($K_{sp} = 10^{-36.2}$). Being an intermediate acceptor between hard and soft acids, copper complexes with nitrogen and sulfur containing ligands. Humic materials in freshwaters bind more than 90% of total-Cu, whereas those in seawater bind only 10% (Mantoura *et al.,* 1978). In the latter case, calcium and magnesium, because of their large concentrations, displace copper from humic materials.

Schnitzer and Kerndorff (1981) studied the solubility of fulvic acid(FA)–metal complexes as a function of pH, and concentrations of the metal ion and FA, for eleven metal ions including Cu^{2+} ion. They showed that (i) FA/metals ratio of >2 favors the formation of soluble metal–FA complexes and (ii) over the environmentally important pH range 6–8, Fe(+3), Cr(+3), Al(+3), Pb(+2), and Cu(+2) form mostly water insoluble complexes with FA which could lead to the accumulation of such complexes in soils, sediments and waters (Figure 5-2). The stability constants of the

Table 5-8. Stability constants of some Cu(+2)–inorganic complexes.[a]

Ligand	Log β_1	β_2	β_3	β_4
OH^-	6.0[b]	14.3[c]	14.2[b]	
CO_3^{2-}	6.1[d]			
Cl^-	0.5	0.3		
SO_4^{2-}	2.3[e]			
NH_3	5.8	10.7	14.7	17.6
HPO_4^{2-}	16.6			
SH^-			26.5	

Sources: [a] Sillén and Martell (1971); [b] Baes and Mesmer (1976); [c] Vuceta and Morgan (1977); [d] Bilinski *et al.* (1976); [e] Davies *et al.* (1957).

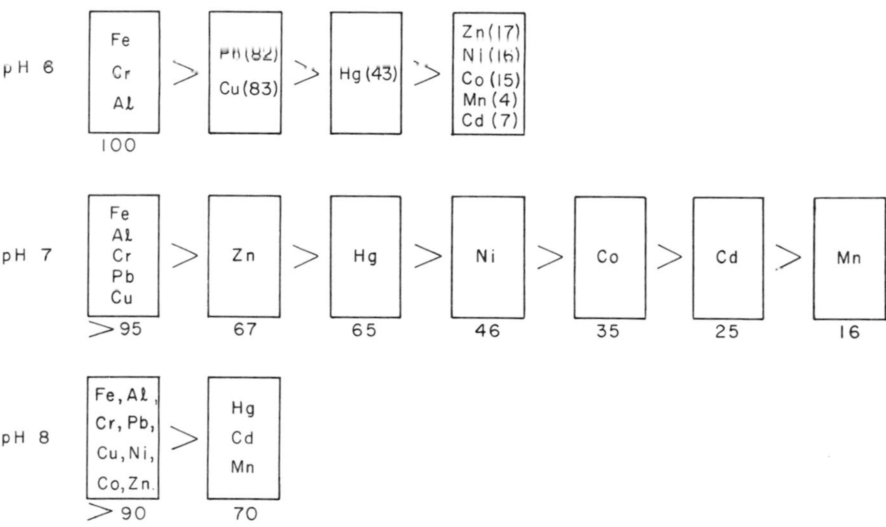

Figure 5-2. The order and percentage precipitations of fulvic acids by metal ions in the pH range 6–8. (From Schnitzer and Kerndorff, 1981.)

complexes of copper with humic materials from different environments followed the sequence:

soil FA < soil HA < peat FA < peat HA < seawater HM < lake HM = river HM < marine sedimentary FA < marine sedimentary HA.

The stabilities of the humic complexes followed the Irving–Williams order:

$$\text{Mg} < \text{Ca} < \text{Cd} \simeq \text{Mn} < \text{Co} < \text{Zn} \simeq \text{Ni} < \text{Cu} < \text{Hg}.$$

The variation in copper binding strength to humic materials from different sources in the environment should be taken into account in speciation studies.

Binding to Particulates. Particulates contain varying fractions (12–97%) of the total copper transported by rivers. Approximately 6.3×10^6 metric tons of copper are transported annually by the rivers to the oceans. Only 1% of this is in the soluble form, 85% in particulate crystalline phases, about 6% bound to metal hydroxide coatings, 4.5% bound to organics, and only 3.5% adsorbed onto the suspended solids.

Transport in Natural Waters

From an eight-year (1970–78) study on the transport and budget of heavy metals in the Ruhr River, Imhoff *et al.* (1980) reported that (i) copper showed a small range of variation in concentration even with changing flows and (ii) the extent of variation was independent of the sampling stations along the river. The relatively small variation in levels was attributed to the equal input from geochemical origin of non-point sources. About 65–81% of the copper was in the soluble fraction, thus minimizing concentration fluctuations. The ratio of the dissolved and total concentration of heavy metals was of great importance in their transport in suface waters.

From size-fractionation studies of the Ottawa River water, Ramamoorthy and Kushner (1975) showed that about 34% of added Cu(+2) was bound to suspended solids (>0.45 μm) and 48% to macromolecules (≤0.45 μm − 45,000 MW fraction>). The rest was bound to size fraction less than 1400 MW. The binding components were identified to be organic compounds, not comparable to the soil-derived fulvic acid. Similarly 40–60% of total copper in estuarine and coastal waters was associated with colloidal matter of organic and inorganic forms (Batley and Gardner, 1978).

Copper-containing minerals of both technological and natural origins have been identified in the suspended particles of the surface and deep waters of the North Atlantic and Pacific oceans (Jedwab, 1979). The fact that the highest frequencies of occurrences are observed in the North Atlantic ocean suggests the obvious technological influence with the possible natural input as well.

Estuaries are considered to be the zone of net deposition of metals. From a study on the St. Lawrence estuary, Bewers and Yeats (1978) reported that 24% of the incoming copper was removed in the estuary. The removal of riverborne particulates is balanced by internal resuspension of sedimentary material which leaves the estuary in mixed outflowing water. The influence of turbidity maximum, formed by particle trapping, periodic settling and resuspension of sedimentary material is confined to the estuary.

Copper in Sediments

Sorption and Desorption. Copper is sorbed rapidly to sediments, resulting in high residue levels (Table 5-9). Rate of sorption varies with the type of clay/sediment, pH, competing cations and the presence of ligands and Fe/Mn oxides. In Lake Ontario, all of the copper in sediments was bound to humic acids (Nriagu and Coker, 1980). Tessier *et al.* (1980), from studies of Yamaska and St. François Rivers (Canada), reported a high percentage of organic-bound copper in the particulates. The fractions from the two rivers were: organic, 31 and 52%; residual, 41 and 22%; Fe–Mn oxide, 20 and 12%; carbonate, 8 and 14%; and exchangeable, 1.0 and 1.2%. These values were different from the Amazon and Yukon Rivers where the organic-Cu was 8–15% and the residual fraction 74–84% (Gibbs, 1977). Similarly Willey and Fitzgerald (1980) reported that about 27% of organic bound-Cu occurred in coarse sediments (≥63 μm) and 62% in the fractions. Copper levels in the St. Lawrence estuary decreased moving seaward coincident with a decline in the rate of deposition of detritus and Fe–Mn oxides.

Desorption from sediments into the bulkwater depends on pH, salinity,

Table 5-9. Total copper levels (mg kg^{-1} dry weight) in freshwater and marine sediments.

Location	Average (range)	Polluting source
Freshwater Sediments		
Coeur d'Alene Lake (USA)[1]	115 (90–150)	metal mining
Natural creek, Montana (USA)[2]	102 (16–590)	abandoned mine
Yellowknife Bay (Canada)[3]	350 (20–1010)	gold mine
Illinois River (USA)[4]	19 (1–82)	industrial, municipal
Restronguet creek (UK)[5]	1830 (1350–2350)	metal mining
Arctic lakes (Canada)[6]	39 (7–62)	natural
Marine Sediments		
Mediterranean (Lebanon)[7]	21 (2–195)	small local industry
Continental Shelf (SE, USA)[8]	≤2	natural
Coastal Waters, North Sea (UK)[9]	8 (2–49)	natural, mixed industry
St. Lawrence Estuary (Canada)[10]	19 (3–76)	natural
Los Angeles harbour (USA)[11]	99 (39–148)	mixed industrial
Baltic Sea[12]	–(1–283)	mixed industrial
Sörfjord (Norway)[13]	2400 (210–12000)	metal mines, smelters

Sources: [1] Maxfield *et al.* 1974); [2] Pagenkopf and Cameron (1979); [3] Moore (1979); [4] Mathis and Cummings (1973); [5] Boyden *et al.* (1979); [6] Moore *et al.* (1979); [7] Shiber (1980); [8] Bothner *et al.* (1980); [9] Taylor (1979); [10] Loring (1978); [11] Emerson *et al.* (1976); [12] Brügmann (1981); [13] Skei *et al.* (1972).

and the presence of natural and/or synthetic chelating agents. Banat *et al.* (1974) showed that copper was solubilized even with relatively low concentrations of NTA, a surfactant used in commercial detergents. Van der Weijden *et al.* (1977) reported the order for decreasing desorption into 1 : 1 diluted seawater and seawater to be Cd > Zn > Mn > Ni > Co > Cu > Cr; no desorption of Fe and Pb was reported.

Residues

Water, Precipitation and Sediments

Soluble copper levels in uncontaminated freshwaters usually range from 0.5 to 1.0 μg L^{-1}, increasing to ≥2 μg L^{-1} in urban areas. Although much higher concentrations (500–2000 μg L^{-1}) occur in the vicinity of some metal mines and during periods of high water (Tyler and Buckney, 1973), there is a gradual decrease in residues moving from coastal marine waters to off-shore areas. Duinker and Nolting (1977) showed that soluble copper exceeded 15 μg L^{-1} in the Rhine Estuary, but decreased to ≤ 0.2 μg L^{-1} in the English Channel. Similarly, concentration of 1–5 μg L^{-1} have been reported for the Mediterranean and Baltic Seas, a reflection of elevated anthropogenic loading, whereas open oceanic waters have residues of ≤ 2 μg L^{-1}.

Precipitation carries significant amounts of copper. Average residues in Belgium and Florence (Italy) were 10–116 and 3–23 μg L^{-1}, respectively (Kretzshmar *et al.,* 1980; Ligittimo *et al.,* 1980). Samples collected over Lake Michigan had 3–5 μg Cu L^{-1} while coal combustion produced residues of 2.4–4.8 μg L^{-1} in precipitation (Wiener, 1979; Eisenreich, 1980).

Elevated copper levels (≥ 1000 mg kg^{-1} dry weight) in sediments are often associated with the discharge of mine wastes (Table 5-9). Although other industrial sources may produce residues of up to 500 mg kg^{-1}, concentrations associated with sewage disposal are generally ≤ 250 mg kg^{-1}. Unpolluted marine and freshwater sediments generally contain levels of ≤ 20 mg kg^{-1} (Table 5-9).

Aquatic Plants

Concentrations in attached species inhabiting polluted waters generally average 10–100 mg kg^{-1} dry weight. However, residues in *Ascophyllum nodosum* collected from a polluted part of Trondheimsfjord (Norway) averaged 166 mg kg^{-1} whereas in a control area they contained 6 mg kg^{-1} (Haug *et al.,* 1974). A mixed population of benthic algae had copper levels of 120–335 mg kg^{-1} throughout a 150 km section of the Neckar River (FRG) (Bartelt and Förstner, 1977). By contrast the differential input of polluted freshwater into coastal waters of the Arabian Sea (India) led to a relatively greater degree of variation in residues (7–80 mg kg^{-1}) in attached Chlorophyta (Agadi *et al.,* 1978). Trollope and Evans (1976) reported levels of 660 mg kg^{-1} in unattached filamentous species inhabiting a Welsh river near a smelter and 140 mg kg^{-1} in a control area.

CF's for copper are lower than those reported for most other heavy metals, including mercury, cadmium, lead, zinc, and nickel. Depending on the extent of ambient pollution, CF's may range from 0.1×10^3 to $\geq 1 \times 10^5$ in both marine and freshwaters. Significant inter-species variability has been recorded. In the Ruhr River (FRG), average CF's for *Nuphar luteum* (Spermatophyta) and *Hygroamblystegium* sp. (Bryophyta) were 78 and 2800, respectively (Dietz, 1973).

Rate of uptake depends on the initial concentration of copper in the media. Although appreciable sorption occurs in the presence of Na^+ and Mg^{2+}, uptake is generally suppressed by H^+. Sorption rates are species dependent, resulting in variable residue levels among different species.

Furthermore, copper may increase the permeability of the cell wall in aquatic plants, thereby increasing susceptibility to other pollutants.

Invertebrates

Invertebrates inhabiting polluted freshwaters generally carry residues of 5–200 mg kg^{-1} dry weight in soft tissues. Manly and George (1977) reported levels of 21–103 mg kg^{-1} in the mussel *Anodonta anatina* found in the River Thames (UK). Similarly, concentrations in herbivorous, omnivorous and carnivorous invertebrates collected from a stream in an industrial area of the USA (Wisconsin) averaged 13.7, 70.9, and 30.5 mg kg^{-1}, respectively (Anderson,1977). Several field studies have shown that there is no accumulation through the food chain. Copper sulfate was added to an irrigation canal in California (USA) to control algae, resulting in water, sediment, and vascular residues of ≤0.010 mg kg^{-1}, 30–60 mg kg^{-1}, and 35 mg kg^{-1} dry weight, respectively (Fuller and Averett, 1975). However, molluscs (*Corbicula* sp.) had average residues of only 13 mg kg^{-1}.

Residues in marine invertebrates are often much higher than those reported for freshwaters. This reflects the presence of bioavailable copper which is then sequestered in tissues. As in freshwater, copper may not be accumulated at higher trophic levels. Some of the highest residues for marine invertebrates have been reported for the Bristol Channel and Severn Estuary (UK). Stenner and Nickless (1974) reported maximum levels of 1750 mg kg^{-1} dry weight for the soft tissues of the mollusc *Nucella lapillus,* whereas the oyster *Crassostrea gigas* contained up to 6480 mg kg^{-1} (Boyden and Romeril, 1974). On the other hand, crustaceans and molluscs from many other parts of Europe and North America bear levels of ≤60 mg kg^{-1}.

Maximum residues are generally found in the internal organs. Though this probably reflects the presence of metal-binding proteins in such tissues, no evidence for binding to metallothioneins was observed in barnacles *Balanus balanoides* (Rainbow *et al.,* 1980). Ireland and Wooton (1977) reported levels of 554 and 61 mg kg^{-1} in the digestive gland/gonad and body of the marine gastropod *Thais lapillus* collected from the coast of Wales while residues in the viscera of shrimp and squid from the Gulf of Mexico were 2–7 times greater than those in muscle (Horowitz and Presley, 1977). Some species, including the gastropods *Helix aspersa* and *Littorina littorea,* show a relatively even distribution of copper throughout their tissues (Coughtrey and Martin, 1976; Ireland and Wootton, 1977).

Rate of sorption by planktonic invertebrates generally depends on copper concentrations in water. In addition, uptake by benthic species is directly related to levels in sediments. Depending on species, low temperatures may reduce the rate of sorption; similarly, uptake varies with salinity and the presence of other metals in solution. While there are many exceptions, concentrations generally increase with the age and size of animal. Watling and Watling (1976) could find no sex related differences in residues in the mussel *Choromytilus meridionalis.*

Fish

Residues in marine fish are generally higher than those in freshwater species, again reflecting the increased bioavailability of $CuCl_2$. In polluted freshwaters, maximum concentrations in muscle tissue seldom exceed 1 mg kg^{-1} wet weight. Although fish collected from polluted marine waters generally contain 0.5–2.0 mg kg^{-1} in muscle tissue, extreme levels of environmental contamination may lead to muscle concentrations of 3–6 mg kg^{-1}. Because muscle residues are generally low, copper does not pose a threat to most fisheries, even those in polluted waters.

There are relatively few food chain studies on the transfer of copper to higher trophic levels. Roth and Hornung (1977) reported that total copper in water, sediments and algae from the Mediterranean coast (Israel) averaged 3.7 $\mu g\ L^{-1}$, 1.6 mg kg^{-1}, and 5.4 mg kg^{-1} dry weight, respectively. Residues in the 12 most common fish were 3.8 mg kg^{-1} dry weight, yielding CF's of 1000 and 2.4 based on water and sediment as the polluting source, respectively. Treatment of two ponds with copper sulfate resulted in water and sediment levels of 15 $\mu g\ L^{-1}$ and 28 mg kg^{-1}, respectively (McIntosh, 1975). Four species of aquatic plants contained average residues of 16–34 mg kg^{-1} dry weight, compared with a mean of 0.6 mg kg^{-1} in green sunfish.

Residues in the liver of eel, whiting, and flounder from the Medway Estuary (UK) were 5–60 times greater than those in muscle (Wharfe and van den Broek, 1977), as also noted for roach and Crucian carp in central Europe (Drbal, 1976), and flounder from the Baltic Sea (von Westernhagen *et al.,* 1981). Although the gut wall is another major site of deposition, concentrations in skin and gills are approximately similar to those in muscle. In some species, ovaries accumulate substantially more copper than testes.

Contaminated food is probably a more important source of copper than water and thus burdens in fish cannot be consistently related to ambient pollution levels in water. Rate of uptake is inversely related to the presence of chelators and inorganic ions in the water, and directly related to exposure period and concentration. Although residues in muscle tissue frequently decline with age and size of fish, McFarlane and Franzin (1980) demonstrated a consistent positive correlation between copper in pike livers and fish age. This implies liver analysis provides a better analysis of the health of fish populations than muscle analysis.

Toxicity

Aquatic Plants

Copper is highly toxic to most aquatic plants. Inhibition of growth generally occurs at ≤ 0.1 mg L^{-1}, regardless of test conditions and species; other effects, such as reduced carbon and silicic acid uptake, occur at copper

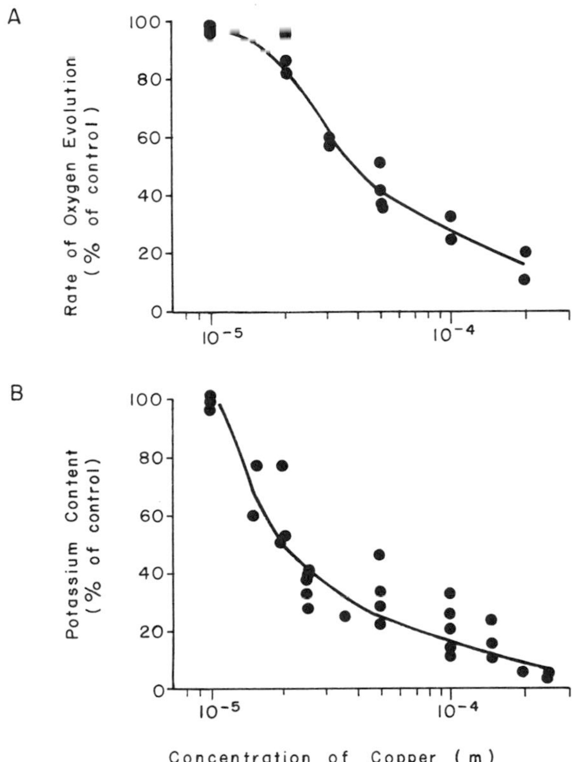

Figure 5-3. Effect of Cu(+2) on light-induced oxygen evolution (A) and potassium content (B) of cells of *Dunaliella tertiolecta.* (From Overnell, 1975.)

concentrations of 0.003–0.03 mg L^{-1}. Mercury is the only metal which is consistently more toxic to aquatic plants than copper.

Sublethal effects of copper intoxication include an initial loss of potassium due to increased permeability of the cell (Figure 5-3). There may also be inhibition of oxygen evolution and assimilation of carbon and decrease in the rate of photosynthesis. Furthermore, cell volumes, particulate carbon and nitrogen levels, and silicic acid uptake may be reduced at concentrations below 0.05 mg L^{-1}. Overnell (1976) showed that, at constant copper concentration, there was a tenfold difference in the degree of inhibition of photosynthesis among seven species of marine algae. In *Chlorella* and presumably other species, copper inhibits electron transport in the photosynthetic system (Cedeno-Maldonado and Swader, 1974). Blue-green algae are particularly susceptible to copper, because of the inhibition of nitrogen fixation.

The presence of complexing substances in water greatly reduces toxicity. Such agents generally originate from decaying matter, though most species of algae also excrete complexing ligands to regulate the concentration of copper in the environment. Additionally, since low pH increases the proportion of free ions in solution, toxicity is greater in acidic waters than in basic waters. Although combinations of copper/lead and copper/cadmium are antagonistic in their effects on algae, combinations of copper/nickel act synergistically. Synergism has also been noted for combinations of copper and flouride, copper and manganese, and copper and zinc.

Many algal species can adapt to high copper levels in water. Tolerant algae accumulate high levels (600 mg kg^{-1} dry weight) of copper while still growing and dividing (Stokes, 1975), thereby increasing the amount of copper in the food chain in polluted waters. Ernst (1975) found no resistant enzymes specific to tolerant populations of mosses and higher plants. He suggested that tolerance depended on the presence of complexing agents which were not organic acids. However, some species can produce a metallothionein-like protein for detoxification.

Invertebrates

Copper is highly toxic to most freshwater and marine invertebrates. LC_{50}'s are generally less than 0.5 mg L^{-1}, though they may range from 0.006 to ⩾225.0 mg L^{-1} under certain conditions. Toxicity is generally greater in freshwater than in marine waters, reflecting the relative proportion of the toxic free copper ion in solution.

Water hardness plays a role in determining toxicity. The LC_{50} for the oligochaete *Tubifex tubifex* increased by a factor of 150 when water hardness increased from 0.1 to 261 mg $CaCO_3$ L^{-1} (Brkovič-Popovič and Popovič, 1977). Similarly, toxicity of *Daphnia magna* was directly related to activities of cationic Cu^{2+}, $CuOH^+$ and $Cu_2(OH)_2^{2+}$ over wide ranges in total copper, dissolved copper, and inorganic chelator concentration (Andrew *et al.,* 1977). Variation in salinity within the range 10–30°/oo may have no substantial effect on the availability of Cu(+2). Although the high buffering capacity of sea water keeps pH above neutrality, toxicity to invertebrates often increases at low pH and salinity (⩽5°/oo) due to the rise in the proportion of free copper.

The presence of organic chelators in solution significantly increases survival. Following exposure to 0.020 mg Cu L^{-1}, mortality in embryos of the Pacific oyster *Crassostrea gigas* exceeded 97% (Knezovich *et al.,* 1981). However, the addition of humic matter and EDTA reduced mortality to 10.8 and 1.4%, respectively. Furthermore sand sorbs copper from solution, thereby decreasing availability of copper to the test organisms.

Some species can adapt to high levels of copper. Brown (1976) reported that the 48h LC_{50} for the isopod *Asellus meridianus* was 2.5 mg L^{-1} for animals collected from a highly polluted river and 1.2 mg L^{-1} for those from less polluted sites. Growth of the tolerant isopods was not influenced by copper levels which restricted non-tolerant animals and their progeny. Similarly, Sudo and Aiba (1973) demonstrated that the mean inhibitory limit to three protozoan species could be increased 1.2–2.2 times, following acclimation for 96h.

In general, sensitivity is inversely related to the age/size of animal. Food ration and population density were directly correlated with LC_{50}'s in the copepod *Acartia tonsa* (Sosnowski *et al.*, 1979). Although this latter study suggested that sublethal effects occur at or slightly above ambient levels in the sea, the ability of wild populations to adapt to copper probably minimizes such effects. Sosnowski and Gentile (1978) concluded that LC_{50} data for wild populations were significantly more variable than those for cultured stocks. While changes in food supply under natural conditions probably account for these differences, the toxicity of copper also depends on the genetic strain of animal under investigation.

Treatment of the polychaete *Eudistylia vancouveri* with 0.010 mg Cu L^{-1} caused a shortening and clubbing of the pinnules on the gills (Young *et al.*, 1981). There was loss of cellular adhesion in the gills and structural derangement that lead to cell necrosis and death. Exposure of the whelk *Busycon canaliculatum* to copper in dilated efferent blood sinuses and blood lacunae in the leaflets of the gills (Betzer and Yevich, 1975). There was a progressive increase in the swelling of the leaflets, followed by necrosis and sloughing of the epithelium of the osphradium and gills. George *et al.* (1978) suggested that oysters *Ostrea edulis* could compartmentalize copper, following sublethal exposures. It was concluded that toxicity is reduced by active uptake from serum into granular amoebocytes. These cells may contain as much as 13,000 mg kg^{-1} copper.

Fish

Copper is usually more toxic to freshwater fish than any other heavy metal except mercury (Figure 5-4). LC_{50}'s range from 0.017 to 1.0 mg L^{-1} under most conditions. However, unusually high water hardness may increase the 96h LC_{50} to 3.0 mg L^{-1}. Copper is much less toxic to marine fish due to the high complexing capacity of salt water. Approximately 30% mortality occurred in mummichog following exposure at 8.0 mg Cu L^{-1} for 96h (Eisler and Gardner, 1973).

Acute toxicity to freshwater fish depends largely on water hardness (Figures 5-4, 5-5). Ionic copper (Cu^{2+}) and ionized hydroxides ($CuOH^{+}$, $Cu_2OH_2^{2+}$) are most toxic, with a combined 96h LC_{50} of 0.00009–0.23 mg L^{-1}. At low water hardness (12 mg L^{-1}) the incipient lethal concentration (ILC) of dissolved Cu to rainbow trout was not affected by a change in alkalinity of 10–50 mg L^{-1} (Miller and Mackay, 1980). However, the same change in alkalinity in hardwater (98 mg L^{-1}) resulted in a 1.8-fold increase in the ILC. Toxicity of combinations of $Cu^{2+}/CuOH^{+}/Cu_2^{2+}$ and H^{+} were antagonistic to rainbow trout at pH <5.4, whereas at pH >5.4 there was synergism between copper toxicity and pH (Miller and Mackay, 1980). Synergistic effects have also been noted for combinations of Cu/Cd/Zn,

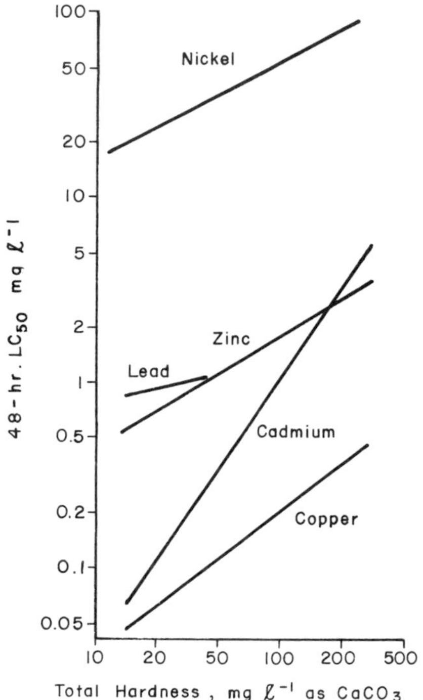

Figure 5-4. Correlation between total hardness of water and the 48h LC_{50} for rainbow trout of nickel, lead, zinc, cadmium, and copper. (From Brown, 1968.)

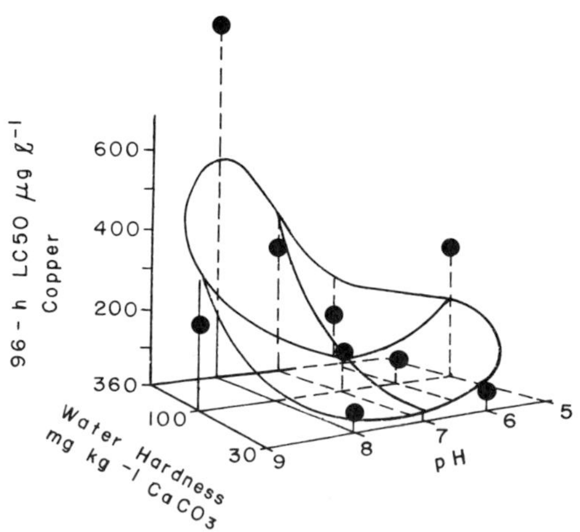

Figure 5-5A. Lethal concentrations of total dissolved copper to rainbow trout at various combinations of water hardness and pH. (From Howarth and Sprague, 1978.)

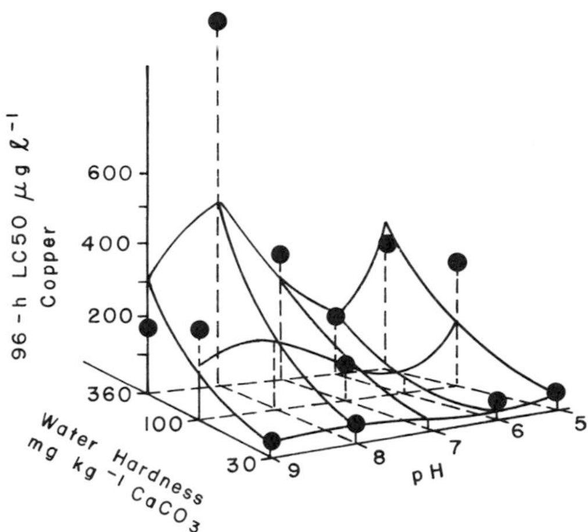

Figure 5-5B. Lethal concentrations of total dissolved copper to rainbow trout for any combination of water hardness from 30 to 360 mg L^{-1} and pH from 5 to 9. (From Howarth and Sprague, 1978.)

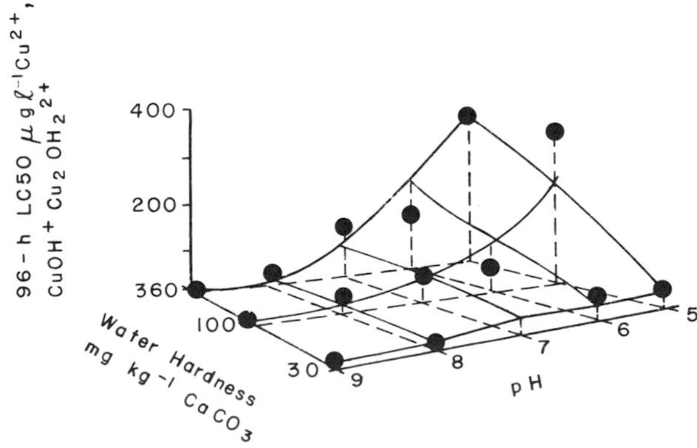

Figure 5-5C. Lethal concentrations of ionic copper (Cu^{2+}) plus $CuOH^{+}$ and $Cu_2OH_2^{2+}$ to rainbow trout at any combination of water hardness from 30 to 360 mg L^{-1} and pH from 5 to 9. (From Howarth and Sprague, 1978).

Cu/Zn, Cu/Zn/Ni, Cu/Zn/phenol, Cu/phenol and Cu/chloroamines/linear alkylsulfonate. Antagonistic effects occur following complexation with organic material, sewage and $CaCO_3$.

Inter-species variability may account for a 30-fold difference in the toxicity of copper under constant test conditions. Some of the more sensitive species include fathead minnows, guppies, and golden shiner (Smith and Heath, 1979; Pickering and Henderson, 1966). Newly-hatched alevins of steelhead trout were less sensitive than the swim-up and parr stages (Chapman, 1978). However, in chinook salmon, alevins and parr showed equal sensitivity to copper. Dixon and Sprague (1981) found that lethal tolerance by rainbow trout increased 60–106% following exposures to 29–59% of the mean incipient lethal limit. However, acclimation to copper was not permanent.

Exposure to chronic/sublethal levels (0.02–0.2 mg L^{-1}) of copper reduces survival, growth, and rate of reproduction in a variety of species. In soft water, fecundity and egg survival may be inhibited at concentrations as low as 0.004 mg L^{-1}. Furthermore, the hematocrit for a range of species increases with copper levels in the environment. Oxygen consumption, blood pH, and energy expenditure may also increase, whereas feeding rate may decline (Waiwood and Beamish, 1978; Lett *et al.,* 1976). Sublethal exposures also result in behavioural changes, such as decreased concealment and ability to orient. This partially reflects depression in olfactory response following treatment with copper. There is also a marked change in preferred temperature, which could in turn produce an immediate decrease in survival. Although not demonstrated, it is likely that salinity preferenda in fish would change, again reducing survival.

Copper ions precipitate gill secretions, causing death by asphyxiation (Tsai, 1979). There is also an impairment of haemopoetic tissue in gill filaments which results in a reduction of oxidative activity. Necrotic kidney cells, fatty degeneration of the liver, and brain hemorrhage have been reported for acutely exposed fish. Eisler and Gardner (1973) found that nuclei of affected squamous epithelial cells of the oral cavity in mummichogs appeared pyknotic and hyperchromatic. There was also a marked increase in the number of dividing cells in the respiratory epithelium. These changes are largely similar to those produced by salts of other heavy metals.

Humans

Copper is not acutely toxic to humans. This is due to the intermediate coordinate character of copper between hard and soft acids. Hence copper seldom interferes with sulfur-containing proteins. By contrast the moderate toxicity to aquatic animals results from its sequestering and precipitating the essential carboxylic acids. In some instances, copper deficiency in humans imitates chronic copper intoxication. There is no indication that copper is carcinogenic or mutagenic to humans.

References

Agadi, V.V., N.B. Bhosle, and A.G. Untawale. 1978. Metal concentration in some seaweeds of Goa (India). *Botanica Marina* **21**:247–250.

Anderson, R.V. 1977. Concentration of cadmium, copper, lead, and zinc in thirty-five genera of freshwater macroinvertebrates from the Fox River, Illinois and Wisconsin. *Bulletin of Environmental Contamination and Toxicology* **18**:345–349.

Andrew, R.W., K.E. Biesinger, and G.E. Glass. 1977. Effects of inorganic complexing on the toxicity of copper to *Daphnia magna. Water Research* **11**:309–315.

Baes, C.F. Jr., and R.E. Mesmer. 1976. *The hydrolysis of cations.* Wiley-Interscience, New York. 489 pp.

Banat, K., U. Förstner, and G. Muller. 1974. Experimental mobilization of metals from aquatic sediments by nitrilotriacetic acid. *Chemical Geology* **14**:199–207.

Bartelt, van R.-D., and U. Förstner. 1977. Schwermetalle im staugeregelten Neckaruntersuchungen and sedimenten, algen und wasserproben. *Jber. Mitt. oberrhein. geol. Ver.* **59**:247–263.

Batley, G.E., and D. Gardner. 1978. A study of copper, lead and cadmium speciation in some estuarine and coastal marine waters. *Estuarine and Coastal Marine Science* **7**:59–70.

Betzer, S.B., and P.P. Ỳevich. 1975. Copper toxicity in *Busycon canaliculatum* L. *Biological Bulletin* **148**:16–25.

Bewers, J.M., and P.A. Yeats. 1978. Trace metals in the waters of a partially mixed estuary. *Estuarine and Coastal Marine Science* **7**:147–162.

Bilinski, H., R. Huston, and W. Stumm. 1976. Determination of the stability constants of some hydroxo and carbonato complexes of Pb(II), Cu(II), Cd(II), and Zn(II) in dilute solutions by anodic stripping voltammetry and differential pulse polarography. *Analytica Chimica Acta* **84**:157.

Bothner, M.H., P.J. Aruscavage, W.M. Ferrebee, and P.A. Baedecker. 1980. Trace metal concentrations in sediment cores from the Continental Shelf off the south-eastern United States. *Estuarine and Coastal Marine Science* **10**:523–541.

Boyden, C.R., and M.G. Romeril. 1974. A trace metal problem in pond oyster culture. *Marine Pollution Bulletin* **5**:74–78.

Boyden, C.R., S.R. Aston, and I. Thornton. 1979. Tidal and seasonal variations of trace elements in two Cornish estuaries. *Estuarine and Coastal Marine Science* **9**:303–317.

Brković-Popović, I., and M. Popović. 1977. Effects of heavy metals on survival and respiration rate of tubificid worms: Part I. Effects on survival. *Environmental Pollution* **13**:65–72.

Brown, B.E. 1976. Observations on the tolerance of the isopod *Asellus meridianus* Rac. to copper and lead. *Water Research* **10**:555–559.

Brown, V.M. 1968. The calculation of the acute toxicity of mixtures of poisons to rainbow trout. *Water Research* **2**:723–733.

Brügmann, L. 1981. Heavy metals in the Baltic Sea. *Marine Pollution Bulletin.* **12**:214–218.

Cedeno-Maldonado, A., and J.A. Swader. 1974. Studies on the mechanism of copper toxicity in *Chlorella. Weed Science* **22**:443–449.

Chapman, G.A. 1978. Toxicities of cadmium, copper and zinc to four juvenile stages of chinook salmon and steelhead. *Transactions of the American Fisheries Society* **107**:841–847.

Coughtrey, P.J., and M.H. Martin. 1976. The distribution of Pb, Zn, Cd, and Cu within the pulmonate mollusc *Helix aspersa* Müller. *Oecologia* **23**:315–322.

Davies, W.G., R.J. Otter, and J.E. Prue. 1957. The dissociation constant of copper sulphate in aqueous solution. *Discussions of the Faraday Society* **24**:103–107.

Dietz, F. 1973. The enrichment of heavy metals in submerged plants. *In*: S.H. Jenkins (Ed.), *Advances in water pollution research.* Pergamon Press, Oxford, pp. 53–62.

Dixon, D.G., and J.B. Sprague. 1981. Acclimation to copper by rainbow trout *(Salmo gairdneri)*—a modifying factor in toxicity. *Canadian Journal of Fisheries and Aquatic Science* **238**:880–888.

Drbal, K. 1976. Relation between copper contents in fish tissues, in plankton and in pond waters. *Zivocisna vyroba* **21**:925–932.

Duinker, J.C., and R.F. Nolting. 1977. Dissolved and particulate trace metals in the Rhine estuary and the Southern Bight. *Marine Pollution Bulletin* **8**:65–71.

Eisenreich, S.J. 1980. Atmospheric input of trace metals to Lake Michigan. *Water, Air, and Soil Pollution* **13**:287–301.

Eisler, R., and G.R. Gardner. 1973. Acute toxicology to an estuarine teleost of mixtures of cadmium, copper, and zinc salts. *Journal of Fish Biology* **5**:131–142.

Emerson, R.R., D.F. Soule, and M. Oguri. 1976. Heavy metal concentrations in marine organisms and sediments collected near an industrial waste outfall. *Proceedings of International Conference on Environmental Sensing and Assessment,* Vol. **1**, Las Vegas, Nevada, Sept. 14-19, 1975. pp. 1–5.

Ernst, W.H.O. 1975. Physiology of heavy metal resistance in plants. *In*: T.C. Hutchinsn (Ed.), *Proceedings of 1st International Conference on Heavy Metals in the Environment,* Vol. **II**, University of Toronto Institute for Environmental Studies, Toronto, Canada, pp. 121–136.

Frieden, E. 1979. Ceruloplasmin: The serum copper transport protein with oxidase activity. *In*: J.O. Nriagu (Ed.), *Copper in the environment, Part 2, Health effects.* Wiley, New York, pp. 242–284.

Fuller, R.H., and R.C. Averett. 1975. Evaluation of copper accumulation in part of the California aqueduct. *Water Resources Bulletin* **11**:946–952.

George, S.G., B.J.S. Pirie, A.R. Cheyne, T.L. Coombs, and P.T. Grant. 1978. Detoxication of metals by marine bivalves: An ultrastructural study of the compartmentation of copper and zinc in the oyster *Ostrea edulis. Marine Biology* **45**:147–156.

Gibbs, R.J. 1977. Transport phases of transition metals in the Amazon and Yukon Rivers. *Geological Society of America Bulletin* **88**:829–843.

Haug, A., S. Melsom, and S. Omang. 1974. Estimation of heavy metal pollution in two Norwegian fjord areas by analysis of the brown alga *Ascophyllum nodosum. Environmental Pollution* **7**:179–192.

Horowitz, A., and B.J. Presley. 1977. Trace metal concentrations and partitioning in zooplankton, neuston, and benthos from the south Texas outer continental shelf. *Archives of Environmental Contamination and Toxicology* **5**:241–255.

Howarth, R.S., and J.B. Sprague. 1978. Copper lethality to rainbow trout in waters of various hardness and pH. *Water Research* **12**:455–462.

Imhoff, K.R., P. Koppe, and F. Dietz. 1980. Heavy metals in the Ruhr River and their budget in the catchment area. *Progress in Water Technology* **12**:735–749.

Ireland, M.P., and R.J. Wootton. 1977. Distribution of lead, zinc, copper and manganese in the marine gastropods *Thais lapillus* and *Littorina littorea* around the coast of Wales. *Environmental Pollution* **12**:27–41.

Jedwab, J. 1979. Copper, zinc and lead minerals suspended in ocean waters. *Geochimica et Cosmochimica Acta* **43**:101–110.

Knezovich, J.P., F.L. Harrison, and J.S. Tucker. 1981. The influence of organic chelators on the toxicity of copper to embryos of the Pacific oyster, *Crassostrea gigas. Archives of Environmental Contamination and Toxicology* **10**:241–249.

Kretzschmar, J.G., I. Delespaul, and Th. de Rijck. 1980. Heavy metal levels in Belgium: a five year survey. *The Science of the Total Environment* **14**:85–97.

Legittimo, P.C., G. Piccardi, and F. Pantani. 1980. Cu, Pb, and Zn determination in rainwater by differential pulse anodic stripping voltammetry. *Water, Air, and Soil Pollution* **14**;435–441.

Lett, P.F., G.J. Farmer, and F.W.H. Beamish. 1976. Effect of copper on some aspects of the bioenergetics of rainbow trout *(Salmo gairdneri). Journal of the Fisheries Research Board of Canada* **33**:1335–1342.

Loring, D.H. 1978. Geochemistry of zinc, copper and lead in the sediments of the estuary and Gulf of St. Lawrence. *Canadian Journal of Earth Sciences* **15**:757–772.

Manly, R., and W.O. George. 1977. The occurrence of some heavy metals in populations of the freshwater mussel *Anodonta anatina* (L.) from the River Thames. *Environmental Pollution* **14**:139–153.

Mantoura, R.F.C., A. Dickson, and J.P. Riley. 1978. The complexation of metals with humic materials in natural waters. *Estuarine and Coastal Marine Science* **6**:387–408.

Mathis, B.J., and T.F. Cummings. 1973. Selected metals in sediments, water, and biota in the Illinois River. *Journal Water Pollution Control Federation* **45**:1573–1583.

Maxfield, D., J.M. Rodriguez, M. Buettner, J. Davis, L. Forbes, R. Kovacs, W. Russel, L. Schultz, R. Smith, J. Stanton, and C.M. Wai. 1974. Heavy metal content in the sediments of the southern part of the Coeur d'Alene lake. *Environmental Pollution* **6**:263–266.

McFarlane, G.A., and W.G. Franzin. 1980. An examination of Cd, Cu, and Hg concentrations in livers of northern pike, *Esox lucius,* and white sucker, *Catostomus commersoni,* from five lakes near a base metal smelter at Flin Flon, Manitoba. *Canadian Journal of Fisheries and Aquatic Sciences* **37**:1573–1578.

McIntosh, A.W. 1975. Fate of copper in ponds. *Pesticides Monitoring Journal* **8**:225–231.

Miller, T.G., and W.C. Mackay. 1980. The effects of hardness, alkalinity and pH of test water on the toxicity of copper to rainbow trout *(Salmo gairdneri). Water Research* **14**:129–133.

Moore, J.W. 1979. Diversity and indicator species as measures of water pollution in a subarctic lake. *Hydrobiologia* **66**:73–80.

Moore, J.W., V.A. Beaubien, and D.J. Sutherland. 1979. Comparative effects of sediment and water contamination on benthic invertebrates in four lakes. *Bulletin of Environmental Contamination and Toxicology* **23**:840–847.

Nriagu, J.O. 1979a. Global inventory of natural and anthropogenic emissions of trace metals to the atmosphere. *Nature* **279**:409–411.

Nriagu, J.O. 1979b. The global copper cycle. *In*: J.O. Nriagu (Ed.), *Copper in the environment, Part I, Ecological cycling*. Wiley, New York, pp. 1–17.

Nriagu, J.O. 1979c. Copper in the atmosphere and precipitation. *In*: J.O. Nriagu (Ed.), *Copper in the environment, Part I, Ecological cycling*. Wiley, New York, pp. 43–75.

Nriagu, J.O., and R.D. Coker. 1980. Trace metals in humic and fulvic acids from Lake Ontario sediments. *Environmental Science and Technology* **4**:443–446.

Overnell, J. 1975. The effect of heavy metals on photosynthesis and loss of cell potassium in two species of marine algae, *Dunaliella tertiolecta* and *Phaeodactylum tricornutum*. *Marine Biology* **29**:99–103.

Overnell, J. 1976. Inhibition of marine algal photosynthesis by heavy metals. *Marine Biology* **38**:335–342.

Pagenkopf, G.K., and D. Cameron. 1979. Deposition of trace metals in stream sediments. *Water, Air, and Soil Pollution* **11**:429–435.

Pickering, Q.H., and C. Henderson. 1966. The acute toxicity of some heavy metals to different species of warmwater fishes. *Air and Water Pollution* **10**:453–463.

Rainbow, P.S., A.G. Scott, E.A. Wiggins, and R.W. Jackson. 1980. Effect of chelating agents on the accumulation of cadmium by the barnacle *Semibalanus balanoides*, and complexation of soluble Cd, Zn, and Cu. *Marine Ecology-Progress Series* **2**:143–152.

Ramamoorthy, S., and D.J. Kushner. 1975. Heavy metal binding components of river water. *Journal of the Fisheries Research Board of Canada* **32**:1755–1766.

Roth, I., and H. Hornung. 1977. Heavy metal concentrations in water, sediments and fish from Mediterranean coastal area, Israel. *Environmental Science and Technology* **11**:265–269.

Schnitzer, M., and H. Kerndorff. 1981. Reactions of fulvic acid with metal ions. *Water, Air, and Soil Pollution* **15**:97–108.

Shiber, J.G. 1980. Metal concentrations in marine sediments from Lebanon. *Water, Air, and Soil Pollution* **13**:35–43.

Sillén, L.G., and A.E. Martell. 1971. *Stability constants of metal-ion complexes. Supplement No. 1.* Special Publication No. 25, The Chemical Society, London, 865 pp.

Skei, J.M., N.B. Price, S.E. Calvert, and H. Holtedahl. 1972. The distribution of heavy metals in sediments of Sorfjord, West Norway. *Water, Air, and Soil Pollution* **1**:452–461.

Smith, M.J., and A.G. Heath. 1979. Acute toxicity of copper, chromate, zinc, and cyanide to freshwater fish: effect of different temperatures. *Bulletin of Environmental Contamination and Toxicology* **22**:113–119.

Sosnowski, S.L., and J.H. Gentile. 1978. Toxicological comparison of natural and cultured populations of *Acartia tonsa* to cadmium, copper, and mercury. *Journal of the Fisheries Research Board of Canada* **35**:1366–1369.

Sosnowski, S.L., D.J. Germond, and J.H. Gentile. 1979. The effect of nutrition on the response of field populations of the calanoid copepod *Acartia tonsa* to copper. *Water Research* **13**:449–452.

Stenner, R.D., and G. Nickless. 1974. Absorption of cadmium, copper, and zinc by dog whelks in the Bristol Channel. *Nature* **247**:198–199.

Stokes, P.M. 1975. Adaptation of green algae to high levels of copper and nickel in aquatic environments. *In*: T.C. Hutchinson (Ed.), *Proceedings of 1st Interna-*

tional Conference on Heavy Metals in the Environment, Vol. **II**, University of Toronto Institute for Environmental Studies, Toronto, Canada, pp. 137–154.

Sudo, R., and S. Aiba. 1973. Effect of copper and hexavalent chromium on the specific growth rate of ciliata isolated from activated-sludge. *Water Research* **7**:1301–1307.

Taylor, D. 1979. The effect of discharges from three industrialized estuaries on the distribution of heavy metals in the coastal sediments of the North Sea. *Estuarine and Coastal Marine Science* **8**:387–393.

Tessier, A., P.G.C. Campbell, and M. Bisson. 1980. Trace metal speciation in the Yamaska and St. François Rivers (Quebec). *Canadian Journal of Earth Sciences* **17**:90–105.

Trollope, D.R., and B. Evans. 1976. Concentrations of copper, iron, lead, nickel, and zinc in freshwater algal blooms. *Environmental Pollution* **11**:109–116.

Tsai, C-F. 1979. Survival, overturning and lethal exposure times for the pearl dace, *Semotilus margaritus* (Cope), exposed to copper solution. *Comparative Biochemistry and Physiology* **64C**:1–6.

Tyler, P.A., and R.T. Buckney. 1973. Pollution of a Tasmanian river by mine effluents. I. Chemical evidence. *Internationale Revue der Gesamten Hydrobiologie* **58**:873–883.

United States Minerals Yearbooks. 1911-1979. Bureau of Mines, US Department of the Interior, Washington, D.C.

Vuceta, J., and J.J. Morgan. 1977. Hydrolysis of Cu (II). *Limnology and Oceanography* **22**:742–746.

van der Weijden, C.H., M.J.H.L. Arnoldus, and C.J. Meurs. 1977. Desorption of metals from suspended material in the Rhine estuary. *Netherlands Journal of Sea Research* **11**:130–145.

von Westernhagen, H., H. Rosenthal, V. Dethlefsen, W. Ernst, U. Harms, and P.D. Hansen. 1981. Bioaccumulating substances and reproductive success in Baltic flounder *Platichthys flesus. Aquatic Toxicology* **1**:85–99.

Waiwood, K.G., and F.W.H. Beamish. 1978. Effects of copper, pH, and hardness on the critical swimming performance of rainbow trout (*Salmo gairdneri* Richardson). *Water Research* **12**:611–619.

Watling, H.R., and R.J. Watling. 1976. Trace metals in *Choromytilus meridionalis. Marine Pollution Bulletin* **7**:91–94.

Wharfe, J.R., and W.L.F. van den Broek. 1977. Heavy metals in macroinvertebrates and fish from the low Medway estuary, Kent. *Marine Pollution Bulletin* **8**:31–34.

Wiener, J.G. 1979. Aerial inputs of cadmium, copper, lead, and manganese into a freshwater pond in the vicinity of a coal-fired power plant. *Water, Air, and Soil Pollution* **12**:343–353.

Willey, J.D., and R.A. Fitzgerald. 1980. Trace metal geochemistry in sediments from the Miramichi estuary, New Brunswick. *Canadian Journal of Earth Sciences* **17**:254–265.

Young, J.S., R.R. Adee, I. Piscopo, and R.L. Buschbom. 1981. Effect of copper on the sabellid polychaete, *Eudistylia vancouveri.* II. Copper accumulation and tissue injury in the branchial crown. *Archives of Environmental Contamination and Toxicology* **10**:87–104.

6
Lead

Chemistry

Lead is a member of the Group IV elements (C, Si, Ge, Sn, and Pb) of the periodic classification. The electropositive character in this group increases with atomic number and lead is truly metallic compared to carbon and silicon. Lead, unlike C and Si, does not bind to another identical atom, shows marked decrease in covalency, and has stable (+2) and (+4) oxidation states. Members of group IV form organo derivatives. Lead forms alkyl and aryl compounds. Tetraethyllead is widely used as an antiknock agent in gasoline. With the exception of nitrate and acetate, most lead(+2) salts are insoluble in water.

Lead is classified as an intermediate acceptor between hard and soft acids in its interaction with ligands. A hard acceptor is characterized by low polarizability, low electronegativity, large positive charge density (high oxidation state and small radius) and formation of ionic bonds. The converse is true for a soft acceptor which forms essentially covalent bonds. Lead resembles the divalent alkaline earth group metals in chemical behavior more than its own Group IVA metals, except in the poor solubility of lead salts such as halides, hydroxides, sulfates and phosphates. Lead resembles Ca in deposition in and remobilization from the skeletal compartment of the body.

Production, Uses, and Discharges

Production and Uses

Global production of lead from both smelter and mining operations has been relatively high throughout this century. Production totaled 9.6×10^6 metric tons during 1900–1909, increasing to 27.7×10^6 metric tons during 1960–1969. Production increased to 34.0×10^6 metric tons during the 1970's and will likely continue to increase slowly for the foreseeable future.

Lead is one of the oldest metals known to man and, since medieval times, has been used in piping, building materials, solders, paint, type metal, ammunition, and castings. In more recent times lead has been used mainly in storage batteries, metal products, chemicals, and pigments (Table 6-1).

Storage Batteries. Lead is mainly used in acid storage batteries. Significant improvements have been made to reduce the unit lead weight in a battery and also to increase the average battery life and performance. This category in 1979 accounted for about 60% of total consumption in the USA (Table 6-1) and it is expected to increase due to rapid growth in the use of electric-powered industrial trucks, particularly fork-lift vehicles. Many governments in Europe and North America have experimental transportation programs involving storage batteries as the power source, substituting the gasoline-operated internal combustion engines.

Metal Products. The two most important products are ammunition and solder followed by casting materials, sheet lead, and others. The consumption of lead for ammunition has steadily increased since 1960: 11.8% (1960), 26.5% (1974), and 22.4% (1979) of the total consumption under metal products category. Because of its unique sound control characteristics, consumption has also increased in the manufacture of sound attenuation material, both as sheets and composition paneling. Lead-coated steel sheeting (terne steel), which combines anti-corrosion and sound-barrier proper-

Table 6-1. Lead consumption (USA) by product class (in metric tons $\times 10^3$).

Product class	1960	1970	1974	1978	1979
Metal products	336.1	319.4	289.6	221.9	237.3
Storage batteries	326.1	539.5	774.4	879.3	814.3
Pigments	89.6	89.8	108.3	91.6	90.8
Chemicals	151.5	253.8	228.3	178.5	186.9
Miscellaneous	30.1	37.1	47.0	61.4	28.9
Total	933.4	1239.6	1447.6	1432.7	1358.2

Source: US Minerals Yearbooks.

ties with the strength of steel, has found several building applications. Lead-asbestos anti-vibration pads are employed in foundations for office buildings, and hotel and apartments exposed to severe vibration from nearby heavy traffic. It is also used in the mounting of various types of equipment including air-conditioning systems, heavy industrial equipment, and commercial laundry machines. Additional applications include cable sheathing, collapsible tubes, caulking materials, and corrosive-liquid containers. Although lead usage for metal sheathing was on the decline in most industrialized countries, it has shown some revival as a result of export, particularly to the Middle East.

Chemicals. Tetraethyllead constitutes an antiknock agent in gasoline. The production of non-leaded gasoline for use in automobiles with emission control devices has sharply reduced the use of tetraethyllead in gasoline since 1972. Lead sterate is used in the production of synthetic polymers. Batteries and gasoline additives accounted for 75% of the total lead consumption in the United States in 1975.

Pigments. The use of lead in decorative paints (such as white lead and colours) has declined significantly. However consumption for anticorrosive and highway traffic safety paints (as red lead and lead chromates) is increasing because of their excellent versatility and relatively low cost.

Miscellaneous Uses. This category includes automotive wheel weights, ship ballast, and various alloys, and as lead-ferrite for permanent magnets in small electric motors. Growing areas of application are leaded-porcelain enamel in coating aluminum and for radiation-shielding against gamma rays in nuclear-powered reactors, nuclear-powered ships, and submarines, and in shipping casks for transporting radioactive materials. Organometallic lead compounds have potential applications as anti-fouling paints, wood and cotton preservatives, lubricant-oil additives, polymethane foam catalysts, molluscicides, antibacterial agents, and rodent repellents.

The three principal grades of refined lead are corroding, chemical, and common desilverized lead. The corroding grade has the highest purity and is used mainly in the manufacture of pigments, battery oxides, and tetraethyllead. Chemical lead is used for cable sheathing because of its superior creep and corrosion resistance. Common lead is used mostly in industrial and home construction.

Discharges

Emissions into the atmosphere have increased sharply during this century, reaching a peak of 4265×10^3 metric tons during the 1970's (Table 6-2). Anthropogenic inputs greatly exceed those from natural sources (Tables 6-2, 6-3). Combustion of oil and gasoline account for $\geqslant 50\%$ of all anthropogenic

Table 6-2. All-time global consumption and anthropogenic emission of lead.

Period	Consumption (10^6 metric tons)	Emission (10^3 metric tons)
Pre-1850	55	2420
1850–1900	25	1100
1901–1910	10.7	471
1911–1920	11.2	493
1921–1930	14.2	1120
1931–1940	14.6	1639
1941–1950	14.9	1672
1951–1960	24	2694
1961–1970	33	3704
1971–1980	38	4265
Total (all-time)	241	19578

Source: Nriagu (1979). Reprinted by permission from *Nature* **279**:409–411.

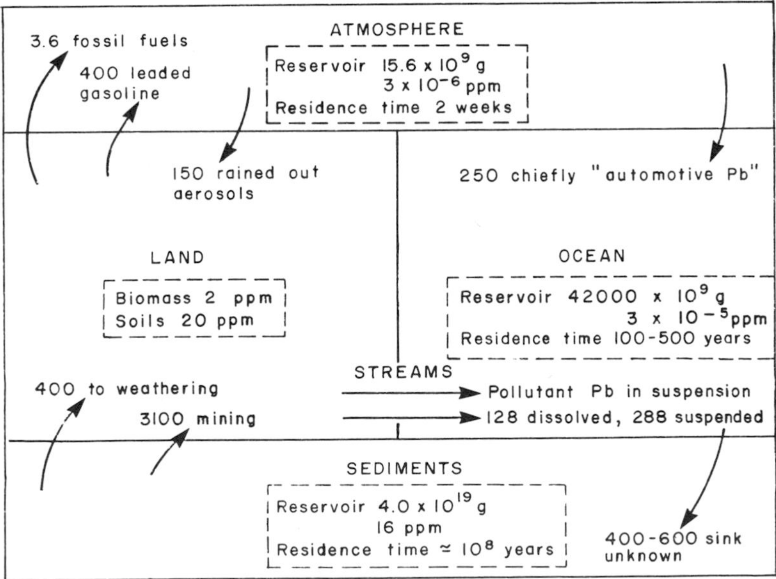

Figure 6-1. Quantitative estimate of lead cycling in the environment (10^6 kg yr^{-1}). From *Chemical Cycles and the Global Environment: Assessing Human Influences,* by R.M. Garrels, F.T. Mackenzie, and C. Hunt.

Table 6-3. Annual global emissions of lead from natural sources.

Source	Global production (10^6 metric tons)*	Global annual emissions (10^3 metric tons)
Wind-blown dusts	500 (6–1100)	16 (0.19–35)
Forest fires	36 (2–200)	0.5 (0.04–2.8)
Volcanogenic particles	10 (6.5–150)	6.4 (4.2–96)
Vegetation	75 (75–1000)	1.6 (1.6–21)
Seasalt sprays	1000 (300–2000)	0.02 (0.01–0.05)
Total **		24.5

Source: Nriagu (1979). Reprinted by permission from *Nature* **279:**409–411. Copyright © 1979, Macmillan Journals Limited.

* Most acceptable values with a range in values from literature in brackets.

** Total values are for the most acceptable production figures.

emissions, and thus form a major component of the global cycle of lead (Figure 6-1). Automobile exhaust accounts for about 50% of the total inorganic lead absorbed by humans, reflecting the high (75%) proportion of inorganic lead in exhaust. Other major anthropogenic inputs include non-

Table 6-4. Annual global anthropogenic lead emissions into the atmosphere in 1975.

Source	Global production (10^6 metric tons)	Emissions (10^3 metric tons)
Mining, non-ferrous metals	16	8.2
Primary non-ferrous metal production		
Copper	7.9	27
Lead	4.0	31
Nickel	0.8	2.5
Zinc	5.6	16
Secondary non-ferrous metal production	4.0	0.77
Iron and steel production	1300	50
Industrial applications	—	7.4
Coal combustion	3100	14
Oil (including gasoline) combustion	2800	273
Wood combustion	640	4.5
Waste incineration	1500	8.9
Phosphate fertilizers manufacture	118	0.05
Miscellaneous	—	5.9
Total		449

Source: Nriagu (1979). Reprinted by permission from *Nature* **279:**409–411. Copyright © 1979, Macmillan Journals Limited.

ferrous metal production and iron/steel production whereas wind-blown dust is the main source of natural emission (Tables 6-3, 6-4).

Mining contributes substantially to lead-containing solid wastes in the environment. In some cases concentrations in solid wastes may approach 20,000 mg kg^{-1}. Although these discharges account for the majority of lead deposited on land, atmospheric fallout is usually the most important source of lead in marine and freshwaters (Figure 6-1).

Lead in Aquatic Systems

Speciation in Natural Waters

The behaviour of lead in natural waters is a combination of precipitation equilibria and complexing with inorganic and organic ligands. The degree of mobility of lead depends on the physicochemical state of the complexes formed.

Binding to Inorganic and Organic Ligands. Hydrolysis of precipitates (lead phosphate and lead sulfides) above pH 6 solubilizes lead as $Pb(OH)^+$ (Table 6-5). Insoluble $Pb(OH)_2$ is not formed till pH 10.0. At pH 8.5, $Pb(OH)^+$ is the only major species in the chloride concentration range 350–56,200 mg L^{-1} (Hahne and Kroontje, 1973). Since sea water contains 20,000 mg L^{-1} of Cl (Klein, 1959), at pH 8.1–8.2 $Pb(OH)^+$ will predominate over the chloride complexes in sea water. The hydroxy ion is common in natural waters and its interaction with heavy metals alters their mobility. Although the Pb(+2) and $Pb(OH)^+$ ions are present in equal concentrations at pH 6, $Pb(OH)^{+1}$ predominates at pH 8. This latter ion has markedly different affinity for sorption sites than Pb(+2).

Lead forms moderately strong chelates with organic ligands containing S, N, and O donor atoms. Lead also binds to different microbial growth media;

Table 6-5. Stability constants and solubility products of some Pb–inorganic complexes.

	Stability constant*				Solubility product	
	Log					
System	β_1	β_2	β_3	β_4	Precipitate	Log K_{sp}
PbCl	0.88	1.49	1.09	0.94	$Pb_3(PO_4)_2$	−42.10
PbOH	7.82	10.88	13.94	16.30	PbS	−28.15
					$Pb(OH)_2$	−19.52

Source: Sillén and Martell (1971).
* β = cumulative formation constant.

Ramamoorthy and Kushner (1975a) showed that the availability of free cations in such media decreased in the order

$$Cd(+2) >> Cu(+2) >> Pb(+2) > Hg(+2).$$

Binding to Particulates. Wilson (1976), reviewing the metal concentration in river waters around the world, reported that lead shows variable levels of binding (15–83%) in association with suspended solids. Ramamoorthy and Kushner (1975b) reported that binding components in Ottawa River water were $<0.45\ \mu m$ in diameter. A significant fraction was bound to organic components of molecular weight (MW) $>45{,}000$ and <1400 and the rest to $<45{,}000 - >16{,}000$ and $<16{,}000 - >1{,}400$. The (α) fraction of total Pb(+2) was 0.38, 0.21 and 0.085, respectively, at 10, 20 and 40 mg L^{-1}, giving an average of 3.8 mg L^{-1} bound. This shows effective metal-buffering by the river water over a wide range of Pb(+2) concentrations. The same study showed that waters from a canal (receiving road-deicing salt discharge) and creek (heavily polluted with sewage) bound all Pb(+2) ion in a component of MW <1400.

Physico-chemical speciation of lead in drinking water indicates little or no free ionic lead. A significant portion of lead is bound to colloids, either hydrous iron oxides or organic macromolecules depending on the composition of the water. A substantial fraction is non-ion-exhangeable.

Transport in Natural Waters

Hart and Davies (1981) studied the speciation of lead in freshwater and estuarine regions of the Yarrah River, Australia. In the freshwater section, about 45% of total lead was present in particulate forms. This value was low compared with those (47–72%) reported from other areas, such as the Susquehanna River, USA (McDuffie *et al.*, 1976) and the Rhine River, FRG (DeGroot *et al.*, 1976). The calculated equilibrium distribution of ion-exchangeable species showed that $PbCO_3$ accounted for about 80% with smaller contributions from Pb(+2) and $PbOH^+$. Pb–organic complexes became significant at ligand concentrations $>10^{-6}$ M. In the estuarine section, 69% of lead was present in the particulate fraction. In the filterable fraction, about 54% was ion-exchangeable. In both fresh and estuarine waters, proportions of dialysable-Pb were similar to the ion-exhangeable-Pb. Organic matter at concentrations $>10^{-5}$ M (~ 5 mg L^{-1} DOC) did not significantly influence speciation. The study also showed that as river water entered the estuary, concentrations of lead increased. This was largely due to the increased particulate fraction in the estuary, reflecting resuspension of sediments, atmospheric fallout and input from urban zone creeks. A strong linear correlation was found between the total Pb and total bound Pb (particulate + bound species in the filterable fraction). Input in the already bound component (irreversible with the aqueous phase) was a possible cause for such a strong correlation.

Boyden *et al.* (1979) arrived at the same conclusion from a study on seasonal changes in the concentrations of trace elements in suspended solids, filtered and bottom waters, and sediments of two Cornish estuaries. At low tide, 58% of total lead in the bottom waters was in the particulate form. All other phases of the cycle showed no significant proportion of particulate form. Runoff of particulate-Pb was the probable cause for the input.

Lead in Sediments

Sorption and Desorption. Sorption of lead by river sediments is correlated to organic content and grain size. In the absence of soluble complexing species, lead is almost totally sorbed as precipitated species at pH > 6.0. In acidic media, humic acid sorbs lead stronger than clays. The trend is reversed at pH $\geqslant 6.5$ where soluble Pb-humate complexes were formed. Clay seems to strongly compete with soluble lead-humates for retention on the solid phase.

About 5 – 10% of lead in Lake Ontario sediments was bound to organic matter and enriched in the humic acid fraction (Nriagu and Coker, 1980). Humic and fulvic acids accounted for 9% and 1 – 4% of the organic matter, respectively. The CF in humic acids (ratio of Pb concentration in humic acids to that in sediments) was 1.4 – 3.0. The order of CF's for several metals studied matches neither the order of stability constants of metal-humic acid complexes nor the order of absorption of metals by humic acids. Hence, humic acids acquire a large fraction of Pb burden from the overlying water through suspended particulates. Additional amounts are deposited after the humic acid is buried in the sediments. Ramamoorthy and Rust (1978) reported the approximate ratios of lead desorption from river sediments by chloride versus NTA was 1 : 10. This is consistent with the ratio of stability constants of PbCl and Pb-NTA : $10^{1.73}$ and $10^{11.47}$, respectively.

Transformations

Organoleads are less stable than their counterparts formed by the lighter members (carbon and silicon) of the same group IVb elements, due to the relative weakness of the lead-carbon bond. The nature and number of organic groups determine the stability of the compound. Generally, aryllead compounds are more stable than the alkyllead compounds and their stability increases with increasing number of organic groups.

Dimethyllead disproportionates irreversibly:

$$2(CH_3)_2PbX_2 \rightarrow (CH_3)_3PbX + PbX_2 + CH_3X \quad \ldots \qquad (1)$$

The stoichiometry is not influenced by the type or concentration of salt added. The reaction is first order, increasing with salt concentration; the order of increasing influence of the anions on the rate of the reaction (1) is

X^-: Acetate, perchlorate, nitrate, chloride, nitrite, bromide, sulfocyanide, iodide

$(CH_3)_3PbX$ disproportionates slowly as follows:

$$3(CH_3)_3PbX \rightarrow 2(CH_3)_4Pb + PbX_2 + CH_3X \ldots \quad (2)$$

The effect of anions on reaction (2) is smaller than on (1).

Biomethylation of Pb(+2) (as lead acetate) by anaerobic cultures to produce tetramethyllead (Me_4Pb) was observed by Schmidt and Huber (1976). The production rate was about 2.5 μg Pb day^{-1}. Biomethylation was reproducible under (i) controlled Pb(+2) concentration (1–10 μg mL^{-1}), (ii) low to medium concentrations of sulfur compounds avoiding precipitation of Pb as PbS and (iii) inoculum not older than 6–7 weeks. Furthermore addition of Me_3PbX to anaerobic cultures (bacteria from the surface of a lake or from anaerobic sediment) increased the rate of the reaction producing less Pb(+2) and more Me_4Pb than expected from the stoichiometry of equation (2). Me_4Pb may be formed on incubation of lead-containing sediments. The addition of Me_3Pb^+ salt and lead nitrate or lead chloride increases Me_4Pb production; pure species of bacterial isolates, however, may not be able to methylate PbX_2 salts.

Jarvie *et al.* (1975) presented evidence for chemical alkylation of lead, denying any microbial methylation in an anaerobic sediment system. Chemical disproportionation explained the formation of Me_4Pb. Huber *et al.* (1978) provided a rough estimate of the ratio of chemically and biologically produced Me_4Pb; about 15–19% of the toal Me_4Pb was produced biologically. The sources of Me_4Pb in an anaerobic microbial system are trimethyllead and inorganic Pb(+2) ion (Figure 6-2).

Similar experiments with triethylead salts (Et_3PbX) showed biomethyla-

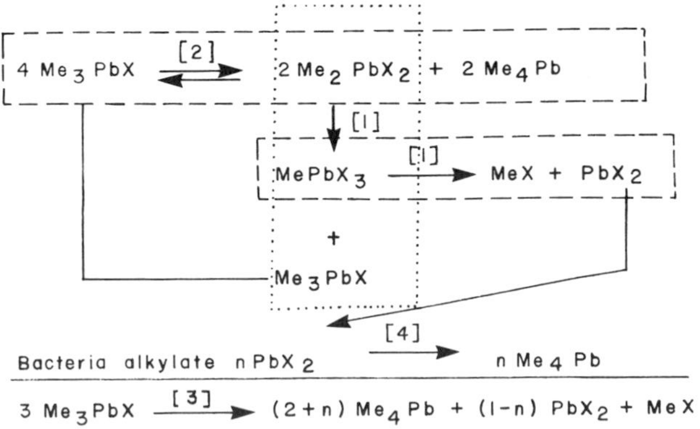

Figure 6-2. Sources of Me_4Pb in an anaerobic bacterial culture. Reprinted with permission from *Organometals and Organometalloids.* Copyright 1978, American Chemical Society.

tion of (i) Et_3PbX directly to Et_3MePb, (ii) biomethylation of the end product Pb(+2) to Me_4Pb, and (iii) chemical disproportionation of Et_3PbX to Et_4Pb (Huber *et al.*, 1978). The chemical and biological modes of production (Et_4Pb chemically, Et_3MePb and Me_4Pb biologically) roughly correspond to the estimate of Me_4Pb production from Me_3PbX.

It should be emphasized that the biomethylation of Pb(+2) has been observed only in laboratory anaerobic conditions. Great care should be exercised in extrapolating data to natural systems.

Residues

Water, Precipitation, and Sediments

Concentrations of soluble lead in uncontaminated freshwaters are generally $\leqslant 3$ μg L^{-1} (Förstner and Wittman, 1979). However much higher levels often occur near highways and cities due to the combustion of gasoline. For example, residues in rivers in southern France and Lake Naini Tal (India) were 3.5–53 and 20–89 μg L^{-1}, respectively (Servant and Delapart, 1979; Pande and Das, 1980). Although a comparable range occurs in major industrial zone rivers, the discharge of liquid mine wastes may produce $\geqslant 500$ μg Pb L^{-1} in receiving waters (Imhoff and Koppe, 1980; van der Veen and Huizenga, 1980). Comparably high levels may occur in estuaries adjacent to mining areas but, in off-shore areas, decline to $\leqslant 0.05$ μg L^{-1}.

Total levels in precipitation generally range from 1–50 μg L^{-1} (Eisenreich, 1980; Legittimo *et al.*, 1980). However, in densely populated areas, residues may exceed 1000 μg L^{-1}, thereby producing significant contamina-

Table 6-6. Total lead levels (mg kg^{-1} dry weight) in freshwater and marine sediments.

Location	Average (range)	Polluting source
Freshwater Sediments		
Coeur d'Alene Lake (USA)[1]	3700 (3000–6300)	metal mining
creek, Montana (USA)[2]	295 (40–1710)	abandoned mine
Illinois River (USA)[3]	28 (3–140)	industrial municipal
Arctic Lakes (Canada)[4]	20 (10–33)	natural
Marine Sediments		
Long Island Sound (USA)[5]	— (200–350)	mixed industrial
Continental Shelf (SE, USA)[6]	$\leqslant 4$	natural
Baltic Sea[7]	— (2–400)	mixed industrial
Sorfjorden (Norway)[8]	11400 (720–70000)	metal mines, smelters

Sources: [1] Maxfield *et al.* (1974); [2] Pagenkopf and Cameron (1979); [3] Mathis and Cummings (1973); [4] Moore *et al.* (1979); [5] Greig *et al.* (1977); [6] Bothner *et al.* (1980); [7] Brügmann (1981); [8] Skei *et al.* (1972).

tion of snow and soils. Hence, Scott (1980) reported concentrations of up to 828,000 μg Pb L^{-1} in snow dumps in Toronto (Canada). By contrast snow in more isolated areas contains only 1 – 100 μg Pb L^{-1}.

Metal mining is the most significant source of lead in sediments (Table 6-6). Concentrations in excess of 6000 mg kg^{-1} dry weight have been reported. Other industrial/municipal sources generally produce residues in sediments of $\leqslant$500 mg kg^{-1}. In uncontaminated areas, concentrations range from 2 – 50 mg kg^{-1}, depending on the nature of the underlying bedrock.

Aquatic Plants

High residues are often reported for attached plants inhabiting polluted waters. Total lead in 6 macroscopic species collected from streams in an industrialized zone of FRG ranged from 100 to 5300 mg kg^{-1} dry weight (Dietz, 1973). In the River Leine (FRG) average residues in 4 species varied from 2 to 49 mg kg^{-1} while aquatic bryophytes growing in polluted mine streams had residues of up to 16,000 mg kg^{-1} (Abo-Rady, 1980; Burton and Peterson, 1979). Although CF's generally range from 5000 to 15,000, phytoplankton in some English lakes exhibited CF's of 100,000 after a 4 month exposure period (Denny and Welsh, 1979). It was suggested that the plankton settled on contaminated bottom sediments during winter and were resuspended during the spring.

Estuarine and coastal marine species generally contain lower residues than those reported for freshwater plants. In the industrialized fjords of Norway, *Fucus vesiculosus* contained up to 202 mg kg^{-1} whereas lead in *Enteromorpha* spp. ranged from 6 to 1200 mg kg^{-1} (Stenner and Nickless, 1974). Coastal species around the UK generally contained $\leqslant$5 mg kg^{-1} (Foster, 1976), and in heavily polluted Raritan Bay (New York City), residues ranged up to 170 mg kg^{-1} in brown algae (Seeliger and Edwards, 1977). The relatively low lead content of marine species is primarily related to the extent of ambient lead pollution in coastal waters.

Because there is often a good correlation between lead levels in water and plant tissues, several major species have been used as biomonitors of environmental contamination. These include freshwater representatives of *Elodea, Cladophora* and *Myriophyllum,* and estuarine *Fucus, Laminaria* and *Ascophyllum.* The use of such species partially overcomes variability in the concentrations of lead in water. There have, however, been few studies on the interactions of sediments and water contamination on tissue residues. Consequently, the use of macrophytes as indicators of water quality has to be viewed with some caution, particularly since lead is generally sorbed to substrate particles rather than dissolved in water.

Relatively little information is available on factors influencing uptake by plants. Sorption rates are species dependent and increase with exposure concentration. Sorption is generally suppressed by H^+ but not by the presence of cations (Na, K). Desorption is considerably faster than that for mercury and cadmium. Myklestad *et al.* (1978) reported a 75–85% reduction in levels in *Ascophyllum* within 2 months of transfer from contaminated to noncontaminated water.

Invertebrates

The majority of benthic and planktonic invertebrates do not concentrate lead from either algal foods or directly from water. In addition, there is often little accumulation at the higher trophic within invertebrate groups. Based on a review of literature on coastal marine species, Bryan (1976) concluded that the average concentration in phytoplankton and seaweed was 4 mg kg^{-1} dry weight, whereas the corresponding values for filter-feeders, omnivores, and carnivores were 3.25, 3.5, and 4.25 mg kg^{-1}, respectively. Similarly in Lake Baldegg, residues in phytoplankton, zooplankton and benthic insects were 420, 20, and 64 $\times 10^5$ moles kg^{-1}, yielding CF's of 0.5, 0.007, and 0.02 $\times 10^5$, respectively (Gächter and Geiger, 1979).

There is apparent biomagnification of lead through the food chain in some polluted waters. In the Severn Estuary (UK), residues in the primary gastropod consumers *Littorina littorea* and *Patella vulgata* averaged 0.2 and 4.3 mg kg^{-1} dry weight, respectively, while the corresponding value in the secondary gastropod consumer *Thais lapillus* was 11.2 mg kg^{-1} (Butterworth *et al.,* 1972). Other studies have demonstrated that significant accumulation may occur directly from water, regardless of exposure history of the test animals. Two species of insects (*Pteronarcys dorsata, Brachycentrus* sp.), one gastropod (*Physa integra*) and one amphipod (*Gammarus pseudolimnaeus*) sorbed $Pb(NO_3)_2$ from water at approximately the same rate, yielding an average CF of 5 (Figure 6-3).

Some of the variability in the sorption of lead from food and water is probably species dependent. In addition, because the feeding habits of invertebrates are diverse under natural conditions, it is difficult to determine the exact source and magnitude of contamination, thereby reducing the accuracy of concentration factors of food. Such determinations also do not consider the sorption of lead by benthic species directly from sediments. Furthermore, it is likely that highly contaminated, disturbed sediments are accidentally ingested by filter-feeders and littoral zooplankton on some occasions.

Residues are normally greater in the exoskeleton of crustaceans than in muscle, whereas the gills, stomach, and shell were the primary sites of accumulation in mussels *Mytilus edulis* collected from Australian waters (Talbot *et al.,* 1976). This latter study also showed that accumulation in muscle tissue was due to the ability of lead to form mercaptides with the SH group of amino acid side chains. Many other molluscs bear maximum residues in the digestive gland and shell. Differential deposition of lead in tissues has several implications for monitoring and assessment investiga-

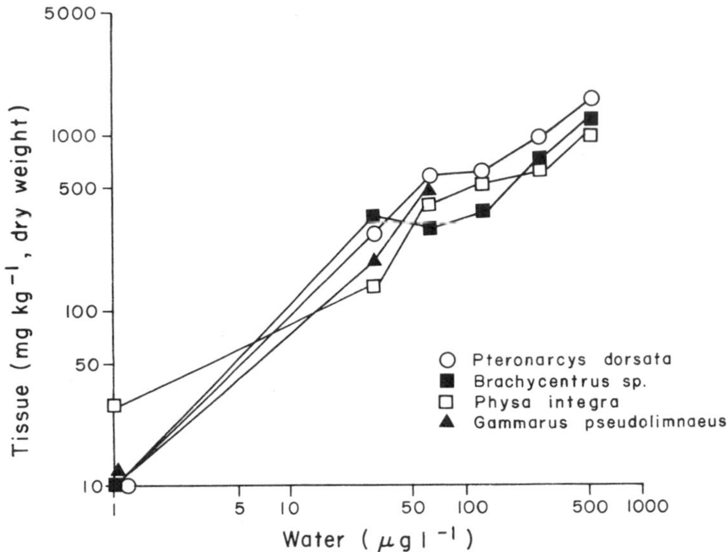

Figure 6-3. Lead residues in aquatic invertebrates exposed to lead for 28 days. (From Spehar *et al.*, 1978.)

tions. These include (i) variability in whole body residues of crustaceans following molting and (ii) inconsistent variability in residue levels due to differential analysis of tissues. In addition, since edible muscle is not necessarily a major site of accumulation, limitation of crustacean and mollusc fisheries should possibly be based on whole body analysis.

Relatively little information is available on factors influencing the uptake and depuration of lead. Phillips (1976) demonstrated that low temperatures had no effect on net uptake by mussels *Mytilus edulis.* There was also no consistent seasonal trend in residues in the barnacle *Balanus balanoides* collected from Welsh coastal waters (Ireland, 1974). However, it is likely that, in many species, low temperatures reduce sorption. There is often no definite relation between the size and age of invertebrates and lead content. In many species, sorption and depuration are linearly dependent on concentrations in either sediments or water. Depuration rates in *M. edulis* were greatest in the digestive gland and adductor muscle (half-life, 20 days) followed by those in kidney, gills, foot and gonad (Schulz-Baldes, 1974).

Fish

There is often little accumulation of lead in marine and freshwater species. Consequently lead is not a threat to fisheries resources except in cases of extreme pollution. Furthermore, there is generally no correlation between residues and feeding habits. In Wintergreen Lake (USA), concentrations

Table 6-7 Distribution of lead in Wintergreen Lake and Illinois River (US).*

Source	Mean	Range
Wintergreen Lake[1]		
Water	0.017	0.015–0.020
Sediment (dry weight)	31.9	7.4–53.8
Macrophytes	1.25	0.40–2.35
Zooplankton	7.9	4.0–18.3
Omnivorous fish (1 species)	0.30	0.22–0.54
Primary carnivores (3 species)	0.36	0.03–0.67
Secondary carnivore (1 species)	0.30	0.17–0.45
Illinois River[2]		
Water	0.002	0.0001–0.018
Sediment (dry weight)	28	3–140
Oligochaetes	17	6–39
Molluscs	3.7	1.8–5.1
Omnivorous fish (5 species)	0.65	0.09–1.78
Carnivorous fish (5 species)	0.62	0.17–1.5

Sources: [1] Mathis and Kevern (1975); [2] Mathis and Cummings (1973).
* Data expressed as mg kg^{-1} wet weight unless otherwise indicated.

averaged 7.9 mg kg^{-1} wet weight in invertebrates, 0.2–0.5 mg kg^{-1} in primary carnivores and 0.3–0.4 in secondary carnivores (Table 6-7). Similarly, omnivorous and carnivorous fish in the Illinois River had average residues of 0.7 and 0.6 mg kg^{-1}, respectively, in muscle tissue, whereas levels in invertebrates ranged up to 39 mg kg^{-1} (Table 6-7). Several studies have also demonstrated that there is often no correlation between lead

Table 6-8. Average lead residues (mg kg^{-1} wet weight) in different fish tissues.

Species	Muscle	Liver	Kidney	Gut Wall	Location
Black Marlin[1]	0.6	0.7	—	—	Pacific Ocean, Australia
Spotted Wolffish[2]	0.06	0.78	0.31	—	Quamarujuk Fjord, Greenland
Eel[3]	0.08	1.91	—	—	Medway Estuary, UK
Whiting[3]	0.32	0.90	—	0.44	Medway Estuary, UK
Flounder[3]	0.34	0.77	—	0.72	Medway Estuary, UK
Plaice[3]	0.94	1.30	—	1.32	Medway Estuary, UK
Atlantic salmon[4]	1.36	5.67	6.82	—	Miramichi River, Canada
Perch[5]	0.048	0.067	—	—	2 lakes, Switzerland

Sources: [1] Mackay *et al.* (1975); [2] Bollingberg and Johansen (1979); [3] Van Den Broek (1979); [4] Ray (1978); [5] Hegi and Geiger (1979).

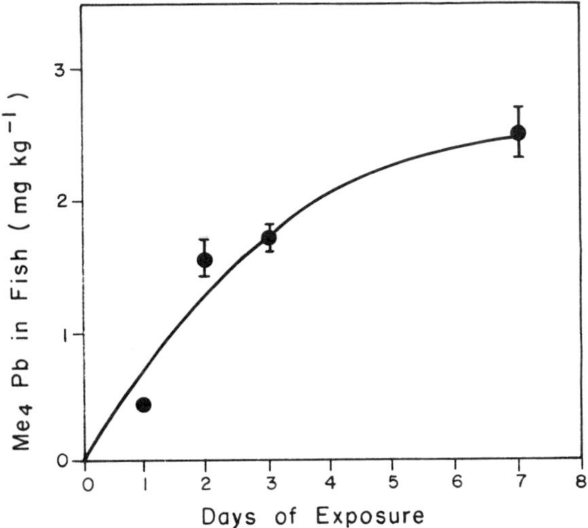

Figure 6-4A. Accumulation of Me_4Pb by fish. (From Wong *et al.*, 1981.)

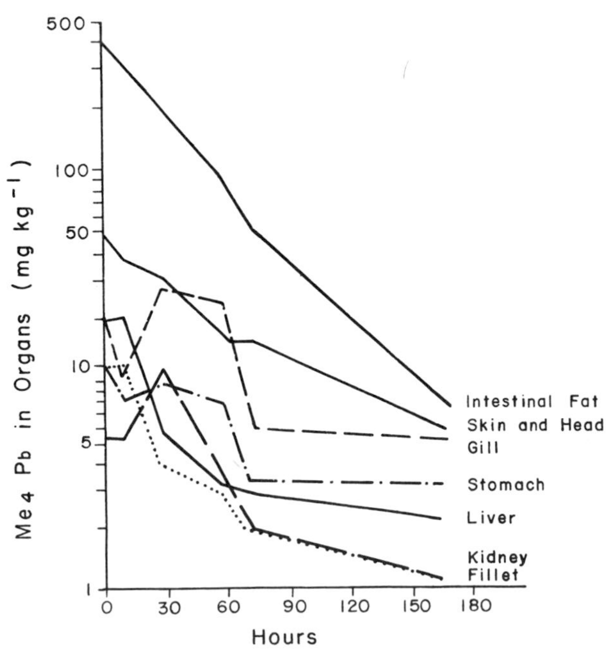

Figure 6-4B. Depuration of Me_4Pb from fish organs. (From Wong *et al.*, 1981.)

content and age or size of fish. Unlike cadmium, residues in muscle tissue are usually only slightly lower than those in organs (Table 6-8). This reflects the relatively low rate of binding to SH groups; in addition the low solubility of lead salts restricts movement across cell membranes. Because of the relatively low CF's in organs, biomonitoring programs for lead do not necessarily have to be based on organ analysis, as is the case with cadmium. This would in turn reduce the cost of analysis or permit additional analysis of edible muscle tissue.

Methylation of lead is rare under natural conditions and consequently organolead is seldom found in fish tissues. These compounds are, however, produced industrially in large quantities and inevitably escape into the environment. Wong *et al.* (1981) demonstrated that tetramethyllead was accumulated rapidly from water by rainbow trout (Figure 6-4A). Highest residues were found in intestinal fat, skin and gills. The half-life of Me_4Pb was short, averaging 30, 45 and 20 h, respectively (Figure 6-4B). Chau *et al.* (1980) reported that 17 of 107 fish collected from the Great Lakes area contained tetraalkyllead. There was no relation between species, feeding habits and size of fish and concentration of tetraalkyllead in tissues. In addition, there were no detectable amounts of tetraalkyllead in water, sediment or vegetation, reflecting the low solubility and high vapour pressure of the compounds. Although currently rare in the aquatic environment, organoleads are highly toxic and may therefore pose an eventual threat to fisheries resources. Hence special consideration should be given to the formulation of guidelines for total lead and organoleads in fishery products.

Toxicity

Aquatic Plants

Inorganic lead is less toxic to aquatic plants than mercurials and copper. Acute/chronic effects generally appear at media concentrations of 0.1–5 mg L^{-1}. While numerous factors influence the toxicity of inorganic lead to plants under laboratory conditions, almost nothing is known of lead toxicity under natural conditions. Manganese and copper may offset lead-induced inhibition, reflecting competition for active sites on enzymes, as also noted for cadmium (Pietilainen, 1975). Several other environmental factors such as temperature, light, and water chemistry likely influence the toxicity of lead to many species. In the euryhaline flagellate *Dunaliella tertiolecta,* however, toxicity was independent of salinity within the range 20–35°/oo (Stewart, 1977). A similar response would also be expected from other euryhaline, eurythermal species, while algae with more limited growth requirements would probably show a toxic response to variable environmental conditions.

Some species appear to be relatively tolerant of lead. Hutchinson (1973) could find no detectable effects of lead nitrate and lead acetate on either

Chlorella or *Chlamydomonas* at concentrations of 5 and 50 mg L^{-1}, respectively. The surprising aspect of such data is that comparable taxa may show a marked sensitivity to lead. These differences in tolerance may reflect differences in experimental conditions and the ability of some species to adapt to high metal levels. In addition, many species form different races, each with specific growth requirements. These races are morphologically indistinct from one another and are therefore described as the same species. As outlined in Chapter 10, physiological differences in response by the same species may limit the usefulness of algae as indicator organisms.

Tetraethyllead (TEL) concentrations of up to 250 μM had no detectable effect on the flagellated alga *Poterioochromonas malhamensis* (Röderer, 1980). Due to its low solubility and high volatility, TEL disappeared rapidly from the culture media; consequently, TEL cannot be considered hazardous to aquatic plants under natural conditions. TEL breaks down to form highly toxic derivatives, which are relatively stable, water-soluble and non-volatile. These include triethyllead and ethyl radicals. Although diethyllead (DiEL) has some TEL activity, DiEL may not be a primary derivative of TEL. DiEL is substatially more toxic to algae than Pb(+2) and less toxic than TEL.

Invertebrates

Lead is less toxic to invertebrates than copper, cadmium, zinc, and mercury, and generally more toxic than nickel, cobalt, and manganese. Acute effects are usually reported at concentrations of 0.1–10 mg L^{-1} (Table 6-9). However, significant mortality may occur within the range 0.002–670 mg L^{-1}. Although data are limited, toxicity appears similar in both marine and freshwaters. Several species, including some isopods and polychaetes, are particularly resistant to lead intoxication (Table 6-9). Chronic effects may or may not appear at concentrations below the LC_{50}. Hence, under natural field conditions, (i) demonstrable impacts of lead discharge may appear abruptly, and (ii) the presence/absence of invertebrates near discharge sites may not give an accurate account of the extent of lead contamination.

An increase in pH reduces the concentration of Pb(+2) in water which in turn decreases toxicity to invertebrates. Furthermore, chelators in water bind Pb(+2), again reducing toxicity. The interaction of temperature and salinity also significantly influences mortality. In mussel (*Mytilus galloprovincialis*) embryos exposed to 0.25 mg L^{-1} total lead, maximum survival (87%) occurred at 32.5‰ salinity and 15°C, whereas at 25‰ salinity and 17.5°C, survival was 0% (Hrs-Brenko *et al.,* 1977).

Toxicity to some invertebrates depends on the ability to develop tolerance to lead. Such adaption is probably based on physiological changes rather than the ameliorating effect of environmental conditions. Although the basis of adaption by invertebrates is not fully understood, it is known that bacteria and algae exposed to lead show changes in genetic material in succeeding generations. These in turn permit the development of physiological adaptations, as also probably occurs in invertebrates.

Table 6-9. Acute toxicity of lead to some freshwater invertebrates.

Species	Toxicity* (mg L^{-1})	Temperature (°C)	pH	Total hardness (mg L^{-1})	Dissolved oxygen (mg L^{-1})
Philodina acuticornis (rotifer)[1]	50.4 (96 h LC_{50})[a]	20	7.4–7.9	NQ	NQ
Lymnaea palustris (gastropod)[2]	0.036 (50% decrease in production)	21	7.8	139	NQ
Daphnia hyalina (cladoceran)[3]	0.055 (48 h LC_{50})[b]	10	7.2	NQ	NQ
Cyclops abyssorum (copepod)[3]	3.8 (48 h LC_{50})[b]	10	7.2	NQ	NQ
Eudiaptomus padanus (copepod)[3]	0.55 (48 h LC_{50})[b]	10	7.2	NQ	NQ
Asellus aquaticus (isopod)[4]	670 (48 h LC_{50})	12	NQ	250	NQ
Tanytarsus dissimilis (dipteran)[5]	0.258 (96 h LC_{50})	22	7.5	46.8	8.7

Sources: [1] Buikema *et al.* (1974); [2] Borgmann *et al.* (1978); [3] Baudouin and Scoppa (1974); [4] Fraser (1980); [5] Anderson *et al.* (1980).

[a] lead chloride

[b] lead acetate.

* Test compound $Pb(NO_3)_2$ unless otherwise indicated. NQ—not quoted.

Fish

The 96 h LC_{50} for total lead generally falls within the range 0.5 – 10 mg L^{-1}. However, in hard waters (≥350 mg L^{-1} $CaCO_3$) the LC_{50} may exceed 400 mg L^{-1}. Chronic effects following long-term exposure to lead nitrate (not found in natural waters) may appear at concentrations as low as 8 μg L^{-1}. Eyed rainbow trout eggs were more sensitive to lead than juveniles (Davies *et al.*, 1976). In addition, if eggs were treated with chronic levels of lead, the resulting fry were more sensitive to intoxication than fish with no previous exposure history. This increase in sensitivity, particularly in the earlier life stages, has also been reported for mercury and cadmium.

Treatment of rainbow trout with lead at concentrations as low as 13 μg L^{-1} caused significant (i) increases in red blood cell (RBC) numbers, (ii) decline in RBC volumes, (iii) decreases in RBC cellular iron content, and (iv) decreases in RBC δ-amino levulinic acid dehydratase activity (Hodson *et al.*, 1978). These changes probably indicated increased erythropoiesis to compensate for reduced hemoglobin production. Holcombe *et al.* (1976) treated three generations of brook trout with total lead concentrations ranging from 0.9 to 474 μg L^{-1}. As a result, all second generation trout exposed to ≥ 235 μg Pb L^{-1} developed severe spinal deformities (scoliosis). Scoliosis was also observed in 21% of the third generation exposed to 119 μg L^{-1}.

Although lead nitrate inhibits several major enzymes, the proportional

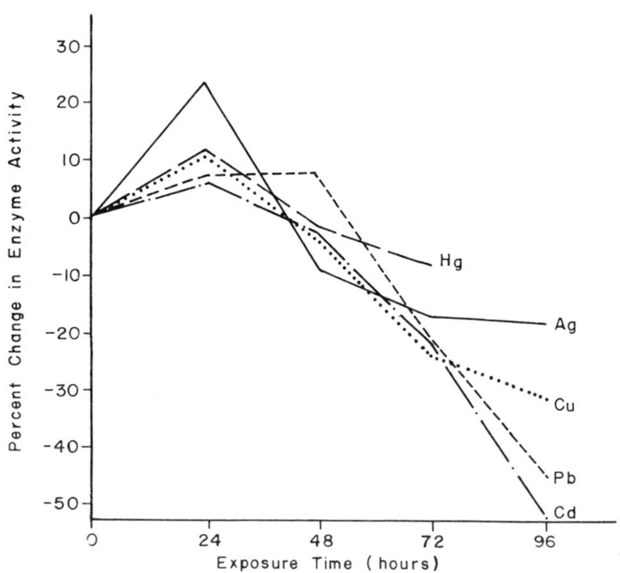

Figure 6-5. Percentage change in the activity of fish liver RNAase exposed to metals. (From Jackim *et al.*, 1970.)

decline in activity depends on the initial concentration of lead, and the duration of exposure (Figure 6-5). Liver enzymes (lipase, RNAase) are particularly susceptible to lead, but enzymes in the stomach remain largely unaffected following treatment (Jackim *et al.,* 1970). Since sulfur-containing chelators can reduce toxicity, lead probably binds to enzyme proteins, particularly the SH group. This in turn results in damage to cell organelles due to decreased synthesis of enzyme proteins. Exposure of fish to sublethal/chronic levels generally results in changes in the structure of specific tissues. There may be fragmentation and reduction in the cristae of erythroblast mitochondria and increase in ribosome density and decrease in polyribosome density. Hence lead induces a slow maturation of erythroblasts but does not inhibit erythroporetic stimulii. Other manifestations of lead toxicosis include blackening of the skin in the caudal region, lordosis, scoliosis, muscle tremors, and necrosis of the sensory and supporting cells of the lateral line.

Humans

Many environmentally important lead compounds such as halides, sulfates, phosphates, and hydroxides are insoluble and thus are of relatively low toxicity in aquatic systems. By contrast, soluble lead compounds are intermediate between hard and soft acids in their interaction towards oxygen- and sulfur-containing ligands. In humans lead resembles calcium in deposition and transport, accounting for the high concentrations of lead in the skeletal compartment. The major source of lead in humans is through the respiratory tract. This reflects the strong association of lead with urban airborne particulates.

Cooper (1976) reported that the incidence of renal tumors and other carcinomas in lead smelter and battery plant workers was not higher than control levels. Furthermore, lead acetate is not mutagenic using the Ames (*Salmonella*) test. By contrast, inorganic lead compounds induce renal carcinoma in rats and mice. Similarly, primary hamster embryo cells treated with lead acetate underwent transformation (Dipaolo *et al.,* 1978). When transformed cells were injected into hamsters, fibrosarcomas were produced.

The embryotoxic effects of lead nitrate in rats are dependent on the day of administration (McClain and Becker, 1975). Injection on day 9 of gestation resulted in urorectocaudal malformations in neonates. If the injections were made on day 16, hydrocephalus and CNS hemorrhages resulted. The teratogenic effects of lead nitrate on various strains of hamsters include fetal resorptions, tail bud abnormalities, hydrocephalus, and skeletal defects. No congenital malformations were evident in mice and rats born of mothers given tetraethyllead or lead acetate at low doses (Kennedy *et al.,* 1975). However, treatment with high concentrations of both compounds produced maternal toxicity, fetal resorption, and general retardation of development observed. It was postulated that neither tetraethyllead nor lead acetate were teratogenic in mice and rats.

References

Abo-Rady, M.D.K. 1980. Aquatic macrophytes as indicator for heavy metal pollution in the River Leine (West Germany). *Archiv fuer Hydrobiologie* **89**:387–404.

Anderson, R.L., C.T. Walbridge, and J.T. Fiandt. 1980. Survival and growth of *Tanytarsus dissimilis* (Chironomidae) exposed to copper, cadmium, zinc, and lead. *Archives of Environmental Contamination and Toxicology* **9**:329–335.

Baudouin, M.F., and P. Scoppa. 1974. Acute toxicity of various metals to freshwater zooplankton. *Bulletin of Environmental Contamination and Toxicology* **12**:745–751.

Bollingberg, H.J., and P. Johansen. 1979. Lead in spotted wolffish, *Anarhichas minor,* near a zinc–lead mine in Greenland. *Journal of the Fisheries Research Board of Canada* **36**:1023–1028.

Borgmann, U., O. Kramar, and C. Loveridge. 1978. Rates of mortality, growth, and biomass production of *Lymnaea palustris* during chronic exposure to lead. *Journal of the Fisheries Research Board of Canada* **35**:1109–1115.

Bothner, M.H., P.J. Aruscavage, W.M. Ferrebee, and P.A. Baedecker. 1980. Trace metal concentrations in sediment cores from the continental shelf off the southeastern United States. *Estuarine and Coastal Marine Science* **10**:523–541.

Boyden, C.R., S.R. Aston, and I. Thornton. 1979. Tidal and seasonal variations of trace elements in two Cornish estuaries. *Estuarine and Coastal Marine Science* **9**:303–317.

Brügmann, L. 1981. Heavy metals in the Baltic Sea. *Marine Pollution Bulletin* **12**:214–218.

Bryan, G.W. 1976. Heavy metal contamination in the sea. *In*: R. Johnston (Ed.), *Marine pollution.* Academic Press, London, pp. 185–291.

Buikema, A.L., Jr., J. Cairns, Jr., and G.W. Sullivan. 1974. Evaluation of *Philodina acuticornis* (Rotifera) as a bioassay organism for heavy metals. *Water Resources Bulletin* **10**:648–661.

Burton, M.A.S., and P.J. Peterson. 1979. Metal accumulation by aquatic bryophytes from polluted mine streams. *Environmental Pollution* **19**:39–46.

Butterworth, J., P. Lester, and G. Nickless. 1972. Distribution of heavy metals in the Severn estuary. *Marine Pollution Bulletin* **3**:72–74.

Chau, Y.K., P.T.S. Wong, O. Kramer, G.A. Bengert, R.B. Cruz, J.O. Kinrade, J. Lye, and J.C. Van Loon. 1980. Occurrence of tetraalkyllead compounds in the aquatic environment. *Bulletin of Environmental Contamination and Toxicology* **24**:265–269.

Cooper, W.C. 1976. Cancer mortality patterns in the lead industry. *Annals of the New York Academy of Sciences* **271**:250–259.

Davies, P.H., J.P. Goettl, Jr., J.R. Sinley, and N.F. Smith. 1976. Acute and chronic toxicity of lead to rainbow trout *Salmo gairdneri,* in hard and soft water. *Water Research* **10**:199–206.

De Groot, A.J., W. Salmons, and E. Allersma. 1976. Processes affecting heavy metals in estuarine sediments. *In*: Burton, J.D., and P.S. Liss (Eds.), *Estuarine chemistry.* Academic Press, London. pp. 131–197.

Denny, P., and R.P. Welsh. 1979. Lead accumulation in plankton blooms from Ullswater, the English Lake District. *Environmental Pollution* **18**:1–9.

Dipaolo, J.A., R.L. Nelson, and B.C. Casto. 1978. *In vitro* neoplastic transformation of syrian hamster cells by lead acetate and its relevance to environmental carcinogenesis. *British Journal of Cancer* **38**:452–455.

Dietz, F. 1973. The enrichment of heavy metals in submerged plants. *In*: S.H. Jenkins (Ed.), *Advances in water pollution research.* Pergamon Press, Oxford, pp. 53–62.

Eisenreich, S.J. 1980. Atmospheric input of trace metals to Lake Michigan. *Water, Air, and Soil Pollution* **13**:287–301.

Förstner, U., and G.T.W. Wittmann. 1979. *Metal pollution in the aquatic environment.* Springer-Verlag, Berlin, Heidelberg, 486 pp.

Foster, P. 1976. Concentrations and concentration factors of heavy metals in brown algae. *Environmental Pollution* **10**:45–53.

Fraser, J. 1980. Acclimation to lead in the freshwater isopod *Asellus aquaticus. Oecologia* **45**:419–420.

Gächter, R., and W. Geiger. 1979. Melimex, an experimental heavy metal pollution study: behavior of heavy metals in an aquatic food chain. *Schweizerische Zeitschrift fuer Hydrologie* **41**:277–290.

Garrels, R.M., F.T. MacKenzie, and C. Hunt. 1975. *Chemical cycles and the global environment.* William Kaufmann Inc., Los Altos, California, 206 pp.

Greig, R.A., R.N. Reid, and D.R. Wenzloff. 1977. Trace metal concentrations in sediments from Long Island Sound. *Marine Pollution Bulletin* **8**:183–188.

Hahne, H.C.H., and W. Kroontje. 1973. Significance of pH and chloride concentration on behavior of heavy metal pollutants: mercury(II), cadmium(II), zinc(II), and lead(II). *Journal of Environmental Quality* **2**:444–450.

Hart, B.T., and S.H.R. Davies. 1981. Trace metal speciation in the freshwater and estuarine regions of the Yarra River, Victoria. *Estuarine, Coastal and Shelf Science* **12**:353–374.

Hegi, V.H.R., and W. Geiger. 1979. Schwermetalle (Hg, Cd, Cu, Pb, Zn) in Lebern und muskulatur des Flussbarsches (*Perca fluviatilis*) aus Bielersee und Walensee. *Schweizerische Zeitschrift fuer Hydrologie* **41**:94–107.

Hodson, P.V., B.R. Blunt, and D.J. Spry. 1978. Chronic toxicity of water-borne and dietary lead to rainbow trout (*Salmo gairdneri*) in Lake Ontario water. *Water Research* **12**:869–878.

Holcombe, G.W., D.A. Benoit, E.N. Leonard, and J.M. McKim. 1976. Long-term effects of lead exposure on three generations of brook trout *Salvelinus fontinalis. Journal of the Fisheries Research Board of Canada* **33**:1731–1741.

Hrs-Brenko, M., C. Claus, and S. Bubić. 1977. Synergistic effects of lead, salinity, and temperature on embryonic development of the mussel *Mytilus galloprovincialis. Marine Biology* **44**:109–115.

Huber, F., U. Schmidt, and H. Kirchmann. 1978. Aqueous chemistry of organolead and organothallium compounds in the presence of microorganisms. *In*: F.E. Brinckman and J.M. Bellama (Eds.), *Organometals and organometalloids.* ACS Symposium Series No. 82, American Chemical Society, Washington, D.C., pp. 65–81.

Hutchinson, T.C. 1973. Comparative studies of the toxicity of heavy metals to phytoplankton and their synergistic interactions. *Water Pollution Research in Canada* **8**:68–90.

Imhoff, K.R., and P. Koppe. 1980. The fate of heavy metals in the Ruhr system and their influence on drinking water quality. *Progress in Water Technology* **13**:211–225.

Ireland, M.P. 1974. Variations in the zinc, copper, manganese and lead content of *Balanus balanoides* in Cardigan Bay, Wales. *Environmental Pollution* **7**:65–75.

Jackim, E., J.M. Hamlin, and S. Sonis. 1970. Effect of metal poisoning on five liver enzymes in the killifish (*Fundulus heteroclitus*). *Journal of the Fisheries Research Board of Canada* **27**:383–390.

Jarvie, A.W.P., R.N. Markall, and H.R. Potter. 1975. Chemical alkylation of lead. *Nature* **255**:217–218.

Kennedy, G.L., D.W. Arnold, and J.C. Calandra. 1975. Teratogenic evaluation of lead compounds in mice and rats. *Food and Cosmetics Toxicology* **13**:629–632.

Klein, L. 1959. River pollution. I. Chemical analysis. Butterworths, London, 206 pp.

Legittimo, P.C., G. Piccardi, and F. Pantani. 1980. Cu, Pb and Zn determination in rainwater by differential pulse anodic stripping voltammetry. *Water, Air, and Soil Pollution* **14**:435–441.

Mackay, N.J., M.N. Kazacos, R.J. Williams, and M.I. Leedow. 1975. Selenium and heavy metals in black marlin. *Marine Pollution Bulletin* **6**:57–60.

Mathis, B.J., and T.F. Cummings. 1973. Selected metals in sediments, water, and biota in the Illinois River. *Journal Water Pollution Control Federation* **45**:1573–1583.

Mathis, B.J., and N.R. Kevern. 1975. Distribution of mercury, cadmium, lead, and thallium in a eutrophic lake. Hydrobiologia **46**:207–222.

McClain, R.M., and B.A. Becker. 1975. *Teratogenicity,* fetal toxicity, and placental transfer of lead nitrate in rats. *Toxicology and Applied Pharmacology* **31**:72–82.

McDuffie, B., I. El-Barbary, G.J. Hollod, and R.D. Tiberio. 1976. Trace metals in rivers—speciation, transport, and role of sediments. *Trace Substances in Environmental Health* **10**:85–95.

Maxfield, D., J.M. Rodriguez, M. Bueltner, J. Davis, L. Forbes, R. Kovacs, W. Russel, L. Schultz, R. Smith, J. Stanton, and C.M. Wai. 1974. Heavy metal content in the sediments of the southern part of the Coeur d'Alene lake. *Environmental Pollution* **6**:263–266.

Moore, J.W., V.A. Beaubien, and D.J. Sutherland. 1979. Comparative effects of sediment and water contamination on benthic invertebrates in four lakes. *Bulletin of Environmental Contamination and Toxicology* **23**:840–847.

Myklestad, S., I. Eide, and S. Melsom. 1978. Exchange of heavy metals in *Ascophyllum nodosum* (L.) Le Jol. *In situ* by means of transplanting experiments. *Environmental Pollution* **16**:277–284.

Nriagu, J.O. 1979. Global inventory of natural and anthropogenic emissions of trace metals to the atmosphere. *Nature* **279**:409–411.

Nriagu, J.O., and R.D. Coker. 1980. Trace metals in humic and fulvic acids from Lake Ontario sediments. *Environmental Science and Technology* **14**:443–446.

Pagenkopf, G.K., and D. Cameron. 1979. Deposition of trace metals in stream sediments. *Water, Air, and Soil Pollution* **11**:429–435.

Pande, J., and S.M. Das. 1980. Metallic contents in water and sediments of Lake Naini Tal, India. *Water, Air, and Soil Pollution* **13**:3–7.

Phillips, D.J.H. 1976. The common mussel *Mytilus edulis* as an indicator of pollution by zinc, cadmium, lead and copper. I. Effects of environmental variables on uptake of metals. *Marine Biology* **38**:59–69.

Pietilainen, K. 1975. Synergistic and antagonistic effects of lead and cadmium on aquatic primary production. *In*: T.C. Hutchinson (Ed.), *Proceedings of 1st International Conference on Heavy Metals in the Environment,* Vol. **II**, Part 2, University of Toronto Institute for Environmental Studies, Toronto, Canada, pp. 861–873.

Ramamoorthy, S., and D.J. Kushner. 1975a. Heavy metal binding sites in river water. *Nature* **256**:399–401.

Ramamoorthy, S., and D.J. Kushner. 1975b. Binding of mercuric and other heavy metal ions by microbial growth media. *Microbial Ecology* **2**:162–176.

Ramamoorthy, S., and B.R. Rust. 1978. Heavy metal exchange processes in sediment water systems. *Environmental Geology* **2**:165–172.

Ray, S. 1978. Bioaccumulation of lead in Atlantic salmon *Salmo salar. Bulletin of Environmental Contamination and Toxicology* **19**:631–636.

Röderer, G. 1980. On the toxic effects of tetraethyl lead and its derivatives on the chrysophyte *Poterioochromonas malhamensis.* I. Tetraethyl lead. *Environmental Research* **23**:371–384.

Schmidt, U., and F. Huber. 1976. Methylation of organolead and lead(II) compounds to $(CH_3)_4Pb$ by microorganisms. *Nature* **259**:157–158.

Schulz-Baldes, M. 1974. Lead uptake from sea water and food, and lead loss in the common mussel *Mytilus edulis. Marine Biology* **25**:177–193.

Scott, W.S. 1980. Occurrence of salt and lead in snow dump sites. *Water, Air, and Soil Pollution* **13**:187–195.

Seeliger, U., and P. Edwards. 1977. Correlation coefficients and concentration factors of copper and lead in seawater and benthic algae. *Marine Pollution Bulletin* **8**:16–19.

Servant, J., and M. Delapart. 1979. Lead and lead-210 in some river waters of the southwestern part of France. Importance of the atmospheric contribution. *Environmental Science and Technology* **13**:105–107.

Sillén, L.G. and A.E. Martell. 1971. *Stability constants of metal-ion complexes. Supplement No. 1.* Special Publication No. 25, The Chemical Society, London, 865 pp.

Skei, J.M., N.B. Price, S.E. Calvert, and H. Holtendahl. 1972. The distribution of heavy metals in sediments of Sorfjord, West Norway. *Water, Air, and Soil Pollution* **1**:452–461.

Spehar, R.L., R.L. Anderson, and J.T. Fiandt. 1978. Toxicity and bioaccumulation of cadmium and lead in aquatic invertebrates. *Environmental Pollution* **15**: 195–208.

Stenner, R.D., and G. Nickless. 1974. Distribution of some heavy metals in organisms in Hardangerfjord and Skjerstadfjord, Norway. *Water, Air, and Soil Pollution* **3**:279–291.

Stewart, J. 1977. Relative sensitivity to lead of a naked green flagellate, *Dunavella tertiolecta. Water, Air, and Soil Pollution* **8**:243–247.

Talbot, V., R.J. Magee, and M. Hussain. 1976. Lead in Port Phillip Bay mussels. *Marine Pollution Bulletin* **7**:234–237.

United States Mineral Yearbooks. 1960–1979. Bureau of Mines, US Government Printing Office, Washington, D.C.

van den Broek, W.L.F. 1979. Seasonal levels of chlorinated hydrocarbons and heavy metals in fish and brown shrimps from the Medway Estuary, Kent. *Environmental Pollution* **19**:21–38.

van der Veen, C., and J. Huizenga. 1980. Combating river pollution taking the Rhine as an example. *Progress in Water Technology* **12**:1035–1059.

Wilson, A.L. 1976. *Concentrations of trace metals in river waters, a review.* Technical Report No. 16., Water Research Centre, Medmenhan Laboratory and Stevenage laboratory, U.K.

Wong, P.T.S., Y.K. Chau, O. Kramar, and G.A. Bengert. 1981. Accumulation and depuration of tetramethyllead by rainbow trout. *Water Research* **15**:621–625.

7
Mercury

Chemistry

Elemental mercury, the only liquid metal at 25°C, is the third member of the group IIb triad of the periodic table of elements. The chemical behavior of mercury is significantly different from that of the other two members of the same triad, zinc and cadmium. Mercury is an enzyme and protein inhibitor whereas zinc plays an active role in protein, lipid, and carbohydrate metabolism in a variety of organisms. The extreme class b character of Hg(+2) and RHg(+1) (R = alkyl or aryl group) gives (i) high affinity towards thiol groups and (ii) enhanced covalency compared to their zinc counterparts, resulting in increased bio-transport, distribution, and toxicity of mercurials. From the toxicological point of view, mercurials are classified into several groups such as elemental mercury, inorganic mercury compounds other than elemental mercury, short-chain alkyl mercurials (methyl and ethyl), and other organomercury compounds (Friberg and Vostal, 1972; Suzuki, 1977). Hg(0), Hg(+1), and Hg(+2) are readily interconverted in the environment (Figure 7-1).

Production, Uses, and Discharges

Production and Uses

Total production of mercury during this century has been about 4.36×10^5 metric tons. Production by decade remained stable from 1900 to 1939 (0.36×10^5 metric tons per decade), but increased to 0.62×10^5 tons during

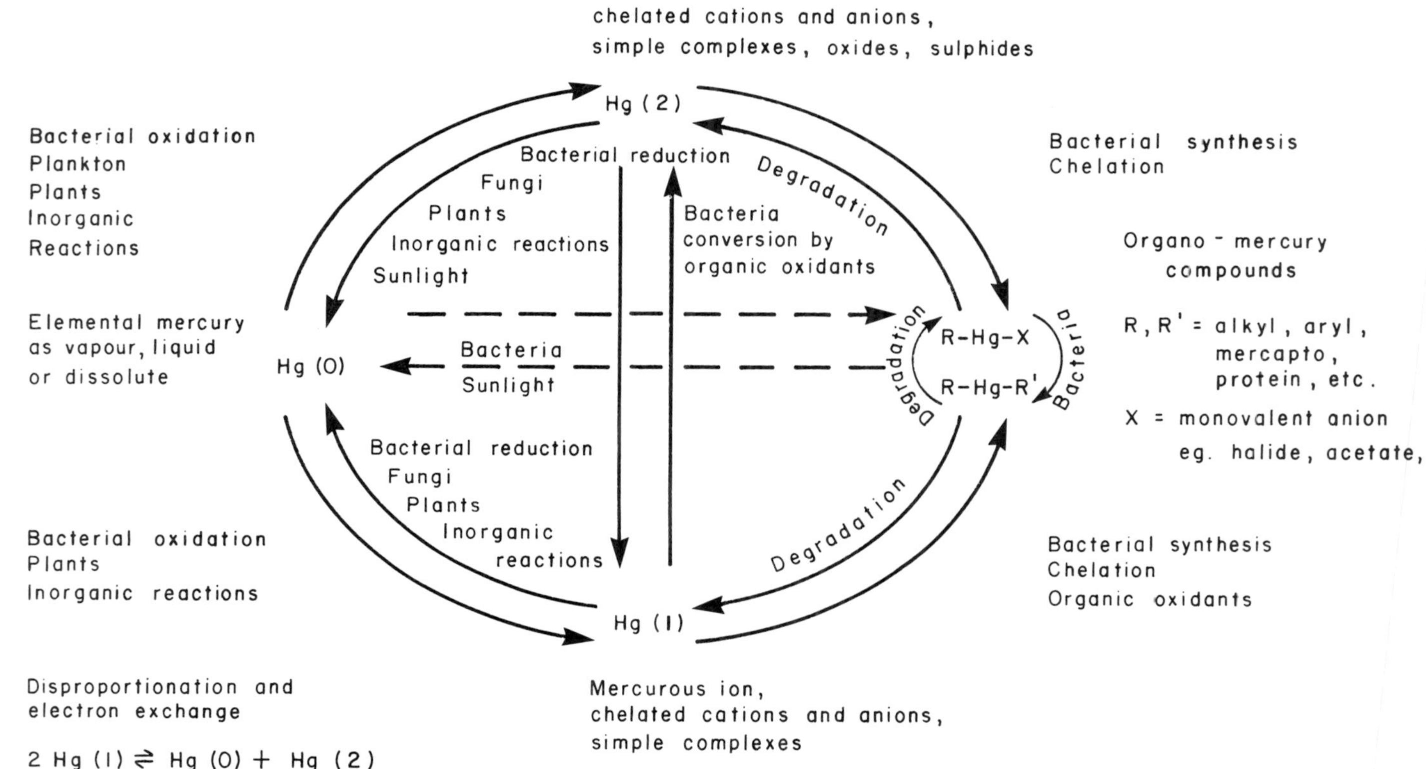

Figure 7-1. Cycle of mercury conversions in nature. (From Jonasson and Boyle, 1972.)

Table 7-1. World production of mercury in the 1970's (in metric tons $\times 10^3$).

	1973	1974	1975	1976	1977	1978	1979
World Production	9.25	8.88	8.70	8.28	6.86	6.07	$<6.0^P$
US Consumption	1.87	2.05	1.75	2.24	2.11	1.68	1.45^P
Canadian Consumption	0.03	0.04	0.03	0.03	0.03	0.03	0.03

Sources: US Minerals Yearbooks; Canadian Minerals Yearbooks.
^{P} = projected.

the 1940's and 0.85×10^5 tons during the 1970's. World production has declined since 1973, falling below 0.06×10^5 tons in 1979. (Table 7-1).

The combination of physico-chemical properties such as liquidity at room temperatures, uniform volume expansion over a wide temperature range, high surface tension (480 dynes cm^{-1} (20°C)), and non-wettability to glass surfaces makes mercury uniquely useful for measuring devices such as thermometers, barometers, and manometers. The low electrical resistivity (95.8×10^{-6} ohm cm^{-1} (20°C)) and high thermal conductivity make mercury an excellent electrical conductor and coolant, respectively. Because of its capacity to absorb neutrons, mercury is employed as a shield against atomic radiation. Mercurials have found widespread use in insecticides, fungicides, bactericides, and pharmaceuticals. Because of their stereospecific nature, several compounds, especially oxides, chlorides, and sulfides have been used as catalysts, especially in the manufacture of synthetic polymers.

Table 7-2. Actual and projected demands (in US) for mercury by industrial categories (metric tons).

Category	1974[a]	1985[b]	2000[b]	2025[b]
Chlorine and caustic soda	582	511	0	0
Paint	235	265	244	355
Pesticides	34	18	21	27
Catalysts	45	46	58	70
Pharmaceuticals	21	27	38	36
Instruments	173	172	142	180
Tubes, switches, and relays	60	82	86	213
Batteries	613	650	787	1028
Lamps	47	68	114	179
Dental	104	89	100	115
Laboratory	16	18	21	27
Other	121	127	159	139
Total	2051	2073	1770	2369
		(1323)[c]	(1153)[c]	(1540)[c]
		(3055)[d]	(2604)[d]	(3531)[d]

Sources: [a]Cammarota (1975); [b]Watson (1979); [c]low estimate; [d]high estimate.

The large hydrogen overpotential of mercury permits its use as a cathode in electrochemical operations (example, chloralkali industry). Amalgams with many metals have a variety of applications in metallurgy and dental fillings. In recent years, two major uses have been in the manufacture of electrical apparatus and in the electrolytic production of chlorine and caustic soda (Table 7-2). These two applications together accounted for 53% of mercury consumed in the United States in 1979.

Discharges

Natural weathering has contributed approximately 1.6×10^{10} metric tons of total mercury into the environment throughout geological times. About 0.1% of this total remains in solution in oceans. Since 97% of surface water is saline, the oceans are the largest sink to mercury in solution. Mercury flux from the earth's surface to the atmosphere through degassing is approximately 49×10^{-6} g m^{-2} year^{-1} for the oceanic shelf and 4.6×10^{-6} g m^{-2} year^{-1} for oceans and polar regions, respectively. These figures are balanced by deposition rates (Nriagu, 1979). The reported estimates on the relative input of anthropogenic mercury into the continental atmosphere vary. Mathematical modeling techniques, allowing for the recycling of mercury in and out of the atmosphere, have shown that a significant portion (~50%) of global mercury cycling is anthropogenic in origin (Miller and Buchanan, 1979). Lower estimates have been reported by other studies (Weiss *et al.*, 1971).

Mercury released in this century through human activities is almost ten times the calculated amount released due to natural weathering (57,000 metric tons). A recent US National Academy of Sciences Report (1978) estimates a global mercury discharge of 1300 metric tons each year to water from natural sources, including 200 metric tons from the USA.

Figures and trends for worldwide mercury discharges have been calculated taking into consideration the economic constraints and/or lack of environmental protection regulations (Watson, 1979). As expected it has been noted that:

1. Most of the increases in mercury discharge into the environment occur in the less developed regions of the world.
2. In the developed regions of the world, discharges into water are either increasing very little or show a real decline.
3. Air discharges through human activities will double between 1975 and 2025, though they might constitute only 25% of natural air discharges.

The consumption and discarding of mercury-containing goods is the largest source of mercury discharge through human activities (Table 7-3). Increasing prices might reduce the uses (and discharge) of mercury where substitutes are available. However, increased fuel burning and materials processing will keep discharge to land and air at relatively high levels. The

Table 7-3. Source discharges of mercury on land through human activities (metric tons).

	1975		1985		2000		2025	
	Quantity	% total	Quantity	% total	Quantity	% total	Quantity	% total
Consumption	3352	69.1	4139	67.4	5466	82.7	7303	79.8
Fuel burning	107	2.2	71	1.2	143	2.2	568	6.2
Mining and smelting	107	2.2	71	1.1	143	2.2	142	1.6
Material processing	142	2.9	214	3.5	286	4.3	568	6.2
Production	1143	23.6	1643	26.8	571	8.6	568	6.2
Total discharge	4851		6138		6609		9149	

* Calculated from Figures in Watson (1979).

amount of mercury removed from municipal sewage plants and appearing in landfills will likely increase in the future. Similarly incineration of municipal sludges will increase the mercury burden in air over time. Though mercury discharges into water are predicted to decline over the next 50 years, the cumulative effect on bottom sediments, especially near sewage disposal sites, will be significant. Hence, frequent monitoring of mercury levels in bottom sediments is a prerequisite for tighter control for clean water legislation.

Mercury in Aquatic Systems

Speciation in Natural Waters

Mercury in natural waters can exist in three oxidation states: elemental mercury(0), the mercurous(+1) state and the mercuric(+2) state. The nature of the species and their distribution will depend upon the pH, redox potential, and nature and concentrations of anions which form stable complexes with mercury. In well-aerated water (Eh $\geqslant$ 0.5 V) mercuric species will predominate whereas, under reducing conditions, elemental mercury should prevail. Presence of enough sulfide ion stabilizes bivalent mercury as hydrosulfide or sulfide complexes, even at very low redox potentials.

Binding to Inorganic Ligands. Among the inorganic anions, Hg(+2) forms the strongest covalent compound with chloride ions, giving a stability constant of $K = 10^{15}$. The mercuric ion hydrolyzes with $Hg(OH)_2$ as the predominant species at pH $\geqslant$4 for aqueous solutions containing less than 10^{-5} M chloride. At concentrations of chloride $\simeq$ 0.01 M, the region of predominance of $Hg(OH)_2$ shifts to pH $\geqslant$6. Well-aerated waters containing traces of stable organic ligands speciate mercury between different chloro and hydroxy mercury complexes as a function of pH and pCl [−log(chloride concentration)] (Figure 7-2).

Binding to Organic Ligands. Mercury forms stable complexes with a variety of organic ligands. The strongest covalent complexes are formed with S-containing ligands such as cysteine, the next strongest with amino acids and hydroxy carboxylic acids.

Dissolved mercury fractionates among compounds of molecular weight ranging from 500–100,000 (Ramamoorthy and Kushner, 1975a). In unpolluted waters, the low molecular weight fraction (<500) predominates (~75%), whereas 66.4% of mercury was found in the >100,000 molecular weight fractions in samples from waters contaminated with industrial effluents. In addition, there was a good correlation between the distribution of mercury and dissolved organic carbon in different molecular weight frac-

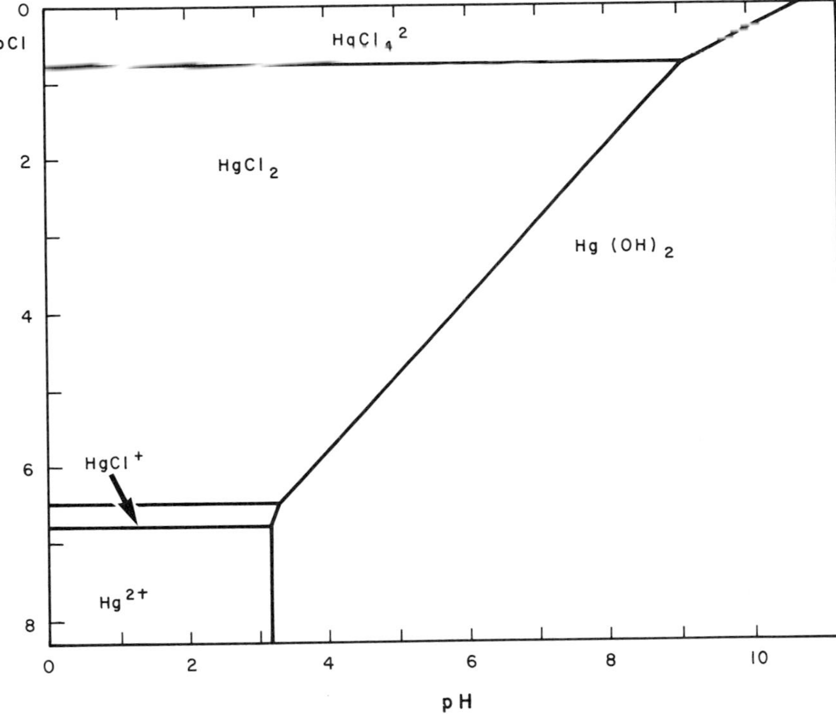

Figure 7-2. Speciation of mercury as a function of chloride and hydrogen ion concentrations. (From Lockwood and Chen, 1973.)

tions. Increasing salinity of the water had a negative effect on the mercury–organic complexing processes.

In the Ottawa River (Canada) low molecular weight organic compounds are primarily responsible for binding mercury (Ramamoorthy and Kushner, 1975a). Mercury also binds strongly to the components of microbial growth media (Figure 7-3). It was suggested that the metal ions might enter bacterial cells as organic complexes or that the cells effectively compete with the components of the growth media for the bound ions (Ramamoorthy and Kushner, 1975b).

Binding to Particulates. Mercury associates strongly with suspended solids in natural waters (Table 7-4). The extent of association is determined by water quality parameters such a pH, salinity, redox potential (Eh), and presence of organic ligands. Partition coefficients for mercury between suspended solids and water have been calculated from field and laboratory data to be $1.34-1.88 \times 10^5$. Suspended solids (⟨20 μm–0.45 μm⟩) partition mercury $\sim 10^5$ times over the surrounding water column, regardless of the nature of the suspended solids. Using these partition coefficients, it is

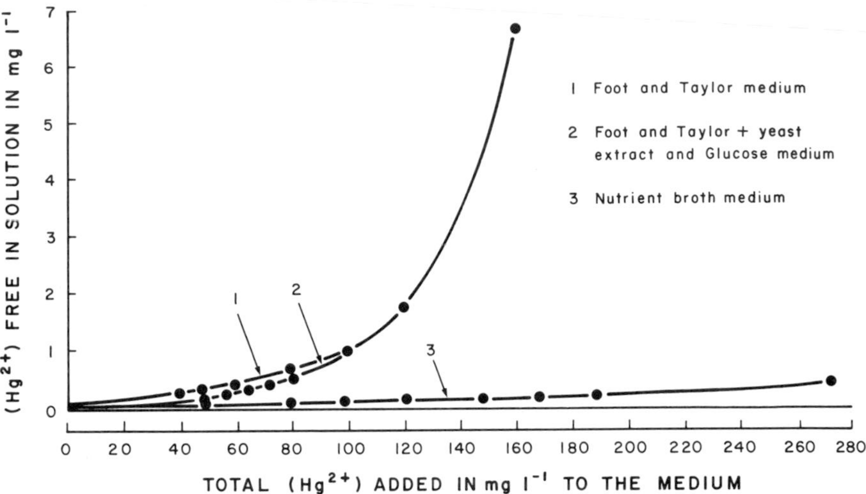

Figure 7-3. Free Hg(+2) remaining after adding Hg(+2) to different growth media. (From Ramamoorthy and Kushner, 1975b.)

possible to estimate the lowest concentration of mercury in the water column by measuring accurately the mercury concentration in suspended solids. This will be more accurate than directly analyzing concentrations of mercury in water. The interaction of trace metals with suspended solids may range from weak van der Waals interaction to strong covalent binding.

The chemical form of dissolved species of mercury determines the mode of association to suspended solids and their residence time in the water column. Sorption–desorption processes on aquatic solids are critical deter-

Table 7-4. Amount of particulate forms of mercury in natural waters.

Type of Water	Percent particulate form	Mercury content (mg kg^{-1})	Method of separation
Lake water[1]	10–13	NR	Centrifugation
River water[2]	83–96*	NR	Filtration
River water[3]	31–58	1.08–1.45	Filtration
River water[4]	87	1.8	Filtration
Estuarine water[4]	96–98	4.3–4.9	Filtration
Sea water[4]	96	9.6	Filtration
Sea water[5]	>60	NR	Filtration
Effluent[4] (Municipal)	82	35.5	Filtration

Sources: [1]Beneš *et al.* (1976); [2]Hinkle and Learned (1969); [3]Ottawa River Project (1977); [4]Smith *et al.* (1971); [5]Carr and Wilkniss (1973).
* Calculated values; NR—not reported.

minants on the fate of mercury in aquatic systems. Sorption to suspended solids and subsequent sedimentation plays an important role in the removal of mercury from the water column. Present knowledge in this area of dynamics in natural waters is fragmentary.

Mercury in Sediments

Sorption. Rate of sorption of mercury depends largely on the physical/chemical characteristics of the sediments. Ramamoorthy and Rust (1976) showed that sorption maximum was correlated to surface area > organic content > cation exchange capacity > grain size, whereas the bonding constant correlated differently: organic content > grain size > cation exchange capacity > surface area. Similarly Kudo and Hart (1974) found that sorption of mercury occurred in the following order: wood chips > clay > sand. They reported that the effective depth of mercury uptake was less than 1 mm for sand bed sediments and also observed no significant difference in uptake rates between aerobic and anaerobic conditions. The association of mercury and heavy metals with sediments can range from weak van der Waals forces to strong covalent bonding, co-precipitation with ferro manganese oxides, and incorporation within crystal lattices. The release of sorbed mercury from sediments into the bulkwater is dependent on partition coefficients, which in turn are related to sediment characteristics and environmental parameters such as pH, Eh, and amount of Cl^- and natural and synthetic chelating agents. Ramamoorthy and Massalski (1979) found that mercury binding in river sediments occurred on sulfur sites but not in a stoichiometric ratio. Not all sites accumulated mercury, possibly because of (i) uneven distribution of binding sites, (ii) inadequate concentration of mercury to saturate all sites, and (iii) desorption of mercury from the binding site by either chemical or biological processes.

Desorption. Desorption is a slow process, posing a long-term problem even after the sources of pollution are eliminated. Reimers and Krenkel (1974) reported negligible desorption of inorganic mercury for all clays, organics, and sands. The order of desorption of $HgCl_2$ and methyl mercuric chloride was 1 : 10 for sands > 1 : 100 for clays > organics (not observable). Desorption was negligible except with illite at high chloride concentrations and pH > 7, and sands with high chloride concentrations. Ramamoorthy and Rust (1976, 1978) from their studies on sand, silt and organic-rich river-bed sediments, reported the cation exchange under various conditions was:

$$Hg > Pb > Cu > Cd.$$

They also observed significant desorption of mercury from leaching sediments with solutions of NaCl and NTA, a surfactant and strong chelating agent in commercial detergents. The two leachates desorbed almost equal amounts of mercury which is consistent with the stability constants for the complexes: $10^{15\cdot15}$ ($HgCl_2$) and $10^{14\cdot60}$ (Hg–NTA). Less than 1% desorption of mercury with fulvic acid leaching was also observed. Varying degrees of desorption of mercury with increasing salinity, depending on the nature and sulfur content of the sorbents, have been reported.

Transformations

Mercury forms in the environment a group of compounds called organic mercurials. These are divided into two categories: (i) those in which mercury is amphiphilic, attached to one organic radical through a covalent bond and to an inorganic anion through an electrovalent bond ($R — Hg^{d+} — X^{d-}$) and (ii) those in which mercury is liphophilic, attached to two organic radicals, both covalently bound, ($R—Hg—R^1$). The first group is characterized by its water and lipid solubility, and persistence in the aquatic system—for example, the methylmercuric ($CH_3 — Hg^+$) ion. The second group is non-polar, almost insoluble in water and extremely volatile.

Mercury is discharged into the aquatic environment mainly in the form of elemental mercury, divalent mercuric ion, Hg(+2), and phenyl mercuric acetate, $C_6H_5Hg(CH_3COO)$. In contaminated waters, almost all mercury in fish is methylmercury. There is substantial evidence for biological and abiological production of methylmercury. Methylation of mercury requires the presence of a free inorganic mercuric ion, Hg(+2), and a methyl donor molecule(s). Many biological end-products commonly found in the aquatic system are potential methylating agents.

Biological methylation could be due to enzymatic or non-enzymatic processes. Enzymatic methylation requires the presence of an actively metabolizing organism and at least three enzyme systems have been identified: (i) methionine synthetase, (ii) acetate synthetase, and (iii) methane synthetase. The non-enzymatic methylation of mercury requires only the methyl donors of an active metabolism. Several studies have been conducted with methyl cobalamine, vitamin B_{12} as methyl donor, and several pathways of non-enzymatic methylation of mercury have been proposed. In bacterial methylation, anaerobes, facultative anaerobes, and aerobes methylate mercury and the efficiency of methylation is dependent upon the metabolic state of the organism and concentration of available mercuric ion which in turn depends on pH, Eh, and the presence of organic complexing agents. Measurements of methylmercury in field and laboratory sediments indicate that $<1\%$ of the total mercury is organic. If fish or some other means of methylmercury clearance are introduced into laboratory experiments, the production of methylmercury is relatively high.

Methyl derivatives of lead, tin, and silicon may be involved in the transmethylation of mercury. Photochemical alkylation of mercury occurs with methyl donors such as acetic acid, propionic acid, methanol, and ethanol. Abiotic methylation of mercury by alkaline extracts of sterile soil samples produces very low yields (0.001%) (Rogers, 1977).

Microbial degradation of methylmercury occurs in mercury-polluted sediments from lakes, rivers, and seas. Hence determination of methylation rates and concentrations of methylmercury in sediments yields a measure of net methylation rather than of the rate of synthesis of methylmercury. Reduction of Hg(+2) to Hg(0) has been reported, as has reduction of phenyl mercuric acetate to Hg(0) by *Pseudomonas.*

The biological and chemical cycling of mercury in the aquatic environment is a complex process involving many pathways and competing reactions depending on the nature of mercury input, physico-chemical composition of the aquatic system, and the metabolic state of various types of biota. Continued multidisciplinary approach to ecological impact studies will certainly reveal valuable insights into the understanding of the critical pathways in the overall dynamics of mercury in the ecosystem.

Residues

Water, Precipitation, and Sediments

Concentrations of dissolved mercury in unpolluted fresh water varies between 0.02 and 0.1 $\mu g\ L^{-1}$ and in ocean water from <0.01 to 0.03 $\mu g\ L^{-1}$ (Table 7-5). A significant fraction of mercury in natural waters is associated with suspended solids and accounts for a substantial portion (next to the water column) of the transported mercury in the aquatic environment. Very little informaton is available on the levels of methylmercury in natural waters. Although non-polluted systems contain less than 0.2 to 1 ng L^{-1} of methylmercury, the biota show relatively higher levels (see following sections). Recent data on the snow cores of Greenland show the absence of any systematic increase in mercury levels due to global activities.

Total mercury levels in precipitation range from <0.01 to >1.0 $\mu g\ L^{-1}$ (National Research Council of Canada, 1979). Maximum concentrations generally occur near industrial sites, such as chlor-alkali plants. There may be efficient mercury removal from air by a heavy rainstorm but, in some areas, significant reduction in atmospheric mercury does not occur. Concentrations reported in precipitation include the input from wind-blown dust, unless operationally defined sampling and analytical procedures are strictly followed. The total atmospheric mercury near the surface is principally composed of several volatile Hg species (Johnson and Braman, 1974). It appears that the rainfall flux of mercury to the oceans is 30×10^8 g $year^{-1}$ (at 10 ng L^{-1} of Hg in oceanic rains). The apparent gaseous nature of mercury species suggests a longer residence time for Hg in the atmospheric reservoir than expected.

Several 'hot spots' of mercury in sediments have been reported around the world posing a potential long-term threat to the aquatic environment even after the sources of mercury discharge have been discontinued (Table 7-6). Levels of up to 9000 mg kg^{-1} were reported in the sediments of Jintsu

Table 7-5. Dissolved mercury levels ($\mu g\ L^{-1}$) in natural waters and effluents.

Source	Average (range)	Source	Average (range)
Freshwaters			
Rhine River (FRG)[1]	0.5(0.2–0.6)	Industrial zone rivers (USA)[2–4]	<0.5(0.0–6.8)
Unpolluted lakes, rivers (Canada)[5]	0.03(0.01–0.1)	Mineral waters (Tuscany)[6]	–(0.01–0.2)
Ice cap (Greenland)[7]	0.01(0.009–0.013)	Ice cap, year 1727–1981 (Greenland)[7]	–(0.002–0.019)
Marine waters			
Japan Sea[8]	0.005(0.004–0.005)	Minamata Bay[9]	(1.6–3.6)
Greenland Sea[10]	0.125(0.016–0.364)	NE Atlantic (0–4030 m)[11]	0.054(0.017–0.142)
NW Atlantic (surface)[12]	0.007(0.002–0.011)	NE Atlantic (surface)[13]	0.015(tr.–0.034)
Mid-Atlantic ridge (bottom)[14]	1.09(0.87–1.42)	NE Pacific (0–5000 m)[15]	0.024(0.012–0.037)
Effluents			
Paper mill[16]	–(2.0–3.4)	Fertilizer plant[16]	–(2.6–4.0)
Smelting plant[16]	–(2.0–4.0)	Chloralkali plant[16]	–(80–2000)
Mixed industrial, sewage[2]	5440(1840–12880)		

Sources: [1]Reichert *et al.* (1972); [2]Cranston and Buckley (1972); [3]Proctor *et al.* (1973); [4]Hinkle and Learned (1969); [5]D'Itri (1972); [6]Dall'Aglio (1968); [7]Applequist *et al.* (1978); [8]Matsunaga *et al.* (1975); [9]Hosohara *et al.* (1961); [10]Carr *et al.* (1972); [11]Leatherland *et al.* (1973); [12]Fitzgerald and Hunt (1974); [13]Gardner (1975); [14]Carr *et al.* (1974); [15]Williams *et al.* (1974); [16]Shiber *et al.* (1978).

Table 7-6. Total mercury residues (mg kg^{-1} dry weight) in surface sediments.

Source	Average (range)	Polluting source
Freshwater sediments		
Lake Dufault (Canada)[1]	0.26(0.07–0.14)	natural
Lake Duparquet (Canada)[1]	0.23(0.19–0.30)	natural
Lake Maggiore (Italy)[2]	7.65(0.24–2.01)	mixed industrial
Lake Erie (USA)[2]	0.61(0.13–7.49)	mixed industrial
Lake Michigan (USA)[2]	0.15(0.03–0.38)	mixed industrial
Lake Ontario (USA)[2]	0.66(0.03–2.10)	mixed industrial
Lake Superior (USA)[2]	0.08(0.004–0.06)	natural
Albegna River (Italy)[3]	–(63.5–688)	industrial, municipal
Thunder Creek (Canada)[4,5]	max. 38.0; max. 0.05*	chloralkali plant
River Seine (France)[3]	–(9.8–15.8)	industrial, municipal
Marine sediments		
Salt marsh (USA)[6]	0.6(0.27–1.70)	chloralkali plant
Howe Sound (Canada)[7]	0.49(0.03–8.8)	pulp and paper
Mersey River estuary (UK)[8]	2.2(0.01–14.3)	mixed industrial sewage
Thames River estuary (UK)[9]	0.1(0.02–0.49)	mixed industrial
Liffey River estuary (Eire)[10]	2.1(0.1–4.6)	pulp and paper
Minamata Bay (Japan)[11]	630(–)	chemical plants
East Pacific rise (Hawaii)[12]	0.06(0.02–0.24)	natural

Sources: [1]Speyer (1980); [2]Damiani and Thomas (1974); [3]Batti *et al.* (1975); [4]Gummer and Fast (1979); [5]Jackson and Woychuk (1979); [6]Gardner *et al.* (1978); [7]MacDonald and Wong (1977); [8]Craig and Morton (1976); [9]Smith *et al.* (1973); [10]Jones and Jordan (1979); [11]Matida and Kumada (1969); [12]Cox and McMurtry (1981).
* Methylmercury.

and Kumano Rivers of Toyama prefectures in Japan in 1970. These were approximately 50% higher than reported in Nigata and were attributed to discharge from a pharmaceutical company on Kumano River. Sediments contaminated from natural sources generally bear levels of <1 mg kg^{-1} (Table 7-6).

Aquatic Plants

Highest residues are usually reported for marine species which live for more than one year (Table 7-7). Concentrations are also relatively high in old, basal tissue of attached plants compared to young leaves. Filamentous and vascular plants with a life span of $\leqslant 1$ year may bear low residues, as also noted for microscopic algae (Table 7-7). In the Neckar River (FRG), maximum combined residues in benthic diatoms and blue-green algae were

Table 7-7. Total mercury residues (mg kg^{-1} dry weight) in some marine and freshwater plants.

Macroscopic/vascular species	Average (range)	Locality	Polluting source
Ascophyllum nodosum[1,2]	0.09(0.05–0.18)	Trondheimsfjord (Norway)	metal smelter, sewage
Ascophyllum nodosum[1,2]	3.1(0.08–20)	Hardangerfjord (Norway)	metal smelter
Vascular Plants (5 species)[3]	0.07(0.02–0.14)	Lake Päijänne (Finland)	pulp and paper mills
Vascular Plants (5 species)[4]	0.18(0.03–0.43)	Ruhr River (FRG)	mixed industrial
Phytoplankton, rotifers[5]	0.21(0.03–0.72)	Lake Päijänne (Finland)	pulp and paper mills
Benthic diatoms[6]	0.6(0.5–0.8)	Neckar River (FRG)	mixed industrial

Sources: [1]Haug *et al.* (1974); [2]Myklestad *et al.* (1978); [3]Särkkä *et al.* (1978a); [4]Dietz (1972); [5]Särkkä *et al.* (1978b); [6]Bartelt and Förstner (1977).

0.8 mg kg^{-1}. Although similarly low levels were reported for planktonic diatoms and blue-greens in heavily contaminated Lake Paijanne (Finland), residues of over 9 mg kg^{-1} were found for benthic algae in the Agano River, Japan.

Approximately 20% of the total amount of mercury found in dividing cells of the diatom *Synedra ulna* was absorbed and about 50% accumulated in inner parts of the cell (Fujita and Hashizume, 1975). Furthermore almost all of the sorbed mercury originated from the 0.45 μm filterable or ionic fraction. Glooschenko (1969) demonstrated that non-dividing cells of the marine diatom *Chaetoceros costatum* accumulated about twice as much ^{203}Hg as dividing cells, presumably by surface absorption. Dividing cells in light accumulated ^{203}Hg longer than did non-dividing cells, indicating the possibility of some active uptake. On the other hand, the rate of sorption of mercury by both live and dead cells of the green alga *Selenastrum capricorum* was similar under light and dark conditions; hence the mechanism of intake appeared to be through initial passive absorption.

Rate of absorption probably varies with algal species. However, it is not possible to consistently correlate sorption with taxonomic classification at the present time. It is also not clear how much is bound inside and on the cell surface, or externally by metabolites. If mercury was internally bound, Hg(+2) residues would likely bioaccumulate through the food chain; in the latter case, external sequestration could remove Hg(+2) from the cells.

Invertebrates

The relative amount of organic mercurials in marine and freshwater invertebrates is extremely variable. Methylmercury accounted for only 3–10% of total mercury in the gastropod *Littorina errorata* collected from a salt marsh, and 43–72% for crabs (*Uca* sp.) from the same area (Gardner *et al.*, 1978). Similarly, 29–83% of mercury in benthic invertebrates in a polluted river (USA) was methylated (Hildebrand *et al.*, 1980). Much of this variability among studies can be related to differences in the trophic position of the consumer organisms. However, the diet of most invertebrates is so diverse that it is difficult to describe any species as truly herbivorous or carnivorous. Trophic position also depends on time of year, locality, and competition for food. Consequently, it may not be possible to consistently predict methylmercury levels in any one species. Physical and chemical factors also influence the rate of methylation in the physical environment. For example, increasing salinity reduces methylation due to interference from chloride ions. Although the relative quantity of methylmercury in freshwater invertebrates should be greater than that in marine species, variability in the existing data base prohibits confirmation of this point.

Total residues in freshwater plankton and benthos generally exceed those reported for marine biota (Table 7-8). This is partly attributed to the smaller

Table 7-8. Total mercury residues* (mg kg^{-1} dry weight) in some marine and freshwater invertebrates.

	Average (range)	Locality	Polluting source
Marine species			
Copepods, tunicates[1]	0.32(0.06–0.69)	Minamata Bay (Japan)	chemical plants
Crustaceans, rotifers, dinoflagellates[2]	6.3(0.5–25.2)	Fjords (Norway)	metal mines
Molluscs (unidentified)[3]	–(11.4–39.0 wet)	Minamata Bay (Japan)	chemical plants
Decapods (unidentified)[3]	–(1.0–36.0 wet)	Minamata Bay (Japan)	chemical plants
Mytilus edulis (mussels)[4]	0.3(0.2–0.65)	Fjords (Norway)	metal mines
Freshwater species			
Crustaceans, rotifers, diatom[5]	0.20(0.02–0.72)	Lake Päijänne (Finland)	pulp and paper mills
Crayfish (*Orconectes virilis*)[6]	0.15(0.06–0.35 wet)	Thunder Creek (Canada)	dredging/sediments
Orconectes virilis[7]	0.15(0.07–0.56 wet)	Wisconsin River (USA)	mixed industrial
Larval dipterans (unidentified)[8]	0.52(0.08–1.43 wet)	N. Fork Holston River (USA)	chloralkali plant
Larval trichopterans (unidentified)[8]	0.88(0.04–3.75 wet)	N. Fork Holston River (USA)	chloralkali plant

Sources: [1]Hirota *et al.* (1974); [2]Skei *et al.* (1976); [3]Takeuchi (1972); [4]Stenner and Nickless (1974); [5]Särkkä *et al.* (1978b); [6]Munro and Gummer (1980); [7]Sheffy (1978); [8]Hildebrand *et al.* (1980).

* Data are for whole body, excluding mollusc shells.

receiving area in freshwater, which leads to relatively high ambient levels. In addition, since organic mercurials are sorbed faster than inorganic forms (see below), the lower rate of methylation in marine waters results in reduced total residues in biota. Despite the large amount of data, it has not been possible to correlate total concentrations with feeding habit or trophic position in either marine or freshwater biota (Table 7-8). Bryan (1976) reviewed 65 papers and found that average residues in marine filter-feeders, omnivores, and carnivores were 0.1 – 0.4, 0.2 – 0.3, and 0.1 – 0.4 mg kg^{-1} dry weight, respectively. This again reflects heterogeneity in diet and differential uptake of organic materials.

Methylmercury is sorbed faster from food and water than inorganic mercury. Accumulation rates increase with temperature and methylmercury residues are eliminated slower than those of the inorganic forms (Ribeyre *et al.,* 1980). In addition, sorption may increase with mercury and chelator concentrations in the water. Although substantial amounts of mercuric chloride and methylmercury chloride can be accumulated directly from the water, food is also a significant source of mercury in some species. Such differences in routes of uptake undoubtedly reflect experimental and species variation.

Fish

Most of the mercury in fish tissues is methylated. Jacobs (1977) found that CH_3Hg accounted for 70 – 98% of total mercury in 9 species in German marine waters. Similarly, methylmercury in edible muscle of canned and unprocessed fish comprised 39.2 – 92.9% of total mercury in 11 marine species and 58.8 – 96.4% in 7 freshwater species (Cappon and Smith, 1981). The current standard (0.5 mg kg^{-1}) used to control the consumption of adulterated tissue assumes that all of the mercury in fish is methylated.

Total residues are usually slightly higher in organs than muscle tissue. For example, the liver : muscle ratio for mercury in black marlin collected from Australia was 1.4 : 1 (Mackay *et al.,* 1975). Average concentrations in the muscle, liver, and kidney of several species from the Mediterranean were 0.30, 0.41 and 0.76 mg kg^{-1} wet weight, respectively (Buggiani and Vannucchi, 1980). By contrast perch inhabiting two Swiss lakes had muscle residues ranging from 0.06 to 0.21 mg kg^{-1}, compared with 0.03 – 0.14 mg kg^{-1} for liver (Hegi and Geiger, 1979).

As in invertebrates, organic mercury enters fish at a faster rate than inorganic forms. The rate of sorption depends primarily on temperature and, to a lesser degree, lipid content (MacLeod and Pessah, 1973). Although this implies that the methylmercury content of fish should vary seasonally, other environmental factors apparently mask potential temperature effects. Methylation of inorganic mercury may also occur *in vitro* in fish livers and intestine. The half-life of elimination of methylmercury is among the longest known for metals; Järvenpää *et al.* (1970) reported values of 640,

780 and 1200 days for flounder, pike and eel, respectively. Rate of elimination appears to be independent of temperature and metabolic rate in fish.

The pathways for mercury uptake by fish have been investigated under a variety of experimental conditions. Kudo (1976) showed that the half-life of total mercury in the unstirred, abiotic sediments of aquaria was 12–20 years. The amount taken up by guppies varied widely (up to 600%) but the biological half-life was 38–75 days. No correlation was found between uptake rate and either size or sex of fish. Although the same species sorbed significant amounts of $^{203}Hg(+2)$ directly from the water of aquaria (Prabhu and Hamdy, 1977), other work has indicated that sediment and algae are the main source of mercury to herbivorous invertebrates and fish.

Håkanson (1980) investigated the interactions of pH, trophic level, and mercury contamination of sediments on methylmercury residues in pike from a series of Scandinavian lakes. In his studies, trophic conditions were expressed as a BioProduction Index (BPI), which was obtained by measuring the organic and nitrogen content of sediments. High BPI (up to 10) implied eutrophic conditions. As shown in Figure 7-4, average mercury content in a 1-kg pike was correlated with Hg levels in sediments and inversely with pH and BPI. Hence, increased acidification of lakes will result in higher residues in fish whereas lime treatment of the water may offer a short term improvement in conditions. Furthermore, reduction in trophic levels in lakes, through improved sewage treatment systems may result in increased methylation in the environment and subsequent uptake by fish.

Natural Sources of Mercury Pollution

Reservoirs. High levels of mercury are often found in fish collected from new impoundments where there is no apparent industrial or municipal waste discharge. Abernathy and Cumbie (1977) reported that total average residues in largemouth bass exceeded 3.1 mg kg^{-1} wet weight in a recently filled reservoir (Table 7-9). However, lower levels (0.5–1.4 mg kg^{-1}) were found for the same species in reservoirs which had been filled 4–12 years prior to the investigation. The initially high rate of uptake was related to the flooding of soil with elevated mercury levels. Cox *et al.* (1979) similarly found residues of up to 1.2 mg kg^{-1} for carnivorous fish collected from a 2 year old reservoir. These authors demonstrated that mercury levels in freshly inundated soil were about 7 times greater than those of the original lake bottom. Immediately after flooding, the rate of methylation increased, thereby increasing uptake to the food chain. It was also suggested that the sediments of older reservoirs developed reducing conditions which increased the rate of binding of mercury to sulfur compounds, in turn decreasing the availability of mercury to the food chain. In American Falls Reservoir (35 years old), residues in fish are generally low, despite the presence of mining activity (Table 7-9).

Geological Sources. Although the overall terrestrial concentration of inorganic mercury is approximately 50 $\mu g\ kg^{-1}$, much higher levels have been found in specific types of ores, minerals, and rocks. These areas of high mercury content have been arranged into belts and generally correspond to zones of instability and volcanic and thermal activity. There are a number of areas which occur outside of the mercuriferous zone, including Eire, Ontario and Arkansas (Figure 7-5).

Pike collected from a river in the mercuriferous belt (Montana) had body burdens averaging approximately 0.5 and 1.35 mg kg^{-1} wet weight at 6 and 10 years of age, respectively (Phillips *et al.*, 1980). Sauger and walleye were also significantly contaminated reaching 0.5 mg kg^{-1} mercury by age 7 and 8, respectively, whereas maximum burdens for white crappie and black crappie were both 0.25 mg kg^{-1}. Although outside of known mercuriferous belts, lakes on Precambrian shield areas may also exhibit significant mercury contamination. Delisle and Demers (1976) found that total residues in pike and walleye inhabiting lakes in arctic/subarctic Quebec generally averaged $\geqslant 1$ mg kg^{-1}. These levels are approximately similar to those found in industrialized parts of the province. There are also many examples of natural contamination in estuarine and coastal marine environments. Koli *et al.* (1978) reported that average residues in shrimp and five species of fish collected from South Carolina ranged from 0.1–0.6 mg kg^{-1} whereas Persian Gulf shrimp contained total body levels of 0.1–0.9 mg kg^{-1} (Parvaneh, 1977).

Toxicity

Aquatic Plants and Invertebrates

All mercurial compounds are highly toxic to aquatic plants. Growth inhibition by mercuric chloride may occur at concentrations of 0.002–0.25 mg L^{-1} (Hollibaugh *et al.*, 1980; Harriss *et al.*, 1970). Under most conditions organic mercurials are significantly more toxic than inorganic forms. Using planktonic algae as test species, Mora and Fábregas (1980) reported that the minimum toxic concentration of $HgSO_4$ was 0.15–0.20 mg Hg L^{-1}, but the corresponding levels for phenyl mercuric acetate (PMA) and methyl mercuric chloride were 0.025–0.050 and 0.025 mg Hg L^{-1}, respectively.

Several factors may decrease the toxicity of mercurials to algae. Mercury and cadmium, and mercury and zinc interact synergistically towards photosynthesis and result in synergism mixed with antagonism depending on metal concentrations (Stratton and Corke, 1979). Pre-treatment with low levels of nickel may protect against mercury toxicity, possibly by reducing active uptake of mercury. Other modifying factors include sequestration by organic chelators and partitioning by suspended solids (Hollibaugh *et al.*, 1980). Toxicity is also often correlated with the pH and phosphate and

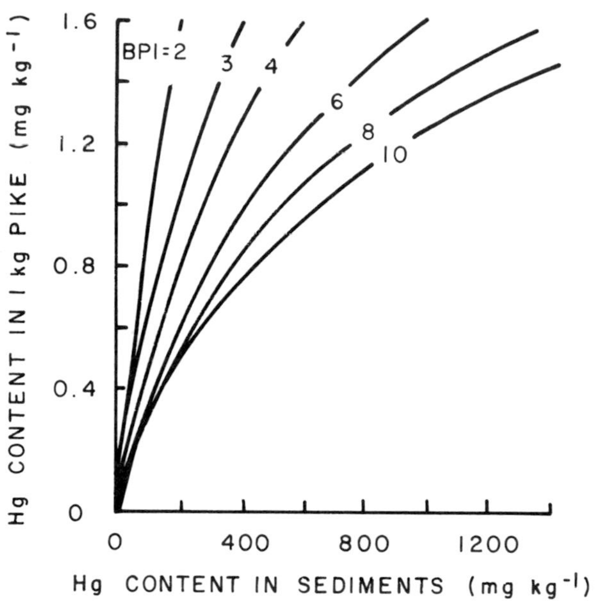

Figure 7-4a. Mercury content in a 1-kg pike as a function of mercury content of sediments and BPI at pH 5. (From Håkanson, 1980.)

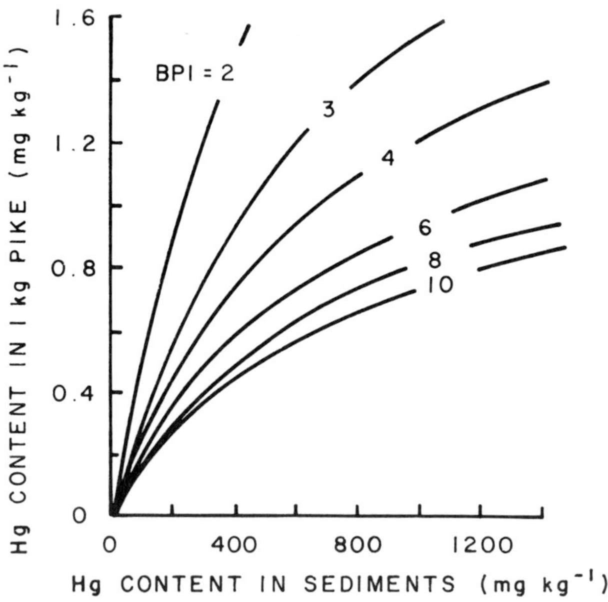

Figure 7-4c. Mercury content in a 1-kg pike as a function of mercury content of sediments and pH for waters of BPI = 4. (From Håkanson, 1980.)

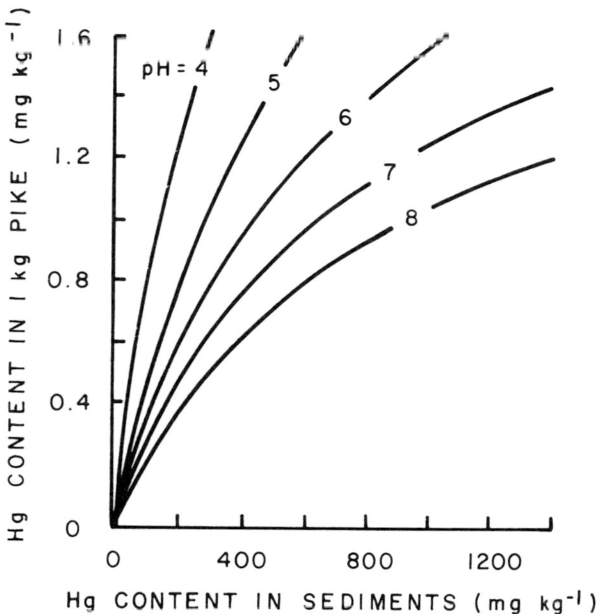

Figure 7-4b. Mercury content in a 1-kg pike as a function of mercury content of sediments and BPI at pH 7. (From Håkanson, 1980.)

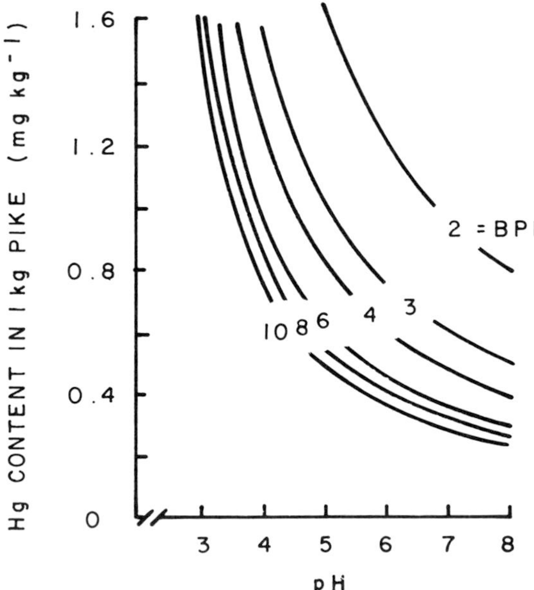

Figure 7-4d. Mercury content in a 1-kg pike as a function of pH and BPI at a constant mean mercury content in sediments 0.2 mg kg^{-1}. (From Håkanson, 1980.)

Table 7-9. Total mercury residues (average, range) in reservoirs.

Reservoir	Years after filling	Water $\mu g\ L^{-1}$	Sediment $mg\ kg^{-1}$ dry weight	Biota $mg\ kg^{-1}$ wet weight
Lake Jocassee, South Carolina, US (1973–74)[1]	1	0.04(0.01–0.06)	0.04(0.03–0.04)	Largemouth bass 3.11(1.87–4.49)
Lake Keowee, South Carolina, US (1973–74)[1]	4	—	—	Largemouth bass 1.39(0.34–3.99)
Lake Hartwell, South Carolina, US (1973–74)[1]	12	—	—	Largemouth bass 0.48(0.38–0.68)
Cedar Lake, Illinois, US (1976)[2]	2	⩽0.01	lake bottom 0.01(0.00–0.02) inundated soil 0.07(0.01–0.15)	Largemouth bass 0.48(0.14–1.20) White crappie 0.59(–) Bluegill ⩽0.20(–)
American Falls Reservoir, Idaho, US (1974)[3]	35	0.87(0.25–1.78)	0.42(0.21–0.95)	Rainbow trout 0.36(0.05–1.20) Yellow perch 0.19(0.11–0.33) Black bullhead 0.17(0.10–0.34) Black crappie 0.37(0.12–0.80)

Sources: [1]Abernathy and Cumbie (1977); [2]Cox *et al.* (1979); [3]Kent and Johnson (1979).

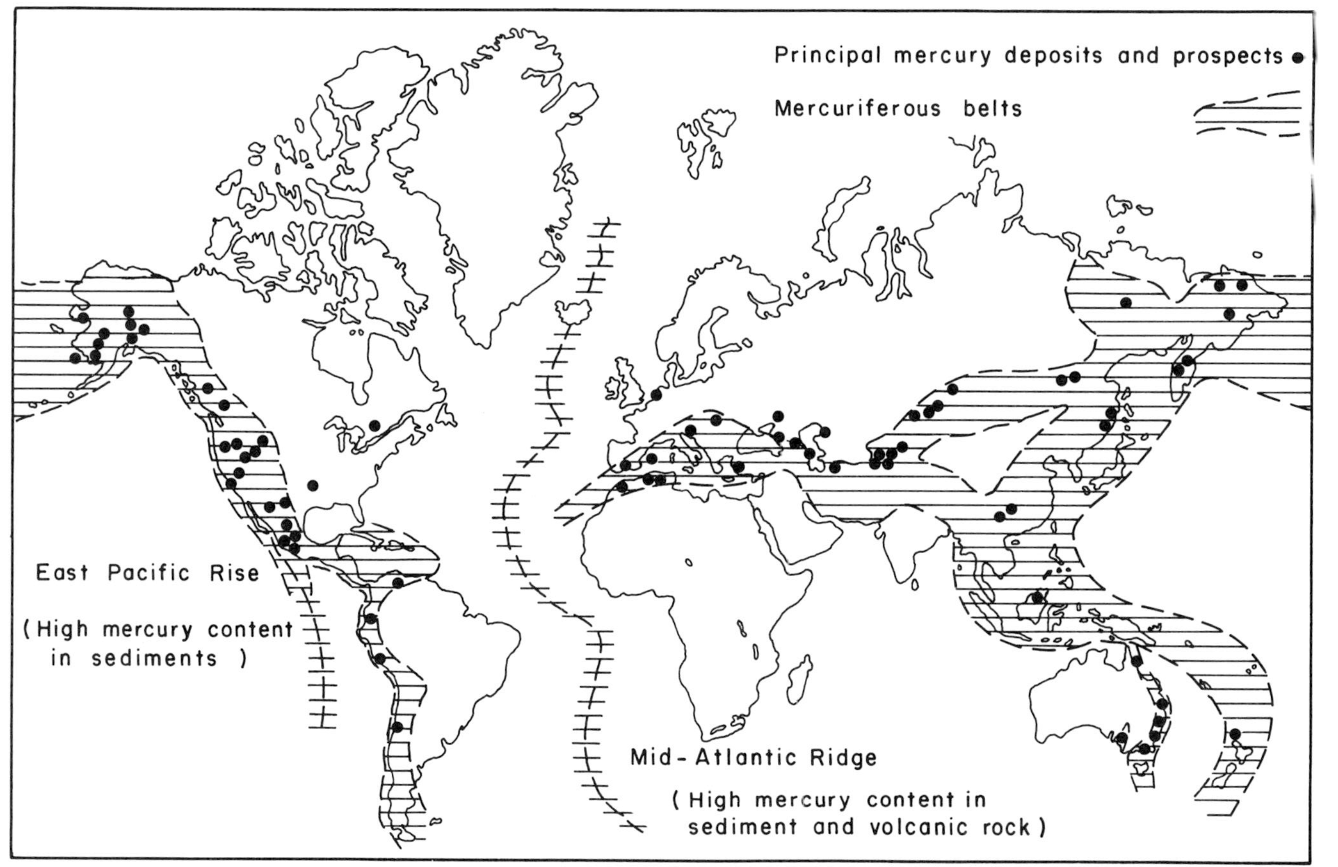

Figure 7-5. Mercuriferous belts of the earth. (From Jonasson and Boyle, 1972.)

calcium content of the water. The presence of a well-developed cell wall in algae may significantly reduce mortality, presumably by reducing rate of sorption.

Acute toxicity of mercuric chloride to marine and freshwater invertebrates depends on species, developmental stage, and environmental conditions (Tables 7-10, 7-11). Microscopic species such as rotifers appear most sensitive, reflecting their large surface area/volume ratio which permits relatively large amounts of dissolved Hg(+2) to enter the cell. Some freshwater oligochaetes and marine polychaetes show considerable tolerance to Hg(+2) under laboratory conditions; this in turn accounts for the presence of such species in highly polluted natural waters. Rehwoldt *et al.* (1973) demonstrated that the eggs of the gastropod *Amnicola* were highly resistant (96h LC_{50} : 2100 $\mu g\ L^{-1}$) to $HgCl_2$ intoxication. Since these eggs have a thick wall to reduce dessication, dissolved pollutants probably pass through the wall at a relatively slow rate. Although data are limited, organic mercurials are more toxic than inorganic forms.

Increasing water hardness reduces the acute toxicity of mercurials to freshwater invertebrates (Table 7-10); however, the degree of antagonism is considerably lower than that reported for many other metals. Similarly, Olson and Harrel (1973) found that the LC_{50} for an estuarine clam increased with salinity to 5.5‰ but declined at 22‰. A number of compounds ameliorate mercury toxicity to invertebrates, including NaCl, SO_3^{2-}, TeO_3^{2-} and TeO_4^{2-}.

Fish

Organic mercury compounds are considerably more toxic than inorganic forms. Wobeser (1975) reported that the 24h median tolerance limit (MTL) of fry and fingerlings of rainbow trout to methyl mercury chloride was 0.084 and 0.125 mg L^{-1}, respectively, whereas the 24h MTL of fingerlings to mercuric chloride was 0.90 mg L^{-1}. Akiyama (1970) found that phenyl mercuric acetate was approximately 7 times more toxic than $HgCl_2$ to the teleost *Oryzias latipes.*

Temperature is an important factor controlling the acute toxicity of mercury compounds. In summer when water temperatures rise, the rate of mortality has been observed to increase without any change in the concentration of pollutant in water (MacLeod and Pessah, 1973). It has also been shown that acute toxicity is inversely related to the dissolved oxygen content of water. This suggests that the high temperatures and low, early morning, DO levels in some waters may significantly contribute to fish mortality. However, air-breathing fish are more resistant than gill breathers to either the direct or indirect effects of mercury poisoning.

In the majority of species studied, the pre-hatch (egg) stage is most susceptible to intoxication. Servizi and Martens (1978) found that the 168h LC_{50} of inorganic mercury for the eggs of sockeye salmon was 4 $\mu g\ L^{-1}$ and

Table 7-10. Acute toxicity of mercury compounds* to some freshwater invertebrates.

Species	Toxicity ($mg\ L^{-1}$)	Temp. (°C)	pH	Total hardness ($mg\ L^{-1}$)	Dissolved oxygen ($mg\ L^{-1}$)
Vorticella convallaria (protozoan)[1]	0.005(50% mortality, 12h)[a]	25	7	NQ	NQ
Paramecium caudata (protozoan)[2]	0.62(75% mortality, 1h)[b]	NQ	NQ	NQ	NQ
Philodina acuticornis (rotifer)[3]	0.7–0.8(96h LC_{50})	20	7.4–7.9	25	NQ
	1.6–2.1(96h LC_{50})	20	7.4–7.9	81	NQ
Tubifex tubifex (oligochaete)[4]	0.058(48h LC_{50})	20	6.3	0.1	NQ
	0.066(48h LC_{50})	20	6.85	34.2	NQ
	0.082(48h LC_{50})	20	7.2	34.2	NQ
	0.100(48h LC_{50})	20	7.3	261	NQ
Nais sp. (oligochaete)[5]	1000(96h LC_{50})	17	7.6	50	6.2
Amnicola sp. (snail)[5]					
egg	2.10(96h LC_{50})	17	7.6	50	6.2
adult	0.08(96h LC_{50})	17	7.6	50	6.2
Gammarus sp. (amphipod)[5]	0.01(96h LC_{50})	17	7.6	50	6.2
Chironomus sp. (dipteran)[5]	0.02(96h LC_{50})	17	7.6	50	6.2

Sources: [1]Sartory and Lloyd (1976); [2]Mills (1976); [3]Buikema *et al.* (1974); [4]Brković-Popović and Popović (1977); [5]Rehwoldt *et al.* (1973).

* Test compound $HgCl_2$ in all cases except: [a]mercuric nitrate; [b]phenylmercuric nitrate. NQ—not quoted.

Table 7-11. Acute toxicity of mercury compounds* to some marine invertebrates.

Species	Toxicity (mg L^{-1})	Temp (°C)	pH	Salinity (⁰/oo)	Dissolved oxygen (mg L^{-1})
Capitella capitata (polychaete)[1]	<0.1(96h LC_{50})	NQ	7.8	NQ	NQ
	0.1(28-day LC_{50})	NQ	7.8	NQ	NQ
Neanthes arenceodentata (polychaete)[1]	0.022(96h LC_{50})	NQ	7.8	NQ	NQ
	0.017(28-day LC_{50})	NQ	7.8	NQ	NQ
Rangia cuneata (clam)[2]	5.1(96h LC_{50})	24	NQ	1	NQ
	10.0(96h LC_{50})	24	NQ	5.5	NQ
	8.7(96h LC_{50})	24	NQ	22	NQ
Mercenaria mercenaria (clam)[3]	0.015(12-day LC_{50})	25	NQ	24	NQ
Crassostrea virginica (oyster)[3]	0.012(12-day LC_{50})	25	NQ	24	NQ
Crassostrea gigas (oyster)[4]	0.006(48h LC_{50})	20	8.1	33.8	6.5–8.0
	0.006(48h LC_{50})	20	8.1	33.8	6.5–8.0
Cancer magister (crab)[4]	0.021(48h LC_{50})[a]	20	8.1	33.8	6.5–8.0
	0.007(48h LC_{50})[a]	20	8.1	33.8	6.5–8.0

Sources: [1]Reish *et al.* (1976); [2]Olson and Harrel (1973); [3]Calabrese *et al.* (1977); [4]Glickstein (1978).

* Test compound $HgCl_2$ except [a]mercuric nitrate. NQ—not quoted.

180–220 $\mu g\ L^{-1}$ for fry and smolt. Exposure of embryos (1–11 days old) of the teleost *Oryzias latipes* to three different mercury compounds resulted in a low rate of survival at 1, 2, 3, and 10 days of age (Figure 7-6). Thus the embryos became increasingly resistant as development proceeded but were also sensitive immediately before hatch. Although increasing hardness of the water probably reduces the toxicity of mercury compounds, Tabata (1969) found no significant change in the onset of toxicosis with hardness.

Chronic/sublethal poisoning may show several other clinical symptoms, including (i) inhibition of enzymes and protein synthesis in liver, kidney, and brain, (ii) structural alterations of fish epidermal mucus, (iii) reduction in sperm viability, embryogenesis, and survival of second generation fry, (iv)

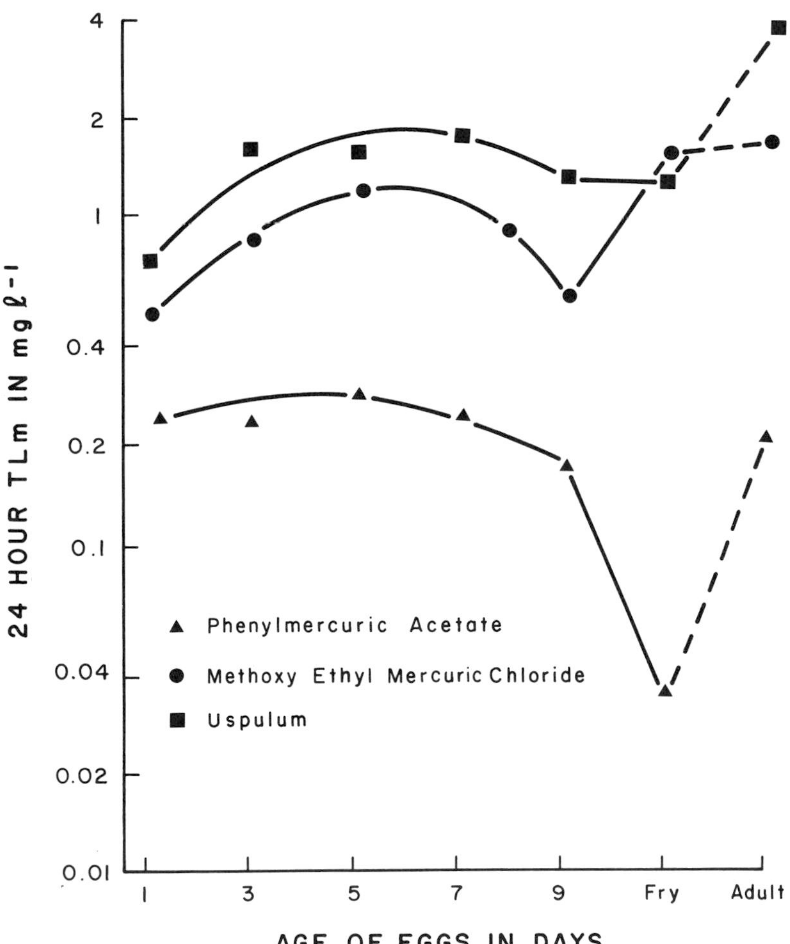

Figure 7-6. The 24-h TLm of *Oryzras latipes* of different developmental stages exposed to different mercury compounds. (From Akiyama, 1970.)

reduction in olfactory response, vision, and respiration, (v) reduction in fin regeneration time in fish and limb regeneration in amphibians, and (vi) decreased ability to osmoregulate. It is apparent from these symptoms that sublethal poisoning may lead to decrease in the ability of fish to survive and reproduce in nature. Fish with reduced olfactory capacity will have difficulty in locating food, thereby reducing growth rate and maximum size of fish. Similarly, retarded fin regeneration will reduce ability to escape predators and catch food while reduced sperm viability would likely result in a decrease in population size. None of these changes would result in drastic or immediate changes in fish stocks.

In rainbow trout, acute poisoning by inorganic mercury is characterized by morphological changes in gills during the initial period of exposure (Daoust, 1981). Branchial changes included an accumulation of cellular debris in the lamellar epithelium, enlarged lamellar epithelial cells, and lamellar fusion (Figure 7-7). Although these changes caused death by asphyxiation, acidosis, hemoconcentration and chloride imbalance may have also contributed to death. After 48 h of exposure, inorganic mercury ceased to cause severe branchial damage; however, lesions were observed in the renal proximal tubules and pancreas.

After 24 h of exposure, acute effects of methyl mercuric chloride in the liver of channel catfish included disorganization and degeneration of exocrine pancreatic tissue, distention of mesenteric vessels, and production of a serous fluid (Kendall, 1977). By 96 h the exocrine pancreatic tissue had severely degenerated and had become necrotic; there was also desquamation of bile duct epithelium into the duct lumen. Exposure of mummichog to 0.5 mg L^{-1} inorganic mercury resulted in lesions in the olfactory and lateral line system (Gardner, 1975). There was severe cytoplasmic and nuclear degeneration of all cellular material. Neurosensory cells of the olfactory organ showed various degrees of degeneration. Although there was no evidence of hyperplasia, the epithelial lining of olfactory pits was necrotic.

Humans

Organic mercurials may induce a toxicity which is commonly referred to as Minamata disease. Clinical signs of intoxication include ataxia, depressed peripheral sensation, and gait and limb reflex dysfunction (National Research Council of Canada, 1979). Sorbed methylmercury is transported by the blood stream, and accumulates in tissues such as liver, kidney, and brain. The whole body half-life of methylmercury is generally <70 and may range from 35–189 days, depending on subject and method of analysis (National Research Council of Canada, 1979). All forms of mercury may be eliminated in the feces and excreted in urine and on hair.

Although mercury compounds are probably not carcinogenic or cocarcinogenic, several studies have demonstrated cytological effects in both plants and animals. Skerfving *et al.* (1970) worked with nine subjects

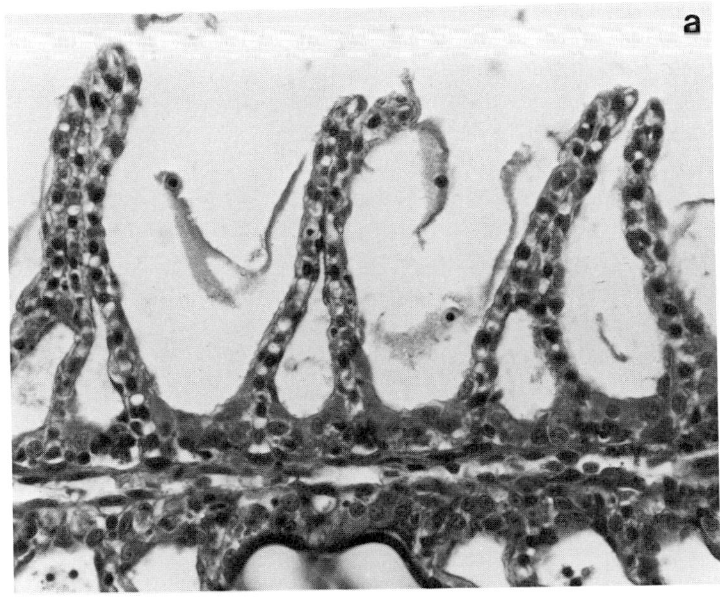

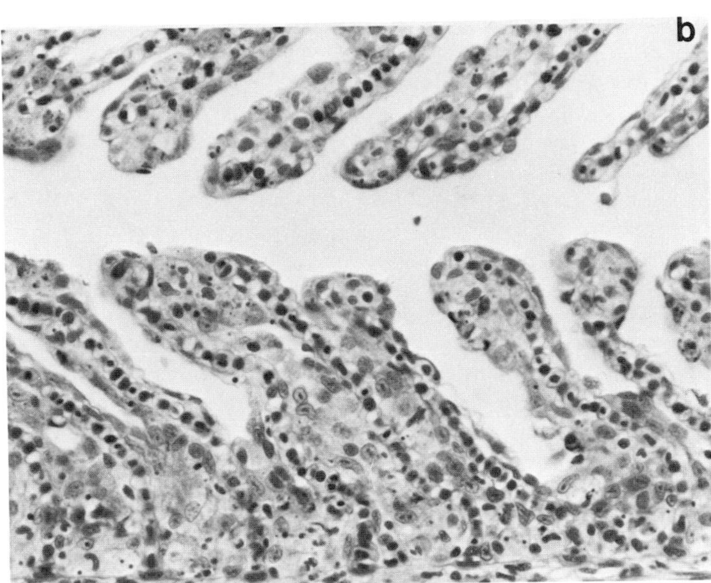

Figure 7-7. Branchial changes in rainbow trout exposed to inorganic mercury. Lamellar fusion (A); enlargement of lamellar epithelial cells and accumulation of cellular debris (B). (From Daoust, 1981.)

showing increased levels of mercury in their red blood cells. These individuals had consumed fish with high levels of methylmercury. It was shown that the frequency of lymphocytes and leucocytes with chromosome breaks was significantly correlated with mercury concentrations in red blood cells. Mercurials may also (i) reduce the rate of DNA replication in Chinese hamsters, (ii) retard leukemic cell multiplication in mice, (iii) inhibit glioma cell mitosis in mice, and (iv) scatter and contract chromosomes in allium roots, resulting in the production of cells with one or more missing or additional chromosomes. Because methylmercury can cross the placenta, there are a number of reported cases of teratogenesis and embryotoxicity. Harada (1978) found that an elevated percentage of offspring of mothers from the Minamata Bay area showed symptoms of cerebral palsy. Gale (1980) demonstrated that in hamsters manifestations of mercury teratogenesis were ventral wall defects, pericardial cavity distension, cleft palate, hydrocephalus, and heart defects.

References

Abernathy, A.R., and P.M. Cumbie. 1977. Mercury accumulation by largemouth bass (*Micropterus salmoides*) in recently impounded reservoirs. *Bulletin of Environmental Contamination and Toxicology* **17**:595–602.

Akiyama, A. 1970. Acute toxicity of two organic mercury compounds to the teleost, *Oryzias latipes* in different stages of development. *Bulletin of the Japanese Society of Scientific Fisheries* **36**:563–570.

Appelquist, H., K.O. Jensen, T. Sevel, and C. Hammer. 1978. Mercury in the Greenland ice sheet. *Nature* **273**:657–659.

Bartelt, von R.D., and U. Förstner. 1977. Schwermetalle im staugeregelten Neckar untersuchungen an sedimenten, algen und wasserproben. *Jber. Mitt. oberrhein. geol. Ver.* **59**:247–263.

Batti, R., R. Magnaval, and E. Lanzola. 1975. Methylmercury in river sediments. *Chemosphere* **1**:13–14.

Beneš, P., E.T. Gjessing, and E. Steinnes. 1976. Interactions between humus and trace elements in fresh water. *Water Research* **10**:711–716.

Brković-Popović, I., and M. Popović. 1977. Effects of heavy metals on survival and respiration rate of tubificid worms: Part I—Effects on survival. *Environmental Pollution* **13**:65–72.

Bryan, G. W. 1976. Some aspects of heavy metal tolerance in aquatic organisms. *In*: A.P.M. Lockwood (Ed.), *Effects of pollutants on aquatic organisms*. Cambridge University Press, Cambridge, pp. 7–34.

Buggiani, S.S., and C. Vannucchi. 1980. Mercury and lead concentrations in some species of fish from the Tuscan coasts (Italy). *Bulletin of Environmental Contamination and Toxicology* **25**:90–92.

Buikema, A.L., Jr., J. Cairns, Jr., and G. W. Sullivan. 1974. Evaluation of *Philodina acuticornis* (Rotifera) as a bioassay organism for heavy metals. *Water Resources Bulletin* **10**:648–661.

Calabrese, A., J.R. MacInnes, D.A. Nelson, and J.E. Miller. 1977. Survival and growth of bivalve larvae under heavy-metal stress. *Marine Biology* **41**:179–184.

Cammarota, V.A. 1975. Mercury. *In: Mineral facts and problems.* Bureau of Mines, Washington, D.C.

Canadian Minerals Yearbooks. 1973–1979. Publishing Center, Department of Supply and Services, Ottawa, Ontario.

Cappon, C.J., and J.C. Smith. 1981. Mercury and selenium content and chemical form in fish muscle. *Archives of Environmental Contamination and Toxicology* **10**:305–319.

Carr, R.A., and P.E. Wilkniss. 1973. Mercury: short-term storage of natural waters. *Environmental Science and Technology* **7**:62–63.

Carr, R.A., J.B. Hoover, and P.E. Wilkniss. 1972. Cold-vapor atomic absorption for mercury in the Greenland Sea. *Deep Sea Research* **19**:747–752.

Carr, R.A., M.M. Jones, and E.R. Russ. 1974. Anomalous mercury in near bottom water of a mid-Atlantic Rift Valley. *Nature* **251**:89–90.

Cox, M.E., and G.M. McMurtry. 1981. Vertical distribution of mercury in sediments from the East Pacific Rise. *Nature* **289**:789–792.

Cox, J.A., J. Carnahan, J. DiNunzio, J. McCoy, and J. Meister. 1979. Source of mercury in fish in new impoundments. *Bulletin of Environmental Contamination and Toxicology* **23**:779–783.

Craig, P.J., and S.F. Morton. 1976. Mercury in Mersey estuary sediments, and the analytical procedure for total mercury. *Nature* **261**:125–126.

Cranston, R.E., and D.E. Buckley. 1972. Mercury pathways in a river and estuary. *Environmental Science and Technology* **6**:274–278.

Dall'Aglio, M. 1968. The abundance of mercury in 300 natural water samples from Tuscany and Latium (Central Italy). *In*: L.H. Ahrens (Ed.), *Origin and distribution of the elements.* Pergamon Press, New York, pp. 1065–1081.

Damiani, V., and R.L. Thomas. 1974. Mercury in the sediments of the Pallanza Basin. Nature 251:696–697.

Daoust, P.Y. 1981. Acute pathological effects of mercury, cadmium and copper in rainbow trout. PhD. thesis. University of Saskatchewan, Canada. 331 pp.

Delisle, C.E., and R.L. Demers. 1976. *Le mercure dans le nord-ouest Québécois (situation actuelle et recommandations) 1976.* Environment Canada, 55 pp.

Dietz, V.F. 1972. Die Anreicherung von Schwermetallen in submersen Pflanzen. *GWF das Gasund Wasserfach, Wasser, Abwasser* **113**:269–273.

D'Itri, F.M. 1972. *The environmental mercury problem.* CRC Press, The Chemical Rubber Company.

Fitzgerald, W.F., and C.D. Hunt. 1974. Distribution of mercury in surface microlayer and in subsurface waters of the Northwest Atlantic. *Journal Rech. Atmos.* **8**:629–637.

Friberg, L., and J. Vostal. 1972. *Mercury in the environment.* CRC Press, Cleveland.

Fujita, M., and K. Hashizume. 1975. Status of uptake of mercury by the fresh water diatom, *Synedra ulna. Water Research* **9**:889–894.

Gale, T.F. 1980. Cardiac and non-cardiac malformations produced by mercury in hamsters. *Bulletin of Environmental Contamination and Toxicology* **25**:726–732.

Gardner, D. 1975. Observations on the distribution of dissolved mercury in the ocean. *Marine Pollution Bulletin* **6**:43–46.

Gardner, G.R. 1975. Chemically induced lesions in estuarine or marine teleosts. *In*:

W.E. Ribelin, and G. Migaki (Eds.), *Pathology of fishes.* University of Wisconsin Press, Madison, pp. 657–693.

Gardner, W.S., D.R. Kendall, R.R. Odom, H.L. Windom, and J.A. Stephens. 1978. The distribution of methyl mercury in a contaminated salt marsh ecosystem. *Environmental Pollution* **15**:243–251.

Glickstein, N. 1978. Acute toxicity of mercury and selenium to *Crassostrea gigas* embryos and *Cancer magister* larvae. *Marine Biology* **49**:113–117.

Glooschenko, W.A. 1969. Accumulation of ^{203}Hg by the marine diatom *Chaetoceros costatum. Journal of Phycology* **5**:224–226.

Gummer, W.D., and D. Fast. 1979. Mercury in the sediments of Thunder Creek and the Moose Jaw river. Unpublished Report prepared by the Moose Jaw River Basin Working Group on Mercury, Water Quality Branch, Environment Canada, Regina, Saskatchewan.

Håkanson, L. 1980. The quantitative impact of pH, bioproduction and Hg-contamination on the Hg-content of fish (pike). *Environmental Pollution* (*Series B*) **1**:286–304.

Harada, M. 1978. Congenital Minamata disease: intrauterine methylmercury poisoning. *Teratology* **18**:285–288.

Harriss, R.C., D.B. White, and R.B. Macfarlane. 1970. Mercury compounds reduce photosynthesis by plankton. *Science* **170**:736–737.

Haug, A., S. Melsom, and S. Omang. 1974. Estimation of heavy metal pollution in two Norwegian fjord areas by analysis of the brown alga *Ascophyllum nodosum. Environmental Pollution* **7**:179–192.

Hegi, H.R., and W. Geiger. 1979. Heavy metals (Hg, Cd, Cu, Pb, Zn) in liver muscle tissue of fresh-water perch (*Perca fluviatilis*) of the Lake of Biel and the Walensee. *Schweizerische Zeitschrift fuer Hydrologie* **41**:94–107.

Hildebrand, S.G., R.H. Strand, and J.W. Huckabee. 1980. Mercury accumulation in fish and invertebrates of the North Fork Holston River, Virginia and Tennessee. *Journal of Environmental Quality* **9**:393–400.

Hinkle, M.E., and R.E. Learned. 1969. Determination of mercury in natural waters by collection on silver screens. *In*: Geological Survey Research. *U.S. Geological Survey Professional Paper 650-D.* pp. D251–D254.

Hirota, R., M. Fujiki, and S. Tajima. 1974. Mercury contents of the plankton collected in Ariake- and Yatsushiro-kai. *Bulletin of the Japanese Society of Scientific Fisheries* **40**:393–397.

Hollibaugh, J.T., D.L.R. Seibert, and W.H. Thomas. 1980. A comparison of the acute toxicities of ten heavy metals to phytoplankton from Saanich Inlet, B.C., Canada. *Estuarine and Coastal Marine Science* **10**:93–105.

Hosohara, K., H. Kozuma, K. Kawasaki, and T. Tsuruta. 1961. Total mercury in sea water. *Nippon Kagaku Zasshi* **82**:1479–1480.

Jackson, T.A., and R.N. Woychuk. 1979. A preliminary report on the geochemistry of mercury in polluted sediments of Thunder Creek, Saskatchewan: Unpublished report, National Water Research Institute, Western and Northern Region, at the Freshwater Institute, Winnipeg, Manitoba.

Jacobs, G. 1977. Total and organically bound mercury content in fishes from German fishing grounds. *Zeitschrift fur Lebensmittel-Untersuchung und-Forschung* **164**:71–76.

Järvenpää T., M. Tillander, and J.K. Miettinen. 1970. Methylmercury: half-time of elimination in flounder, pike and eel. *Suomen Kemistilehti* **B43**:439–442.

Johnson, D.L., and R.S. Braman. 1974. Distribution of atmospheric mercury species near ground. *Environmental Science and Technology* **8**:1003–1009.

Jonasson, I.R., and R.W. Boyle. 1972. Geochemistry of mercury and origins of natural contamination of the environment. *The Canadian Mining and Metallurgical Bulletin,* January , **1972,** pp. 1–8.

Jones, G.B., and M.B. Jordan. 1979. The distribution of organic material and trace metals in sediments from the River Liffey estuary, Dublin. *Estuarine and Coastal Marine Science* **8**:37–47.

Kendall, M.W. 1977. Acute effects of methyl mercury toxicity on channel catfish (*Ictalurus punctatus*) liver. *Bulletin of Environmental Contamination and Toxicology* **18**:143–151.

Kent, J.C. and D.W. Johnson. 1979. Mercury, arsenic, and cadmium in fish, water, and sediment of American Falls Reservoir, Idaho, 1974. *Pesticides Monitoring Journal* **13**:35–40.

Koli, A.K., S.S. Sandhu, W.T. Canty, K.L. Felix, R.J. Reed, and R. Whitmore. 1978. Trace metals in some fish species of South Carolina. *Bulletin of Environmental Contamination and Toxicology* **20**:328–331.

Kudo, A. 1976. Mercury transfer from bed sediments to freshwater fish (guppies). *Journal of Environmental Quality* **5**:427–430.

Kudo, A., and J.S. Hart. 1974. Uptake of inorganic mercury by bed sediments. *Journal of Environmental Quality* **3**:273–278.

Leatherland, T.M., J.D. Burton, F. Culkin, M.J. McCartney, and R.J. Morris. 1973. Concentrations of some trace metals in pelagic organisms and of mercury in Northeast Atlantic Ocean water. *Deep Sea Research* **20**:679–685.

Lockwood, R.A., and K.Y. Chen. 1973. Adsorption of Hg(II) by hydrous manganese oxides. *Environmental Science and Technology* **7**:1028–1034.

Macdonald, R.W., and C.S. Wong. 1977. The distribution of mercury in Howe Sound sediments. Unpublished manuscript, Pacific Marine Science Report 77-22, Department of Fisheries and the Environment, Institute of Ocean Sciences, Patricia Bay, Sidney, British Columbia, Canada, 51 pp.

Mackay, N.J., M.N. Kazacos, R.J. Williams, and M.I. Leedow. 1975. Selenium and heavy metals in black marlin. *Marine Pollution Bulletin* **6**:57–60.

MacLeod, J.C., and E. Pessah. 1973. Temperature effects on mercury accumulation, toxicity, and metabolic rate in rainbow trout (*Salmo gairdneri*). *Journal of the Fisheries Research Board of Canada* **30**:485–492.

Matida, Y., and H. Kumada. 1969. Distribution of mercury in water, bottom sediments and aquatic organisms of Minamata Bay, the River Agano and other water bodies in Japan. *Bulletin of Freshwater Fisheries Research Laboratory* **19**:73.

Matsunaga, K., M. Nishimura, and S. Konishi. 1975. Mercury in the Kuroshio and Oyashio regions and the Japan Sea. *Nature* **258**:224–225.

Miller, D.R. and J.M. Buchanan. 1979. *Atmospheric transport of mercury.* Report Series, Monitoring and Assessment Research Centre, Chelsea, University of London, 1979, 42 pp.

Mills, W.L. 1976. Water quality bioassay using selected protozoa, I. *Journal of Environmental Science and Health* **A11**:491–500.

Mora, B., and J. Fábregas. 1980. The effect of inorganic and organic mercury on growth kinetics of *Nitzschia acicularis* W.Sm. and *Tetraselmis suecica* Butch. *Canadian Journal of Microbiology* **26**:930–937.

Munro, D.J., and W.D. Gummer. 1980. Mercury accumulation in biota of Thunder Creek, Saskatchewan. *Bulletin of Environmental Contamination and Toxicology* **25**:884–890.

Myklestad, S., I. Eide, and S. Melsom. 1978. Exchange of heavy metals in *Ascophyllum nodosum* (L.) Le Jol *in situ* by means of transplanting experiments. *Environmental Pollution* **16**:277–284.

National Research Council of Canada. 1979. *Effects of mercury in the Canadian environment.* Associate Committee on Scientific Criteria for Environmental Quality, Ottawa, Canada. NRC Publication, No. 16739. 290 pp.

Nriagu, J.O. 1979. Global inventory of natural and anthropogenic emissions of trace metals to the atmosphere. *Nature* **279**:409–411.

Olson, K.R., and R.C. Harrel. 1973. Effect of salinity on acute toxicity of mercury, copper, and chromium for *Rangia cuneata* (Pelecypoda, Mactridae). *Contributions in Marine Science* **17**:9–13.

Ottawa River Project. 1977. Distribution and transport of pollutants in flowing water ecosystems. Final Report. University of Ottawa–National Research Council of Canada, Ottawa, 1077 pp.

Parvaneh, V. 1977. A survey on the mercury content of the Persian Gulf shrimp. *Bulletin of Environmental Contamination and Toxicology* **8**:778–782.

Phillips, G.R., T.E. Lenhart, and R.W. Gregory. 1980. Relation between trophic position and mercury accumulation among fishes from the Tongue River Reservoir, Montana. *Environmental Research* **22**:73–80.

Prabhu, N.V., and M.K. Hamdy. 1977. Behavior of mercury in biosystems 1. Uptake and concentration in food chain. *Bulletin of Environmental Contamination and Toxicology* **18**:409–417.

Proctor, P.D., G. Kisvarsanyi, E. Garrison, and A. Williams. 1973. Heavy metal content of surface and ground waters of the Springfield–Joplin areas, Missouri. *In*: D.D. Hemphill (Ed.), *Trace substances in environmental health.* Volume **7**, Columbia, University of Missouri, pp. 63–73.

Ramamoorthy, S., and D.J. Kushner. 1975a. Heavy metal binding sites in river water. *Nature* **256**:399–401.

Ramamoorthy, S., and D.J. Kushner. 1975b. Binding of mercuric and other heavy metal ions by microbial growth media. *Microbial Ecology* **2**:162–176.

Ramamoorthy, S., and A. Massalski. 1979. Analysis of structure-localized mercury in Ottawa River sediments by scanning electron microscopy/energy-dispersive X-ray microanalysis technique. *Environmental Geology* **2**:351–357.

Ramamoorthy, S., and B.R. Rust. 1976. Mercury sorption and desorption characteristics of some Ottawa River sediments. *Canadian Journal of Earth Sciences* **13**:530–536.

Ramamoorthy, S., and B.R. Rust. 1978. Heavy metal exchange processes in sediment-water systems. *Environmental Geology* **2**:165–172.

Rehwoldt, R., L. Lasko, C. Shaw, and E. Wirhowski. 1973. The acute toxicity of some heavy metal ions toward benthic organisms. *Bulletin of Environmental Contamination and Toxicology* **10**:291–294.

Reichert, J., K. Harberer, and S. Normann. 1972. Untersuchungen über das Verhalten von Spurenelementen dei der Trinkwasser. *Vom Wasser,* **39**:137–146.

Reimers, R.S., and P.A. Krenkel. 1974. Kinetics of mercury adsorption and desorption in sediments. *Journal Water Pollution Control Federation* **46**:352–365.

Reish, D.J., J.M. Martin, F.M. Piltz, and J.Q. Word. 1976. The effect of heavy metals on laboratory populations of two polychaetes with comparisons to the water quality conditions and standards in southern California marine waters. *Water Research* **10**:299–302.

Ribeyre, F., A. Delarche, and A. Boudou. 1980. Transfer of methylmercury in an experimental freshwater trophic chain—temperature effects. *Environmental Pollution (Series B)* **1**:259–268.

Rogers, R.D. 1977. Abiological methylation of mercury in soil. U.S. Environmental Protection Agency, EPA-600/3-77-007.

Särkkä, J., M.L. Hattula, J. Janatuinen, and J. Paasivirta. 1978a. Chlorinated hydrocarbons and mercury in aquatic vascular plants of Lake Päijänne, Finland. *Bulletin of Environmental Contamination and Toxicology* **20**:361–368.

Särkkä, J., M.L. Hattula, J. Janatuinen, and J. Paasivirta, 1978b. Mercury and chlorinated hydrocarbons in plankton of Lake Päijänne, Finland. *Environmental Pollution* **16**:41–49.

Sartory, D.P., and B.J. Lloyd. 1976. The toxic effects of selected heavy metals on unadapted populations of *Vorticella convallaria* var. *similis*. *Water Research* **10**:1123–1127.

Servizi, J.A., and D.W. Martens. 1978. Effects of selected heavy metals on early life of sockeye and pink salmon. Progress Report No. 39. International Pacific Salmon Fisheries Commission, New Westminster, B.C., Canada.

Sheffy, T.B. 1978. Mercury burdens in crayfish from the Wisconsin River. *Environmental Pollution* **17**:219–225.

Shiber, J., E. Washburn, and A. Salib. 1978. Lead and mercury concentrations in the coastal waters of North and South Lebanon. *Marine Pollution Bulletin* **9**:109–111.

Skei, J.M., M. Saunders, and N.B. Price. 1976. Mercury in plankton from a polluted Norwegian fjord. *Marine Pollution Bulletin* **7**:34–36.

Skerfving, S., K. Hansson, and J. Lindsten, 1970. Chromosome breakage in humans exposed to methyl mercury through fish consumption. *Archives of Environmental Health* **21**:133–139.

Smith, J.D., R.A. Nicholson, and P.J. Moore. 1973. Mercury in sediments from the Thames estuary. *Environmental Pollution* **4**:153–157.

Smith, S.B., A.Y. Hyndshaw, H.F. Laughlin, and S.C. Maynard. 1971. Mercury pollution control by activated carbon. A review of field experience. Presented at the *44th Annual Water Pollution Control Federation Conference,* San Francisco, California, October 1971.

Speyer, M.R. 1980. Mercury and selenium concentrations in fish, sediments, and water of two northwestern Quebec lakes. *Bulletin of Environmental Contamination and Toxicology* **24**:427–432.

Stenner, R.D., and G. Nickless. 1974. Distribution of some heavy metals in organisms in Hardangerfjord and Skjerstadfjord, Norway. *Water, Air, and Soil Pollution* **3**:279–291.

Stratton, G.W., and C.T. Corke. 1979. The effect of mercuric, cadmium and nickel ion combinations on a blue-green alga. *Chemosphere* **8**:731–740.

Suzuki, T. 1977. Metabolism of mercurial compounds. *In*: R.A. Goyer and M.A. Mehlaman (Eds.), *Toxicology of trace elements.* Halstead Press, New York, pp. 1–39.

Tabata, K. 1969. Studies on the toxicity of heavy metals to aquatic animals and the factors to decrease the toxicity II. The antgonistic action of hardness components in water on the toxicity of heavy metal ions. *Bulletin of the Tokai Regional Fisheries Research Laboratory* **58**:215–232.

Takeuchi, T. 1972. Distribution of mercury in the environment of Minamata Bay and the inland Ariake Sea. *In*: R. Hartung, and B.D. Dinman (Eds.), *Environmental mercury contamination.* Ann Arbor Science Publishers, Ann Arbor, Michigan, pp. 79–81.

Unites States Minerals Yearbooks. 1973–1979. Bureau of Mines, US Department of the Interior, Washington, D.C.

US National Academy of Sciences. 1977. *An assessment of mercury in the environment.* National Academy of Sciences, Washington, D.C.

Watson, W.D. Jr. 1979. Economic considerations in controlling mercury pollution. *In*: J.O. Nriagu (Ed.), *The biogeochemistry of mercury in the environment.* Elsevier-North-Holland Biomedical Press, Amsterdam, pp. 41–77.

Weiss, H.V., M. Koede, and E.D. Goldberg. 1971. Mercury in a Greenland ice sheet: evidence of recent input by man. *Science* **174**:692–694.

Williams, R.M., K.J. Robertson, K. Chew, and H.V. Weiss. 1974. Mercury in the South Polar Seas and in the Northeast Pacific Ocean. *Marine Chemistry* **2**:287–299.

Wobeser, G. 1975. Acute toxicity of methyl mercury chloride and mercuric chloride for rainbow trout (*Salmo gairdneri*) fry and fingerlings. *Journal of the Fisheries Research Board of Canada* **32**:2005–2013.

8
Nickel

Chemistry

Nickel is silvery white, malleable and ductile. It possesses good thermal and electrical conductivity, moderate strength, and hardness. It takes and retains high polish. Although nickel can achieve oxidation states from -1 to $+4$, compounds of the $+2$ state are most common.

Nickel alloys have been known to man since 3500 B.C. Use of alloys for coinage dates back to 327 B.C. in the Bactrian Kingdom (North of Afghanistan). Therapeutic uses were discontinued in the early part of this century after clear demonstration of its acute and chronic toxicity.

Nickel is classified as a borderline element between hard and soft acid acceptors in chemical interactions toward donor atoms. This is reflected in its abundance in the earth's crust as oxides, carbonates, silicates with iron, magnesium, and as sulfides, arsenides and tellurides. Nickel is one of the major elemental constituents of the earth, constituting about 2% by weight. However, it is a minor constituent (0.01%) on the earth's crust.

A mean nickel intake of 200–300 μg day^{-1} is probably adequate for the adult population of western countries. The essentiality of nickel for human health is based on the following considerations:

1. The position of nickel comparable to other well-defined essential elements in the periodic table.
2. Storage of nickel (up to 10 mg) in the human body and the body's ability to regulate absorption, especially during pregnancy.
3. Binding to α-globulin in plasma.

4. Induction of lower activities of many enzymes by nickel deficiency.
5. The demonstrated nutritional essentiality of nickel in chicken and rats.
6. Effect of nickel on the efficiency of iron absorption and consequently blood formation.

It seems that nickel has a central function in metabolism, but, under practical conditions, situations of deficiency in human and animal nutrition are unlikely to occur. Nickel interacts competitively with only five essential elements, Ca, Co, Cu, Fe, and Zn in animals, plants and microorganisms. Competitive interactions are more prevalent with copper than zinc. Nickel, when administered in acute excess, interferes with the detoxification activities and drug metabolism of the liver by altering both the synthesis and the degradation of cellular heme and the liver glutathione levels. Nickel powder and its compounds cause a variety of cancers in rodents and are listed as possible causative agents for occupational or environmental cancer in man. Inhalation and ingestion through diet are the major routes of intake. Percutaneous absorption is important in its desmatopathological effects on humans. In summary, nickel appears to possess a dual status with respect to its role in environmental health.

Production, Uses, and Discharges

Production

The demand for nickel in coinage, electroplating, and alloys has increased strongly since 1850, resulting in an exponential increase in the world mine production of the metal (Table 8-1). The average annual compound growth rate of production from 1900–1975 was 6.3%, which is about five times the rate of increase of world population over the same period (Duke, 1980). Nickel continues to be a vital commodity in every area of industrial activity. It is estimated that the growth in demand will moderate to 3.5–4.5% in the

Table 8-1. Global and Canadian mine production of nickel—10 year annual average (in metric tons × 10^3).

Period	Global	Canada	Period	Global	Canada
1901–1910	7.5	3.0	1951–1960	145.0	111.9
1911–1920	20.5	16.0	1961–1970	321.1	195.0
1921–1930	30.7	25.0	1971–1880	740.0	222.0
1931–1940	54.4	47.2	* 2000	2315.0	390.0
1941–1950	158.3	125.0			

Sources: US Minerals Yearbooks; Canadian Minerals Yearbooks.
* estimate by Department of Energy, Mines and Resources, Canada, 1976.

last two decades of this century due to the awareness of the finite nature of world resources.

Canada continues to be the world's principal producer followed by New Caledonia, the South Pacific, the USSR, Australia, and others. The two major types of nickel ores are sulfide and laterite (silicate or oxide minerals), accounting for about 60% and 40%, respectively, of the world mine production. The depressed demand for nickel in 1975 from a record high consumption in 1974 marked a decrease in production in Canada (Table 8-1). The major producers in the world reduced capacity (about 70%) in 1978 and 1979 to drop excessive inventories and progressive price deterioration. This occurred despite the world record consumption level of 753,640 metric tons of primary nickel.

Seabed resources such as manganese nodules are promising future sources of additional supply of nickel, copper, cobalt, and other metals to meet the growing world consumption. Negotiations are progressing at the ongoing United Nations Conference of the Law of the Sea to formulate a regime that would govern the exploration and exploitation of seabed resources which lie in international waters. Legal, economic, and technological problems including the attainment of cost efficiency of recovery comparable to land-based ores have to be solved before large-scale mining of the seabed nodules can start. The economics of deep-sea bed mining are being worked out by five consortia representing interests in the US, Canada, Japan, Netherlands, West Germany, and Belgium. Three of them completed prototype mining and lift systems during 1978–79. Up to 20% of world primary nickel production could come from deep-sea nodules by the year 2000.

Uses

The physical properties of nickel such as corrosion resistance, high strength and durability over a wide range of temperatures, pleasing appearance, good thermal and electrical conductivity, and alloying ability are the chief advantages in almost all uses of the metal. A finely divided form (Raney nickel) can dissolve about 17 times its volume of hydrogen. This property has led to its extensive use in hydrogenation reactions.

More than 75% of nickel produced is consumed in the production of alloys. There are more than 3000 different alloys including stainless steel and alloy steels, ductile and cast irons, cupronickels and high nickel alloys. Stainless steel is the largest single outlet, followed by plating and high nickel alloys. Addition of nickel to ferrous alloys maximizes the mechanical, physical, and chemical properties of the end product. It also enhances the flexibility of the alloy during fabrication. Stainless steel is primarily used in the field of rapid transit and railway car manufacture, in chemical industries including fertilizer manufacture and food processing machinery, petroleum refining, and in architectural applications. Alloy steels are employed in the

Table 8-2. US consumption of nickel by industrial end use categories in 1970–1979 (in metric tons × 10^3).

Use	1970	1974	1975	1977	1979
Steel:					
Stainless and heat-resisting steels	37.1	67.0	34.7	48.3	63.3
Alloy steels (excluding stainless)	19.0	21.1	18.5	16.2	18.3
Super alloys	10.6	11.8	6.4	10.3	16.0
Nickel–copper alloys	6.0	8.4	8.6	6.3	7.7
Permanent magnet alloys	2.2	4.8	3.0	0.7	0.6
Pure nickel and nickel alloys	32.5	40.1	33.8	29.0	37.3
Cast irons	4.5	4.8	4.0	3.4	4.3
Electroplating	22.3	23.8	17.4	19.8	25.9
Chemicals	0.9	1.8	2.2	2.3	1.1
Other Uses	6.6	5.7	4.6	4.8	3.9
Total	141.7	189.3	133.2	141.1	178.4

Source: US Minerals Yearbooks.

manufacture of heavy machinery, armaments, and tools. Certain alloys are non-brittle even at low temperatures and hence used in cryogenic containers for storage and handling of liquefied natural gas. Ductile and cast irons are widely employed in industrial equipment and in the manufacture of automobiles. Nickel–iron alloys are found in magnetic components in electrical equipment. Marine operations such as piping of salt water, propellers, hulls for ships and boats, and water desalinization plants consume cupronickel alloys because of their exceptional corrosion resistance. High-nickel alloys containing one or more of elements such as chromium, molybdenum, aluminum, titanium, and niobium are used in chemical, electronic, nuclear, and aerospace applications. Nickel-chromium alloys have long been used in heating elements of domestic stoves and industrial furnaces.

Pure nickel is consumed in electroplating and electroforming of products such as trim and bumpers in automobiles, household appliances, and plumbing fixtures. Industrial applications include rejuvenation of worn components and reworking of incorrectly machined parts. Other uses of nickel include: the manufacture of nickel cadmium batteries for stand-by power equipment, and zinc–nickel oxide storage batteries (claimed to have 2.5 times the storage capacity of the conventional lead–acid battery); for bonding ceramics to metal; as a catalyst in the preparation of edible oils; and in solar energy equipment.

US nickel consumption decreased in 1975 by about 20% from a record high in 1974, due to slowdown in the industrial economy. Consumption has increased progressively since then, reaching a near-record level in 1979 (Table 8-2).

Discharges

The nickel burden of the world's oceans is 10^4 times higher than that of freshwaters, 8.4×10^{11} kg and 3.4×10^7 kg, respectively (Nriagu, 1980). The sources of nickel in the ocean are: river transport 1.35×10^9 kg yr^{-1}, atmospheric input 2.5×10^7 kg yr^{-1}, industrial and municipal discharges 3.8×10^6 kg yr^{-1}, and an undetermined amount from leaching from ocean sediments. The residence time of nickel in the oceans is about 2.3×10^4 years, whereas residence time in coastal waters of southern California was found to be only 19 years (Hodge *et al.*, 1978). The accumulation of nickel in oceanic sediments calculated from the hydrogenous precipitation values for manganese and the Mn/Ni ratio in the nodules agrees with the nickel input value from the weathering of continental rocks transported by the rivers, 1.5×10^9 kg yr^{-1} (Nriagu, 1980).

Global emissions of nickel from natural sources and anthropogenic sources are given in Tables 8-3 and 8-4. It should be emphasized that the estimates are approximate due to discrepancies in the reported emission rates and might include retransmission of some anthropogenic nickel. Eroded soil particles account for 77% of all natural emissions, followed by volcanogenic particles (15%). Current anthropogenic emissions exceed the natural rates by 180% (Tables 8-3, 8-4). The automotive combustion of nickel-containing diesel oil represents 57% of anthropogenic emissions. About 25% originates from extraction and industrial applications, followed by waste incineration, wood combustion, iron and steel production, coal combustion, and fertilizer production (Table 8-4). Since 1930, consumption and hence emission of nickel has almost doubled every decade (Table 8-5). About a million metric tons (99% in this century) of nickel have been dispersed into the global ecosystems through man's industrial activities. An enrichment factor of 12 has been calculated for nickel in US urban air particulates compared to the earth's crust.

Table 8-3. Global emissions of nickel from natural sources.

Source	Global production (10^6 metric tons yr^{-1})	Ni emissions (10^3 metric tons yr^{-1})
Windblown dusts	500	20
Volcanogenic particles	10	3.8
Vegetation	75	1.6
Forest fires	36	0.6
Seasalt sprays	1,000	0.04
Total		26.04

Source: Nriagu (1979). Reprinted by permission from *Nature* **279**:409–411. Copyright © 1979, Macmillan Journals Limited.

Table 8-4. Global anthropogenic emissions of nickel in 1975.

Source	Global production/ consumption (10^6 metric tons yr^{-1})	Ni emissions (10^3 metric tons yr^{-1})
Primary non-ferrous metal production		
Copper	7.9	1.5
Lead	4.0	0.34
Nickel	0.8	7.2
Zinc	5.6	0.36
Secondary non-ferrous metal production	4.0	0.2
Iron and steel production	1300	1.2
Industrial application	—	1.9
Coal combustion	3100	0.66
Oil and gasoline combustion	2800	27.0
Wood combustion	640	3.0
Waste incineration	1500	3.4
Phosphate fertilizer production	118	0.6
Total		47.36

Source: Nriagu (1979). Reprinted by permission from *Nature* **279**:409–411. Copyright © 1979, Macmillan Journals Limited.

Table 8-5. All-time global consumption and anthropogenic emissions of nickel.

Period	Global consumption (10^6 metric tons)	Global emissions (10^3 metric tons)
1850–1900	0.2	12
1901–1910	0.14	8.2
1911–1920	0.35	21
1921–1930	0.36	21
1931–1940	0.83	49
1941–1950	1.37	80
1951–1960	2.38	140
1961–1970	4.37	257
1971–1980	7.07	415
Total (all-time)	17.07	1003

Source: Nriagu (1979). Reprinted by permission from *Nature* **279**:409–411. Copyright © 1979, Macmillan Journals Limited.

Nickel in Aquatic Systems

Speciation in Natural Waters

Binding to Inorganic and Organic Ligands. Nickel(+2) forms stable complexes with inorganic and organic ligands. Inorganic complexes include halides, sulfates, phosphates, carbonates and carbonyls. Organic ligands with oxygen, nitrogen, and especially sulfur donor atoms form strong complexes with nickel. Humic and fulvic acids form moderately strong complexes with nickel. Ratios of fulvic acid : nickel greater than 2 favor the formation of soluble nickel–fulvic acid complexes whereas, at lower ratios, nickel tends to form insoluble complexes between pH 8–9. This might lead to its accumulation in suspended matter and eventually in sediments.

Transport in Natural Waters

Nickel binding to particulates in river waters exhibits variable behaviour. Of the nickel contained in the Amazon and Yukon Rivers, 97–98% was found in particulate form whereas some US and German streams showed only 5–30% particulate nickel (Wilson, 1976). The high levels (93%) of dissolved nickel reported for two small streams in Tennessee might be due to sampling during a low flow period resulting in a low suspended solids content (Perhac, 1972).

Nickel transport was studied in two river systems, the Amazon River (a tropical river with a two-fold annual variation in discharge) and the Yukon River (an Arctic river with 35-fold annual variation in discharge) (Gibbs, 1977). Precipitated or coprecipitated coatings and crystalline forms accounted for 80% of nickel transport, followed by organic solids (15%), and soluble and sorbed species (5%). The nickel content in particulates was inversely correlated to grain size, and over 90% of the total nickel in both rivers was transported by particles of size range 0.2–20 μm. Assuming that crystalline-bound nickel is non-bioavailable, the coatings transport 70%, organics 22%, and soluble and sorbed the remaining 8%, of the bioavailable nickel.

Association of nickel with Fe–Mn oxides is an important mode of transport. Relatively low percentages of Fe–Mn oxide fractions (14 and 24%) were reported for the Yamaska and St. François Rivers, Quebec (Tessier *et al.*, 1980) compared to 44 and 48% for the Amazon and Yukon Rivers (Gibbs, 1977); in addition, the crystalline (residual) fraction was 38 and 31% for Amazon and Yukon Rivers and 77 and 66% for Yamaska and St. François Rivers. Particulate nickel concentrations of 7–50 mg kg^{-1} and 26–84 mg kg^{-1} were reported for the epilimnion (5–7 m) and bottom (1 m above lake floor) regions of Lake Michigan (Leland, 1975). Particulate nickel concentrations in the near- and off-shore waters of Lake Ontario are

180 and 34 ng L^{-1}. Up to 40% of the total nickel in the lake water is bound to suspended particulates (Nriagu *et al.*, 1981).

Particulate-bound nickel in estuaries is similar to that in river waters. About 95% of total nickel in the Mississippi River is in the particulate form, fluxing into the Gulf of Mexico, similar to the estimate for the Amazon and Yukon Rivers. Of course, concentrations decrease due to dilution from continental shelf waters towards the sea. Generally, nickel concentrations in polluted estuaries are of the same order of magnitude as in unpolluted estuaries (Snodgrass, 1980).

About 80% of total nickel in the primary effluents of the Los Angeles water treatment plant was soluble (≤ 0.2 μm size fraction) (Chen *et al.*, 1974). Similarly because of low suspended matter content, the secondary effluents contained more than 95% of total nickel in the soluble form. Though the secondary treatment removed less than 40% of the dissolved nickel, it was efficient in removing 90–95% of suspended particulates and thus 80% of particulate nickel. The high levels of nickel in particulate matter of the waste effluents, storm water, and dry weather flows also enriched the coastal sediments in the discharge areas in the southern California region. Studies in other wastewater treatment systems have identified the nickel sources to be mainly industrial followed by municipal and residential sources, runoffs, and inactive industrial and mine tailings.

The concentration and speciation of nickel will depend on competing processes such as coagulation, precipitation, sorption, and complexation/chelation with dissolved organic and inorganic ligands. Although many studies have elucidated basic physicochemical mechanisms, their extrapolation to natural waters is still difficult and controversial due to variability in the nature, composition, and content of sorbates and ligands (organic and inorganic). Laboratory studies using natural water showed that (i) the solubilities of most trace metals including nickel are opposite to that predicted from inorganic solubility considerations, (ii) organic matter complexing is an important process, and (iii) nickel has a range of colloidal properties with only 15% removal by calcium at seawater concentration (Sholkovitz and Copland, 1981).

In summary, the particulates play a vital role in sequestering and transporting nickel in the freshwater environment. Further deposition of particulates occurs at the river–ocean boundary. The release and deposition of particulate-nickel cycle is contained in the estuaries, controlled essentially by Fe and/or Mn content. In oceans, nickel is maintained by a biogeochemical cycle involving particulates.

Nickel in Sediments

Nickel enrichment of sediments, especially nearshore, reflects anthropogenic inputs from industrial (electroplating, battery manufacturing, and photoengraving) and municipal discharges, and other sources such as urban storm water runoff, and sludge disposals. Concentrations of some typical unpolluted and polluted sediments are given in Table 8-6. Low residues in delta sediments compared to river particulates are due to lower oxide phase content by diagenetic loss in the former. Chemical partitioning experiments on sediments of the St. Lawrence estuary and gulf showed 12–23% of total nickel (4–160 mg kg^{-1}) was of non-detrital origin and bioavailable (Loring, 1979). The non-detrital fraction was acetic acid soluble, holding nickel in fine grained organic material, hydrous iron oxides, and ion exchange sites in sediments. Nickel was enriched in the humic and fulvic acids compared to the entrapping sediments and 10% of the total nickel was bound to organic matter (Nriagu and Coker, 1980). Chemical partitioning of nearshore sediments of Los Angeles Harbor identified the following fractions of the total nickel (18–47 mg kg^{-1}): 7–10 μg L^{-1} soluble, 25 μg L^{-1} ion-exchangeable, 37% non-residual (ferromanganese nodules; carbonates, oxides and hydroxides; and organic matter and sulfides), and the rest in the residual (lithogeneous) fraction (Gupta and Chen, 1975). This shows that a large fraction of nickel is removed in the marine environment by the scavenging action of Fe and Mn oxides and hydroxides.

Table 8-6. Total nickel levels (mg kg^{-1} dry weight) in freshwater and marine sediments.

Location	Average (range)	Polluting source
Freshwater Sediments		
Qishon River (Israel)[1]	19 (11–31)	municipal, industrial
Thompson Lake (Canada)[2]	45 (30–85)	gold mine
8 Sudbury rivers (Canada)[3]	102 (13–224)	aerial fallout, nickel smelters
65 Sudbury lakes (Canada)[4]	120 (30–630)	aerial fallout, nickel smelters
28 Sudbury lakes (Canada)[5]	27 (8–2700)	aerial fallout, natural
Illinois River (USA)[6]	27 (3–24)	industrial, municipal
Lake Superior (Canada)[7]	55 (24–70)	natural
Lake Huron (Canada)[7]	59 (30–95)	natural
Marine Sediments		
Baltimore Harbor (USA)[8]	36 (—)	mixed industrial
Ems Estuary (Netherlands)[9]	29 (21–42)	mixed industrial
Rhine Estuary (Netherlands)[9]	41 (19–59)	mixed industrial
Coastal waters (Australia)[10]	8 (9–14)	natural
Bayou Chico Estuary (USA)[11]	7 (1–76)	mixed industrial
Galveston Bay (USA)[12]	22 (0.6–58)	mixed industrial
Houston ship channel (USA)[12]	34 (15–63)	mixed industrial

Sources: [1] Kronfeld and Navrot (1974); [2] Moore (1981); [3] Hutchinson *et al.* (1975); [4] Sempkin (1975); [5] Allan (1975); [6] Mathis and Cummings (1973); [7] Kemp *et al.* (1978); [8] Helz (1976); [9] Salomons and Mook (1977); [10] Knauer (1977); [11] Pilotte *et al.* (1978); [12] Trefry and Presley (1976).

Residues

Water, Precipitation, and Sediments

Dissolved nickel levels in unpolluted freshwater usually range from 1 to 3 μg L^{-1} (Snodgrass, 1980). Input from mixed industrial urban sources may increase these values to 10–50 μg L^{-1} whereas natural intrusions of nickel-bearing rock have produced total residues of 200 μg L^{-1} in overlying water (Agrawal *et al.,* 1978). Aerial emissions from nickel smelters (Canada) resulted in dissolved concentrations of 1–183 μg L^{-1} in nearby lakes (Sempkin, 1975). Although concentrations of dissolved nickel in uncontaminated coastal waters are substantially lower (1.8 μg L^{-1}) (Snodgrass, 1980), anthropogenic inputs may increase levels to 2.5–15 μg L^{-1}. Helz (1976) showed that approximately 73% of the nickel entering Chesapeake Bay (USA) was from rivers, whereas saltwater advection accounted for another 22%. By contrast, direct industrial discharge represented 75% of the total input of nickel into Baltimore Harbor.

Precipitation in remote areas of the world carries detectable levels (0.02–5.0 μg L^{-1}) of nickel, resulting in the mild contamination of snow and ice fields. Concentrations increase substantially over urban areas and, depending on industry type, range from 3 to 100 μg L^{-1}. Under most circumstances, precipitation does not contribute significantly to the flux of nickel in natural waters. However, Sempkin (1975) found that the high concentrations (200–2000 μg L^{-1}) in air near the Sudbury (Canada) nickel smelters were correlated with residues in the sediments of nearby lakes. A similar correlation has been observed for terrestrial soils in other industrialized parts of the world (Anderson *et al.,* 1978).

Nickel is not a significant or widespread contaminant of most freshwater and marine sediments (Table 8-6). Some of the highest residues (>500 mg kg^{-1} dry weight) have been found in lakes near Sudbury smelters. In other industrialized parts of the world concentrations seldom exceed 50–100 mg kg^{-1}. Such values are often comparable to those reported for unpolluted sediments (Table 8-6).

Aquatic Plants

Despite the widespread occurrence of nickel, residues in aquatic plants are generally low. Bartelt and Förstner (1977) reported concentrations ranging from 52 to 74 mg kg^{-1} dry weight in benthic algae from the River Neckar (FRG), while in the River Leine (FRG) and Danube River (Austria), there was a range of 3.0 to 288 mg kg^{-1} (Abo-Rady, 1980; Rehwoldt *et al.,* 1975). Substantially higher levels (150–700 mg kg^{-1}) occurred in algae inhabiting water adjacent to a UK smelter (Trollope and Evans, 1976). Similarly *Potamogeton* sp. from the Sudbury area had burdens of up to 690 mg kg^{-1}

wet weight (Hutchinson *et al.*, 1975). As might be expected, CF's, ranging from 0.25×10^3 to 5×10^3, are lower than those for many other metals.

Marine and estuarine species similarly contain low tissue levels of nickel. Concentrations in *Ascophyllum nodosum* and *Fucus vesiculosus* inhabiting Welsh coastal waters ranged from 4.5 to 8.9 mg kg^{-1}, yielding CF's of $4.6-6.8 \times 10^3$ (Foster, 1976). Slightly higher residues (1–22 mg kg^{-1}) were found in *A. nodosum* from Trondheimsfjorden, Norway (Lande, 1977), while nine species from Indian coastal waters had burdens of 0.5–39.1 mg kg^{-1} (Agadi *et al.*, 1978).

Rate of nickel uptake increases with exposure concentration and the application of phosphate to culture water. Much of the sorbed metal is bound internally, whereas the remainder is loosely attached to the external cell wall. As with other metals, the presence of chelators in water reduces nickel sorption. Although actual uptake mechanisms are not known, it is likely that ion exchange processes occur in most species. These involve the release of calcium and other cations from the cell, coincident with replacement by $Ni(+2)$. The rate of exchange depends on culture pH, and decreases with time as the number of binding sites falls.

Invertebrates

Because concentrations in algae and other foods are generally low, there are few reported cases of significant nickel contamination in freshwater and marine invertebrates. Residues in the soft tissues of the mollusc *Anodonta anatina* from the River Thames (UK) ranged from only 0.1 to 45.9 mg kg^{-1} dry weight (Manly and George, 1977). These values were the highest on record for *Anodonta* in the British Isles. Similarly, maximum levels in molluscs and crustaceans collected near Sudbury were 29–39 mg kg^{-1} wet weight (Hutchinson *et al.*, 1975). Mediterranean coastal waters (Lebanon) yielded molluscs, polychaetes, urchins, and anemones containing soft tissue residues of $<0.1-40$ mg kg^{-1} wet weight (Shiber, 1981) though levels in the bivalve *Scrobicularia plana* inhabiting unpolluted English waters reached 11.9 mg kg^{-1} dry weight (Bryan and Hummerstone, 1978). Several other gastropod and bivalve species, collected from various unpolluted and polluted sites throughout the world, carried residues of $0.3-<9.5$ mg kg^{-1} dry weight (Karbe *et al.*, 1977; Watling and Watling, 1976). Hence, nickel cannot be considered a significant, widespread contaminant of invertebrate tissues, except at site specific points.

Nickel concentrations are generally highest in tissues with high metabolic activity (Table 8-7). These include the kidneys and digestive organs, though the exoskeleton of crustaceans may also carry heavy nickel burdens. In most cases, muscle residues are lower than those of any other tissue; thus, biomonitoring and impact assessment programs should not be based on muscle residue data.

Table 8-7. Average concentrations (mg kg^{-1} dry weight) of nickel in different invertebrate tissue.

Species	Location	Tissue	Concentration
Anodonta anatina[1] (bivalve)	River Thames (UK)	Mantle	35.0
		Ctenidia	37.8
		Digestive gland	18.1
		Kidney	138.8
		Gonad	24.3
		Foot	9.4
		Adductor muscle	6.6
Scrobicularia plana[2] (bivalve)	Estuaries (UK)	Mantle and siphons	3.2
		Digestive gland	43.1*
		Foot and gonad	3.9
Squid[3]	Gulf of Mexico (USA)	Muscle	2.5
		Viscera	1.7
		Pens	0.8
Penaeus aztecus[3] (brown shrimp)	Gulf of Mexico (USA)	Muscle	1.4
		Exoskeleton	6.2
		Viscera	5.7
Carcinus maenus[4] (green crab)	Estuaries (UK)	Muscle	6.2†
		Gill	6.7
		Hepatopancreas	9.0
		Haemolymph	12.3

Sources: [1] Manly and George (1977); [2] Bryan and Hummerstone (1978); [3] Horowitz and Presley (1977); [4] Wright (1976).
* Indicates maximum value.
† Wet weight.

Total concentrations in invertebrates vary seasonally in response to reproductive condition and changing water temperature. Karbe *et al.* (1977) observed maximum residues ($\sim$ 4 mg kg^{-1} dry weight) in the mussel *Mytilus edulis* during April and May whereas much lower values ($\leq$0.5 mg kg^{-1}) occurred during the fall. Similarly, the maturation of female mussels *Choromytilus meridionalis* produced soft tissue levels which were about twice as high as those of mature males (Watling and Watling, 1976). Although there are many exceptions, residues generally either increase, or remain constant, with increasing age and size of animal. Hence uptake is more rapid in juveniles, reflecting their relatively high rate of metabolism. Bryan and Hummerstone (1978) demonstrated that the rate of sorption of nickel by the bivalve *Scorbicularia plana* was substantially slower than that recorded for lead, zinc, and cadmium, and comparable to that for silver, copper, and chromium. The primary route of uptake of nickel may be either food or water depending on species.

Fish

One of the few significant cases of nickel contamination in freshwater fish was reported by Hutchinson *et al.* (1975) for lakes and rivers near Sudbury. Average concentrations in the six species investigated were approximately similar, ranging from 9.5 – 13.8 mg kg^{-1} wet weight (Table 8-8). Residues in fish from other industrialized parts are always much lower than these, seldom exceeding 1 mg kg^{-1}. Although the same may be said about most marine fish, Wright (1976) recorded concentrations of 0.5 – 7.2 mg kg^{-1} for several species off the NE coast of England. Because there was no apparent source of industrial contamination, these levels must have been derived from natural sources. There are a number of similar cases in other coastal waters, such as near Israel (Roth and Hornung, 1977) and Los Angeles harbor (Emerson *et al.*, 1976). However, in most areas, residues are <1.0 mg kg^{-1}. Concentrations in muscle are generally slightly lower than those in liver, kidneys, and gills (Table 8-8). Although data are limited, gonads also appear to carry more nickel than muscle, whereas residues in the fat body are generally less than those in muscle.

Nickel does not accumulate through the food chain. In the Illinois River (USA) average concentrations in sediments, invertebrates, and in the muscle of omnivorous and carnivorous fish were 27 mg kg^{-1} (dry weight), 11,

Table 8-8. Average nickel residues (mg kg^{-1} wet weight) in different tissues of fish collected near the Sudbury nickel mines (Canada) and coastal waters of the UK and New Zealand.

Species	Muscle	Liver	Kidney	Gill	Gonad	Fat body
Sudbury[1]						
Brown bullhead	9.5	10.7	11.8	11.1	—	—
Pike	13.3	15.4	27.7	17.3	—	—
Rock bass	12.5	17.0	17.3	31.7	—	—
Shorthead redhorse	12.9	14.5	14.1	12.2	—	—
Walleye	13.8	14.4	51.6	16.0	—	—
White sucker	13.2	16.5	14.0	12.6	—	—
Coastal waters (UK)[2]						
Cod	2.3	2.5	4.6	3.3	4.2	1.6
Dab	7.1	2.4	2.0	—	—	—
Haddock	0.5	1.7	3.6	3.4	—	0.5
Lumpsucker	3.4	4.0	5.5	—	—	3.2
Plaice	2.8	10.8	6.0	—	9.0	4.2
Sprat	7.2	2.1	3.8	4.5	—	2.0
Coastal waters (New Zealand)[3]						
8 species	0.03	0.26	0.35	0.56	0.20	—

Sources: [1] Hutchinson *et al.* (1975); [2] Wright (1976); [3] Brooks and Rumsey (1974).

0.18 and 0.13 mg kg^{-1} wet weight, respectively (Mathis and Cummings, 1973). Similarly residues in aquatic plants, invertebrates and fish from the Sudbury area were 3–690, 4–39 and 9.5–13.8 mg kg^{-1} wet weight, respectively (Hutchinson *et al.,* 1975). There is usually no correlation between nickel residues in muscle and feeding habits. However, levels in kidney of northern pike and walleye (carnivores) were substantially higher than those of omnivores (Table 8-8). Although there are few data relating the age/size of fish to nickel levels in kidney, muscle residues either increase or remain constant as the fish grows. Detailed information on the mechanisms of nickel uptake in fish is also not available.

Toxicity

Aquatic Plants

Under most test conditions, nickel is less toxic to aquatic plants than mercury, copper, cadmium, silver, and thallium, but more toxic than lead and zinc. Significant reductions in growth and photosynthesis generally occur at 0.1–0.5 mg Ni L^{-1}. However, Patrick *et al.* (1975) noted changes in the species composition of benthic algae at 0.002 mg L^{-1} whereas concentrations of 723 mg L^{-1} were needed to reduce photosynthetic oxygen evolution in *Elodea canadensis* by 50% (Brown and Rattigan, 1979). Much of this variability probably reflects the availability of the free nickel ion and the presence of organic/inorganic chelators in the culture media. Hence it is important to fully document experimental conditions in order to clarify differential toxic responses. Susceptibility to nickel poisoning is also largely species dependent. Following a 12-day exposure period at 0.10 mg Ni L^{-1}, the density of *Scenedesmus acuminata* fell by approximately 95% but the three other test species were not affected at 0.10 mg L^{-1} (Figure 8-1).

Combinations of nickel/mercury, and nickel/copper act synergistically towards many species. Antagonism often occurs between mixtures of nickel/cadmium, nickel/zinc, and nickel/cadmium/mercury. Pre-treatment of algae with either nickel or mercury reduces the toxicity of cadmium. As with other metals, high water hardness and the presence of chelators also reduce toxicity.

Several algal species can adapt to high nickel levels. Growth of unadapted strains of *Scenedesmus acutiformis* stopped at 0.5 mg Ni L^{-1} (Stokes *et al.,* 1973). However, the same species collected from polluted lakes continued to divide at 1.5 mg L^{-1}. Unlike copper, pre-exposure to trace levels of nickel was not required to develop tolerance. In addition, adapted strains sorbed much more metal than non-adapted strains, while continuing to divide.

Invertebrates

Nickel is one of the least toxic priority heavy metals. The 48–96 h LC_{50} for freshwater invertebrates generally ranges from 0.5 to 20 mg L^{-1}. Hence

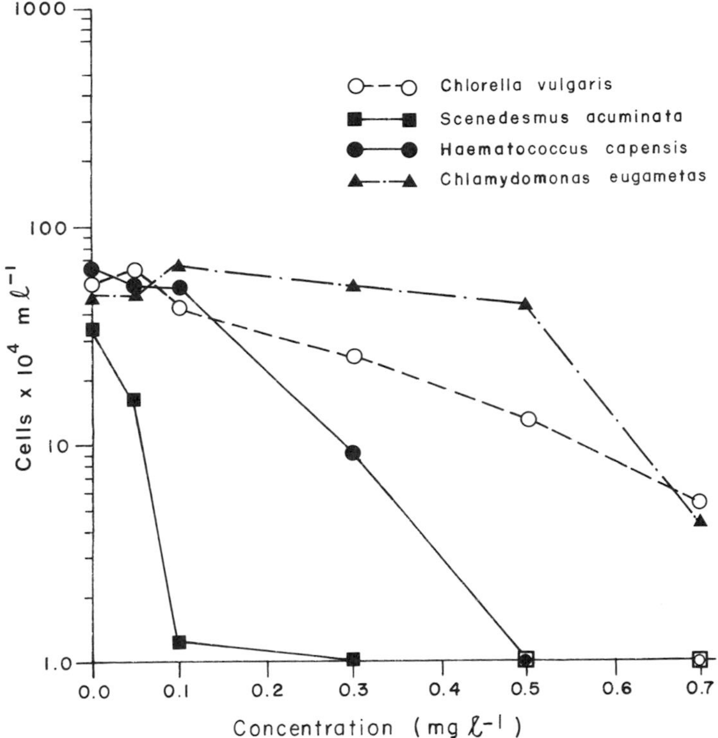

Figure 8-1. Growth of four algal species at different nickel concentrations. Growth measured after 12 days. (From Hutchinson, 1973).

nickel is usually less toxic than mercury, cadmium, copper, lead, and silver; depending on test conditions, it may also be less hazardous than zinc, chromium, and arsenic. A comparable sequence of toxicity can also be applied to marine and estuarine invertebrates. Although LC_{50}'s generally vary from 0.5 to 10 mg L^{-1}, values of 50–500 mg l^{-1} have been reported for some species. This range in toxicity is due to differences in the availability of the free nickel ion, which in turn reflects variable test conditions. Susceptibility to intoxication cannot be consistently related to taxonomic position or life stage of invertebrate. Hence biological monitors for nickel contamination should consist of a number of species and life stages.

Acute effects of nickel intoxication can be suppressed through the addition of Ca, Mg, and other cations. The 48 h LC_{50} for the oligochaete *Tubifex tubifex* increased 750 times when water hardness increased from 0.1 to 261 mg L^{-1} (Brković-Popović and Popović, 1977). This change in sensitivity was greater than that observed for zinc, chromium, cadmium, copper, and mercury. While a comparable response would likely occur in marine waters, there are few comprehensive data to confirm this point. Antagonistic/synergistic effects have also not been fully investigated; however, it is likely that those interactions reported for aquatic plants can also be applied to invertebrates.

Exposure to sublethal/chronic levels of nickel generally results in a reduction in growth. Calabrese *et al.* (1977) reported that the growth of oysters *Crassostrea virginica* was not influenced at the LC_5 value, but was markedly reduced at the LC_{50}. A comparable respone was recorded for the cladoceran *Daphnia magna,* whereas gastrulation in sea urchin (*Lytechinus pictus*) embryos did not occur below 10^{-3} M $NiCl_2$ (Biesinger and Christensen, 1972; Timourian and Watchmaker, 1972). As might be expected, moderate nickel concentrations (0.03–0.05 mg L^{-1}) reduce fecundity, reproduction, and respiration in several species (Biesinger and Christensen, 1972; Timourian and Watchmaker, 1972).

Fish

Nickel is generally less toxic to fish than copper, mercury, lead, zinc, cadmium, silver, chromium, and arsenic. The 48–96 h LC_{50} for adults ranges from 5 to 100 mg L^{-1}. However, lower values have been reported for immature stages of several species. In carp, the 72 h LC_{50}'s for freshly fertilized eggs and one day old larvae were 6.1 and 6.2 mg L^{-1}, respectively (Blaylock and Frank, 1979).

Acute toxicity of nickel is significantly reduced in seawater through competitive interactions with cations. Similarly, toxicity to rainbow trout decreased by a factor of 5, when total water hardness increased from 10 to 200 mg $CaCO_3$ L^{-1} (Brown, 1968). Pre-treatment of fish with high concentrations of cations also reduces the acute effects of nickel. Accordingly, sensitve life stages require a greater degree of treatment than adults. Additive effects on toxicity have been reported for mixtures of nickel and copper, and nickel, copper, and zinc.

Although treatment of fathead minnows with 1.6 mg Ni L^{-1} had no immediate effect on survival or growth, there was a 50% reduction in fecundity and egg survival in these fish, 5 months after exposure (Pickering, 1974). Additionally Blaylock and Frank (1979) reported that nickel concentrations of 3–4 mg L^{-1} did not influence the survival of carp eggs. However, the percentage of teratic larvae in these experiments increased from 23 to 100% at nominal nickel levels of 3 and 7 mg L^{-1}. Comparable results were obtained for goldfish treated with 0.1–10 mg Ni L^{-1} (Birge and Black, 1980). Other sublethal chronic effects have not been fully investigated. While the same may be said about most potential toxic mechanisms, nickel causes a reduction in the number of small lymphocytes moving to the brain in a freshwater teleost, *Colisa fasciatus* (Agrawal *et al.,* 1979). There was a corresponding increase in the haematocrit and haemoglobin values and retardation of the erythrocyte sedimentation rate. Nickel also reduces the diffusion capacity of gills, and promotes an increase in the thickness of lamellar membranes. This leads to death by asphyxiation and is consistent with the high nickel residues in gills. Although effects on gills generally occur rapidly after initial treatment, rainbow trout can recover when placed in clean water, providing exposure to nickel is brief.

Humans

Nickel, being a borderline element, is essential at trace levels for human health. Acute toxicity arises from competitive interaction with five major essential elements, calcium, cobalt, copper, iron, and zinc. Nickel interference with the synthesis and degradation of cellular heme is due to the combination of its strong binding to sulfur and the rigid geometrical structures of the resultant complexes. Elemental nickel is highly surface active and thus adheres to breathable air-borne particulates. This can constitute a major source of nickel uptake by the urban population. Epidemiological studies conducted on refinery workers indicate that nickel compounds induce nasal, laryngeal, and lung cancers (Pedersen *et al.,* 1978; Lessard *et al.,* 1978). Malignant renal tumours can also be induced in rats following intrarenal injections. Teratogenic effects, such as exencephaly, fused ribs, and cleft palate, occur in mammalian and avian embryos exposed to a variety of nickel compounds (Gilani and Marano, 1980; Sunderman *et al.,* 1980).

References

Abo-Rady, M.D.K. 1980. Aquatic macrophytes as indicator for heavy metal pollution in the River Leine (West Germany). *Archiv fuer Hydrobiologie* **89**:387–404.

Agadi, V.V., N.B. Bhosle, and A.G. Untawale. 1978. Metal concentration in some seaweeds of Goa (India). *Botanica Marina* **21**:247–250.

Agrawal, S.J., A.K. Srivastava, and H.S. Chaudhry. 1979. Haematological effects of nickel toxicity on a freshwater teleost, *Colisa fasciatus. Acta Pharmacologica et Toxicologica* **45**:215–217.

Agrawal, Y.K., K.P.S. Raj, and M.R. Patel. 1978. Metal contents in the drinking water of Cambay. *Water, Air, and Soil Pollution* **9**:429–431.

Allan, R.J. 1975. Natural versus unnatural heavy metal concentrations in lake sediments in Canada. *In*: T.C. Hutchinson (Ed.), *Proceedings of 1st International Conference on Heavy Metals in the Environment,* Volume II, Part 2, University of Toronto Institute for Environmental Studies, Toronto, Canada, pp. 785–808.

Anderson, J.W., R.G. Riley and R.M. Bean. 1978. Recruitment of benthic animals as a function of petroleum hydrocarbon concentrations in the sediment. *Journal of the Fisheries Research Board of Canada* **35**:776–790.

Bartelt, R.D., and U. Förstner. 1977. Schwermetalle im staugeregelten Neckar. Untersuchungen an sedimenten, algen und wasserproben. *Jahresber. Mitt. Oberrheinischen Geol. Ver.* **59**:247–263.

Biesinger, K.E., and G.M. Christensen. 1972. Effects of various metals on survival, growth, reproduction, and metabolism of *Daphnia magna. Journal of the Fisheries Research Board of Canada* **29**:1691–1700.

Birge, W.J., and J.A. Black. 1980. Aquatic toxicology of nickel. *In*: J.O. Nriagu (Ed.), *Nickel in the environment.* Wiley, New York, pp. 349–366.

Blaylock, B.G., and M.L. Frank. 1979. A comparison of the toxicity of nickel to the developing eggs and larvae of carp (*Cyprinus carpio*). *Bulletin of Environmental Contamination and Toxicology* **21**:604–611.

Brković-Popović, I., and M. Popović. 1977. Effects of heavy metals on survival and respiration rate of tubificid worms: Part I—Effects on survival. *Environmental Pollution* **13**:65–72.

Brooks, R.R., and D. Rumsey. 1974. Heavy metals in some New Zealand commercial sea fishes. *New Zealand Journal of Marine and Freshwater Research* **8**:155–166.

Brown, B.T., and B.M. Rattigan. 1979. Toxicity of soluble copper and other metal ions to *Elodea canadensis. Environmental Pollution* **20**:303–314.

Brown, V.M. 1968. The calculation of the acute toxicity of mixtures of poisons to rainbow trout. *Water Research* **2**:723–733.

Bryan, G.W., and L.G. Hummerstone. 1978. Heavy metals in the burrowing bivalve *Scrobicularia plana* from contaminated and uncontaminated estuaries. *Journal of the Marine Biological Association of the United Kingdom* **58**:401–419.

Calabrese, A., J.R. MacInnes, D.A. Nelson, and J.E. Miller. 1977. Survival and growth of bivalve larvae under heavy-metal stress. *Marine Biology* **41**:179–184.

Canadian Minerals Yearbooks. 1900–1979. Publishing Center, Department of Supply and Services, Ottawa, Ontario.

Chen, K.Y., C.S. Young, T.K. Jan, and N. Rohatgi. 1974. Trace metals in wastewater effluents. *Journal Water Pollution Control Federation* **46**:2663–2675.

Duke, J.M. 1980. Nickel in rocks and ores. *In*: J.O. Nriagu (Ed.), *Nickel in the Environment.* Wiley, New York, pp. 27–50.

Emerson, R.R., D.F. Soule, and M. Oguri. 1976. Heavy metal concentrations in marine organisms and sediments collected near an industrial waste outfall. In: *Proceedings of International Conference on Environmental Sensing and Assessment,* Volume **1**, Las Vegas, Nevada, September 14–19, 1975, pp. 1–5.

Foster, P. 1976. Concentrations and concentration factors of heavy metals in brown algae. *Environmental Pollution* **10**:45–53.

Gibbs, R.J. 1977. Transport phases of transition metals in the Amazon and Yukon rivers. *Geological Society of America Bulletin* **88**:829–843.

Gilani, S.H., and M. Marano. 1980. Congenital abnormalities in nickel poisoning in chick embryos. *Archives of Environmental Contamination and Toxicology* **9**:17–22.

Gupta, S.K., and K.Y. Chen. 1975. Partitioning of trace metals in selective chemical fractions of nearshore sediments. *Environmental Letters* **10**:129–158.

Helz, G.R. 1976. Trace element inventory for the northern Chesapeake Bay with emphasis on the influence of man. *Geochimica et Cosmochimica Acta* **40**:573–580.

Hodge, V., S.R. Johnson, and E.D. Goldberg. 1978. Influence of atmospherically transported aerosols on surface ocean water composition. *Geochemical Journal* **12**:7–20.

Horowitz, A., and B.J. Presley. 1977. Trace metal concentrations and partitioning in zooplankton, neuston, and benthos from the south Texas outer continental shelf. *Archives of Environmental Contamination and Toxicology* **5**:241–255.

Hutchinson, T.C. 1973. Comparative studies of the toxicity of heavy metals to phytoplankton and their synergistic interactions. *Water Pollution Research in Canada* **8**:68–90.

Hutchinson, T.C., A. Fedorenko, J. Fitchko, A. Kuja, J. Van Loon, and J. Lichwa. 1975. Movement and compartmentation of nickel and copper in an aquatic ecosystem. *In*: D.D. Hemphill (Ed.), *Trace substances in environmental health—IX. A symposium.* University of Missouri Press, Columbia, pp. 89–105.

Karbe, L., C.H. Schnier, and H.O. Siewers. 1977. Trace elements in mussels (*Mytilus edulis*) from coastal areas of the North Sea and the Baltic. Multielement analyses using instrumental neutron activation analysis. *Journal of Radioanalytical Chemistry* **37**:927–943.

Kemp, A.L.W., J.D.H. Williams, R.L. Thomas, and M.L. Gregory. 1978. Impact of man's activities on the chemical composition of the sediments of Lake Superior and Huron. *Water, Air and Soil Pollution* **10**:381–402.

Knauer, G.A. 1977. Immediate industrial effects on sediment metals in a clean coastal environment. *Marine Pollution Bulletin* **8**:249–254.

Kronfeld, J., and J.Navrot. 1974. Transition metal contamination in the Qishon River system, Israel. *Environmental Pollution* **6**:281–288.

Lande, E. 1977. Heavy metal pollution in Trondheimsfjorden, Norway, and the recorded effects on the fauna and flora. *Environmental Pollution* **12**:187–198.

Leland, H.V. 1975. Distribution of solute and particulate trace elements in southern Lake Michigan. *In*: T.C. Hutchinson (Ed.), *Proceedings of 1st International Conference on Heavy Metals in the Environment,* Volume II, University of Toronto Institute for Environmental Studies, Toronto, Canada, pp. 709–730.

Lessard, R., D. Reed, B. Maheux, and J. Lambert. 1978. Lung cancer in New Caledonia, a nickel smelting island. *Journal of Occupational Medicine* **20**:815–817.

Loring, D.H. 1979. Geochemistry of cobalt, nickel, chromium, and vanadium in the sediments of the estuary and open Gulf of St. Lawrence. *Canadian Journal of Earth Sciences* **16**:1196–1209.

Manly, R., and W.D. George. 1977. The occurrence of some heavy metals in populations of the freshwater mussel *Anodonta anatina* (L.) from the River Thames. *Environmental Pollution* **14**:139–154.

Mathis, B.J., and T.F. Cummings. 1973. Selected metals in sediments, water, and biota in the Illinois River. *Journal Water Pollution Control Federation* **45**:1573–1583.

Moore, J.W. 1981. Epipelic algal communities in a eutrophic northern lake contaminated with mine wastes. *Water Research* **15**:97–105.

Nriagu, J.O. 1979. Global inventory of natural and anthropogenic emissions of trace metals to the atmosphere. *Nature* **279**:409–411.

Nriagu, J.O. 1980. Global cycle and properties of nickel. *In*: J.O. Nriagu (Ed.), *Nickel in the environment.* Wiley, New York, pp. 1–26.

Nriagu, J.O., and R.D. Coker. 1980. Trace metals in humic and fulvic acids from Lake Ontario sediments. *Environmental Science and Technology* **14**:443–446.

Nriagu, J.O., H.K.T. Wong, and R.D. Coker. 1981. Particulate and dissolved trace metals in Lake Ontario. *Water Research* **15**:91–96.

Patrick, R.T., T. Bott, and R. Larson. 1975. *The role of trace elements in management of nuisance growths.* U.S. Environmental Protection Agency, Publication No. EPA-660/2-75-008, Corvallis, Oregon, 250 pp.

Pedersen, E., A. Anderson, and A. Høgetveit. 1978. Second study of the incidence and mortality of cancer of respiratory organs among workers at a nickel refinery. *Annals of Clinical and Laboratory Science (Abstract)* **8**:503.

Perhac, R.M. 1972. Distribution of Cd, Co, Cu, Fe, Mn, Ni, Pb, and Zn in dissolved and particulate solids from two streams in Tennessee. *Journal of Hydrology* **15**:177–186.

Pickering, Q.H. 1974. Chronic toxicity of nickel to the fathead minnow. *Journal Water Pollution Control Federation* **46**:760–765.

Pilotte, J.O., J.W. Winchester, and R.C. Glassen. 1978. Detection of heavy metal pollution in estuarine sediments. *Water, Air, and Soil Pollution* **9**:363–368.

Rehwoldt, R., D. Karimian-Teherani, and H. Altmann. 1975. Measurement and distribution of various heavy metals in the Danube River and Danube Canal aquatic communities in the vicinity of Vienna, Austria. *The Science of the Total Environment* **3**:341–348.

Roth, I., and H. Hornung. 1977. Heavy metal concentrations in water, sediments, and fish from Mediterranean coastal area, Israel. *Environmental Science and Technology* **11**:265–269.

Salomons, W., and W.G. Mook. 1977. Trace metal concentrations in estuarine sediments: mobilization, mixing or precipitation. *Netherlands Journal of Sea Research* **11**:119–129.

Sempkin, R.G. 1975. A limnogeochemical study of Sudbury area lakes. M.Sc. Thesis, McMaster University, Hamilton, Ontario, 248 pp.

Shiber, J.G. 1981. Metal concentrations in certain coastal organisms from Beirut. *Hydrobiologia* **83**:181–195.

Sholkovitz, E.R., and D. Copland. 1981. The coagulation, solubility and adsorption properties of Fe, Mn, Cu, Ni, Cd, Co, and humic acids in a river water. *Geochimica et Cosmochimica Acta* **45**:181–189.

Snodgrass, W.J. 1980. Distribution and behavior of nickel in the aquatic environment. *In*: J.O. Nriagu (Ed.), *Nickel in the environment.* Wiley, New York, pp. 203–274.

Stokes, P.M., T.C. Hutchinson, and K. Krauter. 1973. Heavy metal tolerance in algae isolated from polluted lakes near the Sudbury, Ontario smelters. *Water Pollution Research in Canada* **8**:178–187.

Sunderman, F.W., Jr., S.K. Shen, M.C. Reid, and P.R. Allpass. 1980. Teratogenicity and embryotoxicity of nickel carbonyl in Syrian hamsters. *Teratogenesis, Carcinogenesis, and Mutagenesis* **1**:223–233.

Tessier, A., P.G.C. Campbell, and M. Bisson. 1980. Trace metal speciation in the Yamaska and St. François rivers (Quebec). *Canadian Journal of Earth Sciences* **17**:90–105.

Timourian, H., and G. Watchmaker. 1972. Nickel uptake by sea urchin embryos and their subsequent development. *Journal of Experimental Zoology* **182**:379–388.

Trefry, J.H., and B.J. Presley. 1976. Heavy metals in sediments from San Antonio Bay and the northwest Gulf of Mexico. *Environmental Geology* **1**:283–294.

Trollope, D.R., and B. Evans. 1976. Concentrations of copper, iron, lead, nickel and zinc in freshwater algal blooms. *Environmental Pollution* **11**:109–116.

United States Minerals Yearbooks. 1900–1979. Bureau of Mines, US Department of the Interior, Washington, D.C.

Watling, H.R., and R.J. Watling. 1976. Trace metals in *Choromytilus meridionalis. Marine Pollution Bulletin* **7**:91–94.

Wilson, A.L. 1976. Concentrations of trace metals in river waters, a review. Technical Report No 16, Water Research Centre, Medmenham Laboratory and Stevenage Laboratory, U.K.

Wright, D.A. 1976. Heavy metals in animals from the north east coast. *Marine Pollution Bulletin* 7:36–38.

9
Zinc

Chemistry

Zinc is relatively rare in nature but has a long history of use because of its occurrence in localized deposits and ease of extraction from ores. Zinc occurs in a number of minerals—zinc blends, ZnS, smithsonite, $ZnCO_3$, willemite, Zn_2SiO_4, zincite, ZnO, and others. Commercially important ores are mainly those of carbonate and sulfide. Although zinc metallurgy is at least 1000 years old, zinc (described as false silver) has been known for some 2000 years. Smelting technology was brought to Europe from India and China around the eighteenth century and today zinc is mined and produced in over 30 countries. It ranks fourth among the metals next to steel, aluminum, and copper in annual global consumption.

Zinc is a member of the group IIb triad of the periodic classification of elements, along with cadmium and mercury. It does not exhibit multiple valency and is softer and lower melting than the neighbouring transition metal triad, Cu, Ag, and Au (Group Ib). However, zinc resembles transition metals in its ability to form complexes, particularly with ammonia, amines, halide ions, and cyanide.

Zinc is intermediate between hard and soft acceptors in its chemical interaction with ligands, whereas Cd(+2) and Hg(+2) are typical class b (soft acid) acceptors. Thus, zinc forms complexes with both hard (oxygen donors) bases and soft (sulfur donors) bases. This is reflected in the occurrence of zinc in nature, both as sulfide and carbonate ores.

Zinc is an essential element for mammals. More than twenty different zinc metalloenzymes have been identified. These include carbonic anhydrase, alkaline phosphatase, and alcohol dehydrogenase. Zinc plays a vital role in the biosynthesis of nucleic acids, RNA polymerases, and DNA polymerases. Thus zinc is involved in the healing processes of tissues in the body. A number of other physiological processes, including hormone metabolism, immune response, and stabilization of ribosomes and membranes, require zinc.

Production, Uses, and Discharges

Production

The principal ore, ZnS, occurs worldwide along with lead deposits. ZnS oxidizes readily to yield a number of secondary minerals such as $ZnCO_3$. Trace metals (e.g., Cd, Ge, and Ga) associated with the ZnS ores are recovered during extraction. Many of the major zinc–lead deposits of the world are classified as strata-bound deposits in carbonate rocks and commonly occur along and within dolomite and calcite minerals. Most zinc production in the world originates from ZnS minerals. After mining and milling the ore, zinc sulfide concentrates are converted into metallic zinc by either pyrometallurgical or a combination of pyrometallurgical and electrolytical processes.

Global production of zinc has been increasing steadily during this century and has almost doubled during the last decade (Table 9-1). By contrast, mine output and production through smelters remained relatively steady between 1970 and 1974 (Table 9-2). In 1975–76, smelter and mine production declined to reduce stocks. Canada is the world's largest producer and trader of zinc; about 25% of all zinc consumed in the western world originates from Canadian mines (Table 9-3).

Table 9-1. All-time global zinc production* (in metric tons $\times 10^3$).

Period	Production	Period	Production
Pre-1800	**	1931–1940	13,847
1801–1900	11,558+	1941–1950	18,059
1901–1910	6,742	1951–1960	24,965
1911–1920	7,574	1961–1970	39,260
1921–1930	10,877	1971–1980++	63,245
		Total (all-time)	196,127

* *Source:* US Minerals Yearbooks unless otherwise specified. ** estimated to be only a few thousand tons. + *Source:* Cammarota, 1980. ++ 1980 figures estimated.

Table 9-2. World mine and smelter production of zinc during 1970–1979 (in metric tons × 10³).

Year	Mine Production	Smelter Production	Year	Mine Production	Smelter Production
1970	5476	4837	1975	5850	5013
1971	5527	4754	1976	5690	5362
1972	5447	5142	1977	5906	5527
1973	5721	5342	1978	5878	5614
1974	5792	5621	1979	5998	5998

Source: US Minerals Yearbooks.

Table 9-3. Production and consumption of zinc by Canada (in metric tons × 10³).

	Production (Canada)		Consumption	
Year	Mine	Metal	US	Canada
1970	1226	418	1429	98
1971	1273	373	1501	110
1972	1274	477	1676	122
1973	1360	534	1756	121
1974	1240	437	1521	118
1975	1229	427	1117	98
1976	1145	472	1394	99
1977	1300	495	1368	105
1978	1245	495	1442	121
1979	1204[P]	580[P]	1394[P]	124[P]

Sources: Canada Minerals Yearbooks, US Minerals Yearbooks. [P] projected.

Uses

The wide industrial application of zinc stems from its chemical and metallurgical properties. The largest use of zinc is in galvanizing iron and steel products (Table 9-4). This provides a corrosion-resistant coating which can be finished with electroplated metal coating or organic coatings. Such products are used in construction, automobile and building industries for roofing, siding, appliance casings, office equipment, heating and ventilation ducts, automobile door panels, and underbody parts. Zinc is substituted by copper and plastic products in residential plumbing systems. New alloys such as zinc–aluminum have been developed as protective coatings.

Table 9-4. Zinc consumption by industrial categories in US (in metric tons × 10^3).

Industry	1970	1972	1974	1976	1978	1979
Galvanizing	431	471	476	393	454	453
Brass products	116	174	165	151	141	141
Zinc-base alloys	421	527	400	387	354	314
Rolled zinc	37	41	36	27	25	22
Zinc oxide	40	47	59	35	37	35
Other uses	33	28	34	35	39	35
Total consumption	1078	1288	1170	1028	1050	1000

Source: US Minerals Yearbooks.

Zinc diecast products (zinc-base alloys), the second largest consumer, are used in trim pieces, grills, door and window handles, carburetors, pumps, door locks, and other mechanical components in automobiles (Table 9-4). The US automobile industry consumes approximately two-thirds of the total production of zinc diecastings. Diecastings are also employed in small appliances, business equipment, and light engineering industries.

The trend to produce smaller and lighter cars in North America is reflected in the decline of zinc diecast consumption by auto industries. These are being replaced by light-weight aluminum, magnesium, and plastics, because of the development of metal-on-plastic plating techniques. Increasing acceptance of thin wall zinc diecast parts will be responsible for continued consumption. The development of nickel–zinc batteries for use in electric vehicles will provide a significant new market in the coming years.

The third major consumption category, brass, is employed in a variety of applications from decorative hardware to plumbing and heat exchange units. Brass combines good physical, electrical, thermal, and corrosion-resistant qualities amenable to a variety of treatment processes. Rolled zinc is required for dry battery production, photo engraving, lithographic printing plates, roofing (mostly in Europe), and rain water gutters and pipes.

More than half of zinc oxide produced is employed as a catalyst in the vulcanization of natural and synthetic rubber. Zinc oxide is also required for paints, and other end-products such as photocopy paper, agricultural products, and cosmetic and medicinal products.

Zinc dust, a finely divided form of the metal, is used in the printing and dyeing of textiles, purifying fats, and precipitating silver and gold from cyanide solutions. Weather-resistant paints based on zinc oxide and zinc dust provide one of the most effective and durable coatings on outside surfaces. Other uses of zinc are: ZnS in pigment, $ZnSO_4$ in rayon fibre manufacture, and $ZnCl_2$ as a wood preservative.

Discharges

Eroded soil particles account for ~ 58% of zinc originating from natural sources (Table 9-5). Emissions from vegetation account for another 20%, though sea salt sprays are not a significant source of zinc, despite high production. Current anthropogenic emission exceeds natural rates by 700% (Table 9-6). Non-ferrous metal production and use account for 43% of

Table 9-5. Worldwide emissions of zinc from natural sources.

Source	Global production (10^9 kg yr^{-1})	Worldwide emissions (10^6 kg yr^{-1})
Windblown dusts	500	25
Forest fires	36	2.1
Volcanic particles	10	7.0
Vegetation	75	9.4
Seasalt sprays	1000	0.01
Total		43.5

Source: Nriagu (1979). Reprinted by permission from *Nature* **279**:409–411. Copyright © 1979, Macmillan Journals Limited.

Table 9-6. Worldwide anthropogenic emissions of zinc.

Source	Global production (10^9 kg yr^{-1})	Worldwide Zn emissions (10^6 kg yr^{-1})
Mining, non-ferrous	16	1.6
Primary metal production		
Copper	7.9	6.6
Lead	4.0	0.44
Nickel	0.8	0.68
Zinc	5.6	99
Secondary metal production	4.0	9.5
Iron and steel production	1300	35
Industrial applications	—	26
Coal combustion	3100	15
Oil (including gasoline) combustion	2800	0.07
Wood combustion	640	75
Waste incineration	1500	37
Rubber tire wear	5×10^8 tires	2.2
Phosphate fertilizers	118	1.8
Miscellaneous	—	4.5
Total		314

Source: Nriagu (1979). Reprinted by permission from *Nature* **279**:409–411. Copyright © 1979, Macmillan Journals Limited.

Table 9-7. All-time global consumption and anthropogenic emissions of zinc.

Period	Consumption (10^9 kg)	Emissions (10^6 kg)
Pre-1850	50	2804
1850–1900	15	841
1901–1910	7.0	392
1911–1920	8.8	493
1921–1930	11.1	622
1931–1940	13.3	746
1941–1950	17.1	959
1951–1960	27.0	1514
1961–1970	42.3	2372
1971–1980	58.0	3252
Total (all-time)	250	13,995

Source: Nriagu (1979). Reprinted by permission from *Nature* **279**:409–411. Copyright © 1979, Macmillan Journals Limited.

anthropogenic release to the atmosphere. Other important sources include wood combustion and waste incineration.

Sharp increases in man-induced emissions have occurred during this century (Table 9-7). The ratios of emissions for the decades 1901–1910/1971–1980 are consistent with rates of deposition in polar caps of the Northern Hemisphere. The fact that the bulk of man-made emissions occur in the Northern Hemisphere is evidenced by reports of the lack of significant elevations of trace metals in polar caps of the Southern Hemisphere.

Zinc in Aquatic Systems

Speciation in Natural Waters

Zinc is classified as a borderline element, according to HSAB (Hard-Soft Acid Base) classification, in binding to ligands. *In vivo,* zinc in carbonic anhydrase binds halide ions $I^- > Br^- > Cl^- > F^-$, behaving as a soft acid metal, whereas in aqueous solution, $Zn(+2)_{aq}$ binds $F^- > Cl^- > Br^- > I^-$ behaving as a hard acid metal. The borderline character of zinc is reflected in its ability to form bonds with oxygen as well as nitrogen and sulfur donor atoms.

Binding to Inorganic Ligands. Zinc begins to hydrolyze at pH 7–7.5 forming relatively stable $Zn(OH)_2$ at pH > 8.0 (Table 9-8). At pH 6.7, zinc

Table 9-8. Stability constants and solubility products of some Zn–inorganic compounds.

	Stability constants (log)				Solubility product	
Compound	β_1	β_2	β_3	β_4	Precipitate	Log K_{sp}
ZnCl	0.43	0.61	0.53	0.20	$Zn_3(PO_4)_2$	−32.04
ZnOH	4.40	12.89	15.86	15.95	ZnS	−25.15
					$Zn(OH)_2$	−15.50

Source: Sillén and Martell (1971).

is present as divalent zinc, available for sorption onto suspended mineral colloids and complexation with organic matter. Calculation of intrinsic solubility of $Zn(OH)_2$ shows that up to 160 mg Zn(+2) L^{-1} as $Zn(OH)_2$ can exist in solution (Hahne and Kroontje, 1973). Under natural environmental conditions, the hydroxide may even promote mobilization of zinc due to reduction in the charge of the cation and increase in the solubility of sparingly soluble salts. The zinc–chloride complexes do not form until chloride concentration is $\geqslant$0.4 M (89,000 mg L^{-1}).

Binding to Organic Ligands. Stability of zinc-organic complexes are enhanced by the presence of nitrogen and sulfur donor atoms in the ligand. The conditional stability constants of zinc complexes with soil-derived fulvic acid are 1.73 (pH 3.5) and 2.34 (pH 5.0). Humic substances from various natural environments show a range of overall stability constants (K_o) for zinc: (Mantoura *et al.*, 1978).

	Peat	Lakes	Rivers	Sediment	Seawater	Soil
log K_o	4.83	5.05–5.31	5.36–5.41	4.99–5.87	5.27–5.31	2.34–3.7

Therefore it is important when modeling any particular water system to use K_o values relevant to that system.

Binding to Particulates. Zinc shows variable behaviour in binding to particulates depending on the physico-chemical characteristics of the aquatic system. From data collected for rivers around the world, Wilson (1976) showed that particulate-zinc varied from 10–78% of total zinc (3–60 μg L^{-1}). Nriagu *et al.* (1981) reported a significant variation in particulate-zinc concentrations of the nearshore (1.69 μg L^{-1}) and offshore waters (0.23 μg L^{-1}) of Lake Ontario. Similarly, substantial surface enrichment (about 68%) of zinc in the particulate phase was found in Lake Michigan (Elzerman *et al.*, 1979).

Transport in Natural Waters. Benes and Steines (1974) reported that about 90% of zinc in river systems was dialyzable whereas 80% was ion-exchangeable. About 84% of particulate-Zn was reported in the Guanajibo River decreasing to 40% from the head of the river to the coastal zone (Montgomery and Santiago, 1978). It has also been shown that humic acid of marine origin binds trace metals stronger than terrestrial humic acids. However, complexation of zinc and other trace metals by humic materials at the concentrations present in the open sea is insignificant. In estuarine waters, humic acid contributions become significant due to lower concentrations of competing cations such as Ca and Mg. Speciation of zinc in oligotrophic hard water was about 55% chelated by organic materials of different molecular weights (Steinberg, 1980). Zinc was preferentially bound by oligomeric substances of 580–750 daltons.

The majority (75%) of zinc in the Yarrah River (Australia) was transported in the filterable fractions (Hart and Davies, 1981). About 40–50% of filterable zinc was ion-exchangeable, whereas the remainder was in a bound form as part of the non-dialyzable fraction. A higher percentage (36%) of particulate-zinc was observed in the estuarine section of the Yarrah, and about 47% of zinc of the filterable fraction was ion-exchangeable.

Zinc in Sediments

Sorption and Desorption. Zinc in sediments collected from two rivers in Canada was found predominantly in soluble forms (Tessier *et al.*, 1980). The fractions were: Fe–Mn oxide (41 and 39%), carbonate bound (21 and 24%), organic matter (5 and 6%), residual (33 and 29%), and exchangeable fraction (0.9 and 2.2%). These results are consistent with the known ability of Fe–Mn oxides to scavenge zinc from solution. From a study on the Fraser River Estuary, Grieve and Fletcher (1977) reported that the percent of total zinc associated with Fe–Mn oxides increased from about 20 in the fresh water zone to >80 in the brackish waters. The bioavailability of metals generally followed the order: exchangeable fraction > carbonate-bound fraction ≃ Fe–Mn oxide-bound fraction ≃ organic-bound fraction >> residual fraction. The release of metals depended on the redox potential, pH, and presence of leaching ligands of both natural and synthetic origin.

Release of zinc and other heavy metals from polluted river sediments was studied in the presence of added NTA (Banat *et al.*, 1974). The release of zinc was positively correlated with NTA concentrations and negatively with the time of shaking. The latter was explained as possible biodegradation of the Zn–NTA chelate. The enrichment of zinc in humic and fulvic acids compared to the associated sediments is much less than for Cu, Pb, Ni, and Cr. For example, enrichment factors for Zn and Cu are 0.61–0.81 and 21–25, respectively (Nriagu and Coker, 1980). It was estimated that <5% of Zn in sediments was bound to the organic matter.

Residues

Water, Precipitation, and Sediments

Although dissolved zinc levels in unpolluted freshwaters generally range from 0.5 to 15 μg L^{-1}, much higher residues have been recorded for industrial areas. For example, aerial emissions from a lead–zinc smelter at Flin Flon (Canada) resulted in concentrations of $\geqslant$100 μg L^{-1} in nearby lakes (Van Loon and Beamish, 1977). Similarly, metal mining and flooding of industrial zone rivers may produce total residues of $\geqslant$3000 μg L^{-1} in receiving waters. Dissolved zinc levels in offshore marine areas range from 0.4 to 5 μg L^{-1}. Enrichment often occurs below depths of 1 km due to the downward drift of decaying plankton. Similarly anthropogenic inputs may increase levels to $>$20 μg L^{-1} in coastal waters.

Large quantities of zinc are transported and deposited by precipitation. Legittimo *et al.* (1980) reported concentrations of 8–330 μg L^{-1} in rain-water in Florence (Italy). Snow in southern Norway contained 10–205 μg Zn L^{-1}, whereas in the Antarctic, an average concentration of 0.06 μg L^{-1} was reported (Elgmork *et al.*, 1973; Landy and Peel, 1981). Depending on location, aerial inputs may account for $>$50% of total zinc deposited in receiving waters (Peyton *et al.*, 1976).

Total zinc levels in freshwater sediments often exceed 1000 mg kg^{-1} dry weight in the vicinity of metal mines (Table 9-9). Lower values are generally recorded for rivers flowing through urban areas and, in uncontaminated areas, concentrations are $<$50 mg kg^{-1}. A similar range in values has been reported for estuarine and coastal waters (Table 9-9).

Table 9-9. Total zinc levels (mg kg^{-1} dry weight) in freshwater and marine sediments.

Location	Average (range)	Polluting source
Freshwater sediments		
Coeur d'Alene Lake (USA)[1]	3800(3200–4700)	metal mining
Natural creek, Montana (USA)[2]	1045(80–9110)	abandoned mine
Derwent Reservoir (UK)[3]	1035(—)	fluorspar mine
Yellowknife Bay (Canada)[4]	200(25–1005)	gold mine
Illinois River (USA)[5]	81(6–339)	industrial, municipal
Marine sediments		
Mediterranean (Lebanon)[6]	60(13–155)	small local industry
Continental Shelf (SE, USA)[7]	6(3–10)	natural
Los Angeles Harbour (USA)[8]	202(98–325)	mixed industrial
Baltic Sea[9]	–(6–2090)	mixed industrial
Sørfjord (Norway)[10]	20000(830–118000)	metal mines, smelters

Sources: [1] Maxfield *et al.* (1974); [2] Pagenkopf and Cameron (1979); [3] Harding and Whitton (1978); [4] Moore (1979); [5] Mathis and Cummings (1973); [6] Shiber (1980); [7] Bothner *et al.* (1980); [8] Emerson *et al.* (1976); [9] Brugmann (1981); [10] Skei *et al.* (1972).

Aquatic Plants

Zinc residues in plants collected from polluted freshwaters generally range from 100 to 500 mg kg^{-1} dry weight. However, considerably higher levels have been found in several European rivers contaminated with metal mine wastes. Harding and Whitton (1981) reported maximum residues of 2700 mg kg^{-1} in *Lemanea fluviatilis* inhabiting Afon Craffnant (Wales), 1250 mg kg^{-1} in plants from the River Lot (France), and 2290 in those from the River Nent (England). Comparable or higher levels (up to 12,300 mg kg^{-1}) occur in various aquatic macrophytes collected from mining areas of eastern and northern Canada (Franzin and McFarlane, 1980). Although concentrations of up to 45,900 mg kg^{-1} occurred in algae growing next to zinc smelter waste (UK), residues in plants from uncontaminated parts of the world are generally below 50 mg kg^{-1} (Trollope and Evans, 1976).

Zinc levels in marine plants are generally lower than those reported for freshwater species, reflecting the greater level of zinc pollution in freshwaters. Some of the highest residues in marine plants have been associated with the deposition of mine wastes into Norwegian fjords. Concentrations of 3000–4000 mg kg^{-1} were reported for *Ascophyllum nodosum* from Trondheimsfjord and Sørfjorden whereas *Chorda filum* contained average residues of 7350 mg kg^{-1} in Sørfjorden (Eide *et al.*, 1980; Melhuus *et al.*, 1978). Zinc mining also led to contamination (800 mg kg^{-1}) of *Fucus vesiculosus* in the Severn Estuary (Butterworth *et al.*, 1972). In the absence of metal mines/smelters, levels in plants are generally <100 mg kg^{-1}.

Uptake is generally slow and depends on metabolic/photosynthetic rate, temperature, light, and the concentrations of zinc in the environment

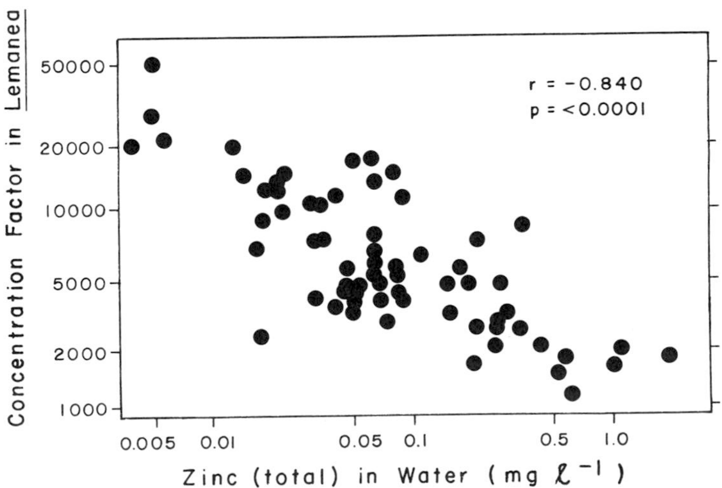

Figure 9-1. Relationship between concentration factor for zinc in *Lemanea fluviatilis* and total zinc in river water. (From Harding and Whitton, 1981.)

(Figure 9-1). Although the presence of other metals such as magnesium often reduces uptake through competitive interactions, calcium did not influence residues in *Lemanea fluviatilis* collected from European rivers (Harding and Whitton, 1981). Sorption is generally suppressed by H^+, chelators, and Na^+. Skipnes *et al.* (1975) suggested that only a small fraction of uptake in living algae was due to ion exchange with intracellular polysaccharides. Sorption was much more rapid in dead algae, indicating that membrane integrity controls the movement of zinc into the cell.

The rate of loss of zinc from plant tissues is slow, approaching that of mercury. Myklestad *et al.* (1978) reported a reduction of only 50% in residues in *Ascophyllum nodosum* over two months in an uncontaminated Norwegian fjord. There is often no measurable desorption in winter due to low temperatures whereas, during the summer, residues may decline significantly. Zinc may be bound to organic moleucles such as phenols and excreted from plants.

Invertebrates

Although zinc is transferred to freshwater invertebrates, accumulation does not occur in most species. Occasionally high levels are reported for animals from polluted waters, reflecting even higher levels in food and sediment. Specimens of the mollusc *Anodonta anatina* collected from the River Thames (UK) contained up to 3500 mg kg^{-1} dry weight in soft tissues (Manly and George, 1977). Comparable values (2000 mg kg^{-1}) were reported for omnivorous Trichoptera larvae in the River Hayle (UK), whereas the corresponding levels for omnivorous Plecoptera nymphs and carnivorous Odonata nymphs were 780 and 1200 mg kg^{-1}, respectively (Brown, 1977).

Some marine species appear to concentrate zinc. In the Tamar River (Tasmania), maximum residues in Pacific oysters *Crassostrea gigas* and bottom sediment were 14,000 and 500 mg kg^{-1}, respectively (Ayling, 1974). Similarly, zinc in water, sediments, and gastropods from the Severn estuary reached 0.052 mg L^{-1}, 540 mg kg^{-1} and 3100 mg kg^{-1}, respectively (Butterworth *et al.*, 1972). Although these data imply that zinc is accumulated in some species, it is possible that much higher levels occurred in food, thereby leading to elevated tissue burdens. In unpolluted coastal waters, residues in benthic invertebrates range from 20 to 200 mg kg^{-1}.

Maximum levels in molluscs are generally associated with the digestive gland and gonad (Table 9-10). Residues in the shell are normally low, as is also the case with the exoskeleton of crustaceans. Although specific data are lacking, it appears that organs in this latter group also concentrate zinc (Table 9-10).

Whole body concentrations generally increase with the size and age of animal; in addition, there is often a significant correlation between levels in specific organs and organ weight. Under natural conditions, whole body

Table 9-10. Concentration (mg kg^{-1} dry weight) of zinc in tissues of different invertebrates.

Species	Location	Organ	Concentration
Thais lapillus[1]	Wales	Digestive gland/gonad	559–761
(gastropod)		Body	359–578
Littorina littorea[1]	Wales	Digestive gland/gonad	111–134
(gastropod)		Body	84–87
Mytilus edulis[2]	California	Digestive gland	3.0–240
(mussel)		Gonad	4.1–360
		Muscle	2.7–210
Squid[3]	Gulf of Mexico	Viscera	80–110
		Muscle	41–56
Decapods[3]	Gulf of Mexico	Viscera	84–136
		Muscle	20–62
		Exoskeleton	8–42
Orconectes virilis[4]	US	Viscera	82.5
(crayfish)		Muscle	52.5
		Exoskeleton	34.8

Sources: [1] Ireland and Wootton (1977); [2] Young *et al.* (1978); [3] Horowitz and Presley (1977); [4] Anderson and Brower (1978).

residues may be greatest in the winter and spring, or summer, or show no consistent seasonal trend. These differences reflect variation in reproductive condition, and effect on temperature on rate of uptake. Rate of elimination from tissues is highly variable. Renfro (1973) estimated the half-life in the polychaete *Hermione hystrix* to be 52–197 days. In the crayfish *Procambarus acutus* rate of elimination of zinc derived from water was almost twice as fast as that from food (Giesy *et al.,* 1980).

Fish

Residues in freshwater and marine fish are generally much lower than those found in algae and invertebrates. Thus there is little evidence for accumulation in many species. Zinc in muscle tissue from 15 species of omnivorous and carnivorous fish collected from industrial and agricultural areas of the lower Great Lakes were 16–82 and 3–9 mg kg^{-1} wet weight, respectively (Brown and Chow, 1977). Similarly, yellow perch, bluegill, and black crappie inhabiting recreational and industrial zone rivers in the USA had average muscle burdens of 106, 108, and 103, and 100, 109, and 101 mg kg^{-1} dry weight, respectively (Vinikour *et al.,* 1980; Adams *et al.,* 1980). Marine fish also have low but variable zinc residues in muscle. Of the 11 species collected from the Mediterranean coast (Israel), 10 showed little evidence of contamination (0.5–33 mg kg^{-1} dry weight), but levels in one pelagic species ranged up to 84 mg kg^{-1} dry weight (Roth and Hornung,

1977). Residues in black marlin from Australia were always low, 5.7–14.6 mg kg^{-1} wet weight (Mackay *et al.*, 1975), as was the case with eight species of New Zealand sea fish (Brooks and Rumsey, 1974). Overall, therefore, zinc does not pose a threat to marine and freshwater fisheries rcsources.

As with invertebrates, zinc residues are greatest in specific organs of all species. Mackay *et al.* (1975) reported that the ratio of zinc in the liver : muscle of black marlin was 5.5 : 1. Similarly, average concentrations in the muscle, liver, kidney, heart, gonad, spleen, and gills of eight species of New Zealand sea fish were 8, 76, 78, 24, 93, 73 and 22 mg kg^{-1}, respectively (Brooks and Rumsey, 1974). Exposure of dogfish to 15 mg Zn L^{-1} in water resulted in average levels in muscle, liver, kidney, spleen and pancreas of 68, 243, 93, 147 and 620 mg kg^{-1} dry weight, respectively (Flos *et al.*, 1979).

In most cases, zinc concentrations cannot be related to feeding habits. Although there is often no correlation between residues in muscle and age/size of fish, uptake of zinc by young fish is generally greater than that of old fish. Most species show a seasonal variation in zinc levels in muscle and organs. Whole body residues in five-bearded rockling from the Severn Estuary (UK) peaked during the spring and early summer and declined in August (Badsha and Sainsbury, 1978). This decrease was possibly related to a change in diet during the summer. Migratory whiting from the same estuary showed maximum and minimum concentrations during November and January, respectively (Badsha and Sainsbury, 1977). It was suggested that the rapid growth of whiting during the winter, at a time when uptake was constant, lead to the decline in residues.

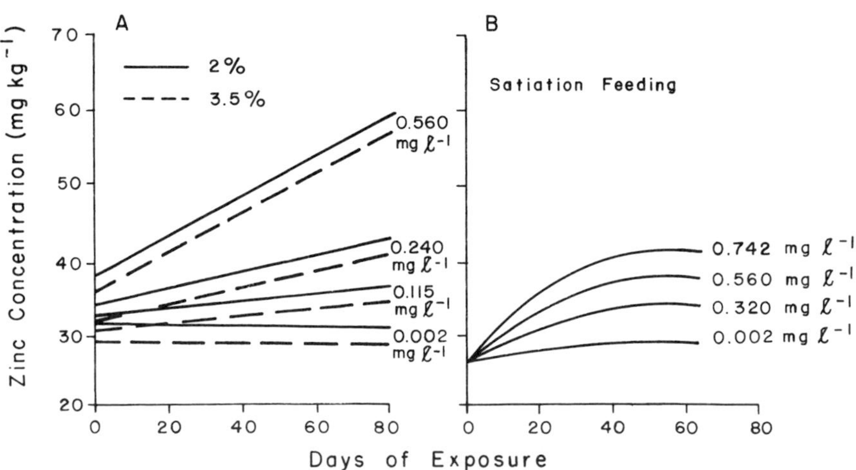

Figure 9-2A. Influence of ambient zinc levels and ration size (% dry body weight per day) on zinc residues in Atlantic salmon. (From Farmer *et al.* 1979.)
Figure 9-2B. Influence of ambient zinc levels on zinc residues in Atlantic salmon fed to satiation. (From Farmer *et al.* 1979.)

With the exception of non-feeding periods, fish normally obtain the majority of zinc from dietary sources rather than from water. Farmer *et al.* (1979) showed that zinc in well-fed Atlantic salmon increased during the first two months of exposure before reaching a maximum (Figure 9-2). However, concentrations in salmon offered smaller rations increased linearly during a three-month period. The attainment of an equilibrium state for metal residues occurs in many species and often reflects a balance between growth and metal uptake. Hence, the majority of fish cannot be used as biological indicators of zinc contamination in the environment. In the absence of contaminated food, rate of uptake depends directly on exposure duration and level in water. In addition, the presence of organic and inorganic chelators in solution may significantly reduce sorption by fish.

Toxicity

Aquatic Plants

Toxicity of zinc to aquatic plants is highly variable, with LC_{50}'s ranging from 0.0075 to >50 mg L^{-1}. Part of this extreme inconsistency is due to the ability of many species to adapt to high zinc levels in water; in addition, physico-chemical factors may significantly influence the bioavailability of zinc. Under most circumstances, mercury and copper are more toxic to aquatic plants than zinc whereas chromium, cadmium, nickel, lead, and arsenic may be more or less toxic, depending on conditions.

Some of the most sensitive species appear to be euglenoids and other flagellated forms. Mills (1976) reported that concentrations of 0.0075 and 0.075 mg Zn(+2) L^{-1} caused reductions of 20 and 100%, respectively, in cell numbers of *Euglena gracilis* within 48 h. Part of this sensitivity is probably related to the absence of thick cell walls, which permit the movement of zinc into the cell interior. In addition, since flagellated forms often have a relatively high metabolic rate, uptake of metals would be rapid. Some green algae, such as *Stigeoclonium, Ulothrix, Hormidium* and *Microspora* can adapt to high zinc levels in water. The TIC (Tolerance Index Concentration) for tolerant forms of *Stigeoclonium tenue* ranged up to 14.1 mg L^{-1} compared with a maximum of 1.0 mg L^{-1} for non-tolerant forms (Harding and Whitton, 1977). The same species has also been found growing in a river in the FRG at 20 mg L^{-1}. Development of resistance is accompanied by a reduction in the number of exchange sites available in the cell walls of algae. Tolerance is therefore characterized by the development of exclusion mechanisms.

An increase in the calcium and magnesium content of water reduces toxicity to most species. Similarly additions of phosphate to culture media also significantly decrease toxic effects and, in some instances, increase growth beyond control levels. Toxicity depends on pH of the water, which controls the concentration of Zn(+2) in solution. A comparable effect can be achieved by the addition of organic chelators in solution. Some species of blue-green algae produce chelating agents which in turn protect the cells from zinc. Combinations of zinc and cadmium, and zinc and copper act as synergists in inhibiting growth. On the other hand, mixtures of zinc, copper, and cadmium may be comparable in toxicity to equal concentrations of zinc. Competition for uptake sites or routes is probably one mechanism for the interaction of zinc with other metals. This response is species specific and influenced by the presence of non-toxic metals, such as manganese.

Invertebrates

Acute toxicity of zinc to freshwater invertebrates is relatively low. Under most conditions, mercury, cadmium, copper, chromium, nickel, and arsenic are more toxic than zinc. Although the 48–96 h LC_{50}'s generally range from 0.5 to 5 mg L^{-1}, some species are particularly sensitive to zinc. Under similar environmental conditions, the 48 h LC_{50} for the cladoceran *Daphnia hyalina* was 0.055 mg L^{-1}, and 5.5 mg L^{-1} for the copepod *Cyclops abyssorum* (Baudouin and Scoppa, 1974). On the other hand, the LC_{50} for a number of insect and crustacean species may exceed 55 mg L^{-1}, possibly reflecting ability to adapt to high zinc levels. In general, immature stages are more sensitive to zinc than adults. Toxicity of zinc to marine invertebrates is comparable to that for freshwater species with LC_{50}'s ranging from 0.2 to 3.5 mg L^{-1}. There are few examples of extreme sensitivity or resistance to zinc in marine invertebrates. This partially reflects consistence in the availability of Zn species due to stable pH conditions in salt water (Young *et al.*, 1980).

As with other metals, increasing water hardness decreases toxicity to invertebrates. For example LC_{50}'s of 0.1 and 60.2 mg L^{-1} were obtained when the oligochaete *Tubifex tubifex* was exposed to 0.1 and 261 mg $CaCO_3$ L^{-1} (Brković-Popović and Popović, 1977). Calcium ions appear to offer better protection from intoxication than magnesium ions (Tabata, 1969). Hence, the ionic constituents of receiving waters should be routinely monitored to help assess potential industrial impacts. Zinc and cadmium may interact additively or antagonistically, depending on species. Zinc and flouride were antagonists to the rotifer *Philodina acuticornis* but zinc and chlorine acted additively (Buikema *et al.,* 1977).

Fish

Mercury and copper are consistently more toxic to fish than zinc. Under most conditions, cadmium is also more toxic, while nickel, lead, and other metals are less toxic. Although the 48–96 h LC_{50} for zinc usually falls within 0.5–5.0 mg L^{-1}, physico-chemical and biological factors may extend the range from 0.09 to >100 mg L^{-1}. Susceptibility to zinc is largely species dependent. Smith and Heath (1979) found that goldfish were particularly tolerant with a 24 h LC_{50} of 110 mg L^{-1}, whereas the LC_{50} for rainbow trout was only 5 mg L^{-1}. Some species have the ability to develop a tolerance to zinc. Although the actual mechanism for adaptation is not clearly understood, fish previously exposed to water of high total hardness are more tolerant of zinc than those exposed to low total hardness levels. Hence, competitive inhibition may constitute part of the adaptive mechanism. Stage of development also significantly influences toxicity. Average 96 h LC_{50}'s for alevins of steelhead trout and chinook salmon were 10–15 times greater than those for swim-ups (Chapman, 1978). Juvenile rainbow trout were about 3 times more resistant to zinc than eyed-eggs (Sinley *et al.,* 1974). This implies that the development of biomonitoring methods must carefully consider both age and species of principal organisms.

Temperature significantly influences toxicity of zinc to many species. For example the incipient lethal level to juvenile Atlantic salmon inhabiting the Miramichi River (Canada) varied seasonally from 0.15 to 1.10 mg L^{-1} (Figure 9-3). The most sensitive period occurred between March and August, a reflection of developmental stage of the fish and environmental conditions (Zitko and Carson, 1977). Based on laboratory experiments acute toxicity of zinc may also depend on acclimation temperature (Figure 9-4). Temperature stress, combined with low oxygen levels, probably increases the susceptibility of most species of fish to zinc. Hence, the decision to discharge metal wastes should be made considering seasonal temperature and oxygen cycles, and their subsequent effects on fish survival.

Treatment of fish with zinc results in substantial gill damage (Figure 9-5). There is initial separation of epithelium, followed by occlusion of central blood spaces and enlargement of central and marginal channels. Lamellar

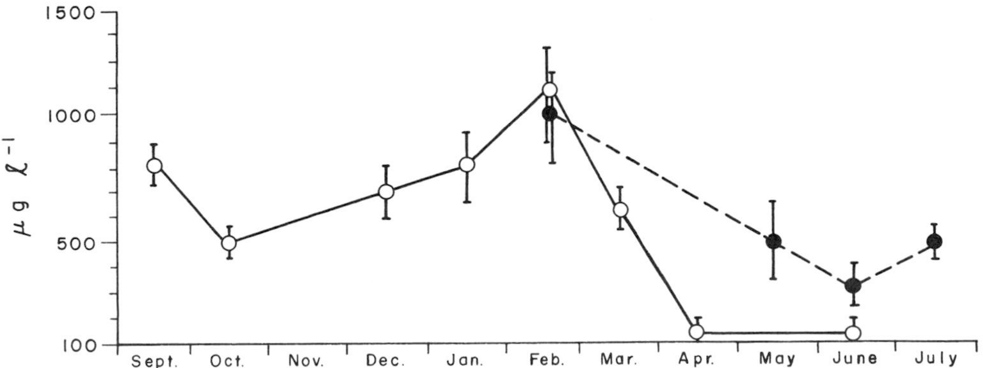

Figure 9-3. Variation in incipient lethal level of zinc to Atlantic salmon. — fish kept at 10°C, --- fish kept at ambient temperature. (From Zitko and Carson, *Journal of the Fisheries Research Board of Canada,* 1977.)

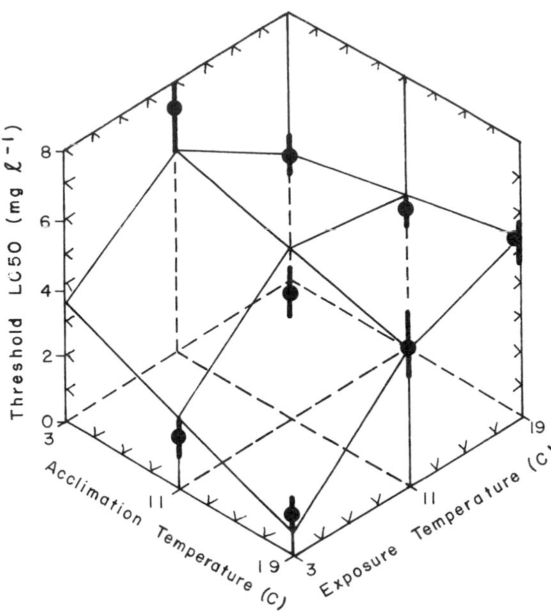

Figure 9-4. Variation in lethal threshold of zinc to Atlantic salmon with acclimation and exposure temperature. (From Hodson and Sprague, *Journal of the Fisheries Research Board of Canada,* 1975.)

height is progressively reduced and, ultimately, central blood spaces are completely occluded. These changes result in a decrease in oxygen consumption and ability to transport ions across the gill and increase in hypoxia, opercular amplitude, buccal amplitude, ventilation frequency, and coughing frequency. Numerous other physical and biochemical changes have been reported for intoxicated fish. These include (i) increase in production of lactic acid and pyruvic acid, thereby decreasing blood pH, (ii) dysfunction of kidney tissue and enzymes, (iii) decrease in growth, maximum size and fecundity, and (iv) alteration in schooling and reproductive behaviour.

Humans

Although classified as an intermediate element between hard and soft acceptors in its interaction with ligands, zinc seldom interferes with sulfur or sulfhydryl groups in biological systems. Thus zinc is an essential element mediating a variety of metalloenzymes and the biosynthesis of nucleic acids and polypeptides. The rare toxicity of zinc arises from its synergistic/antagonistic interaction with other heavy metals, particularly its homologue cad-

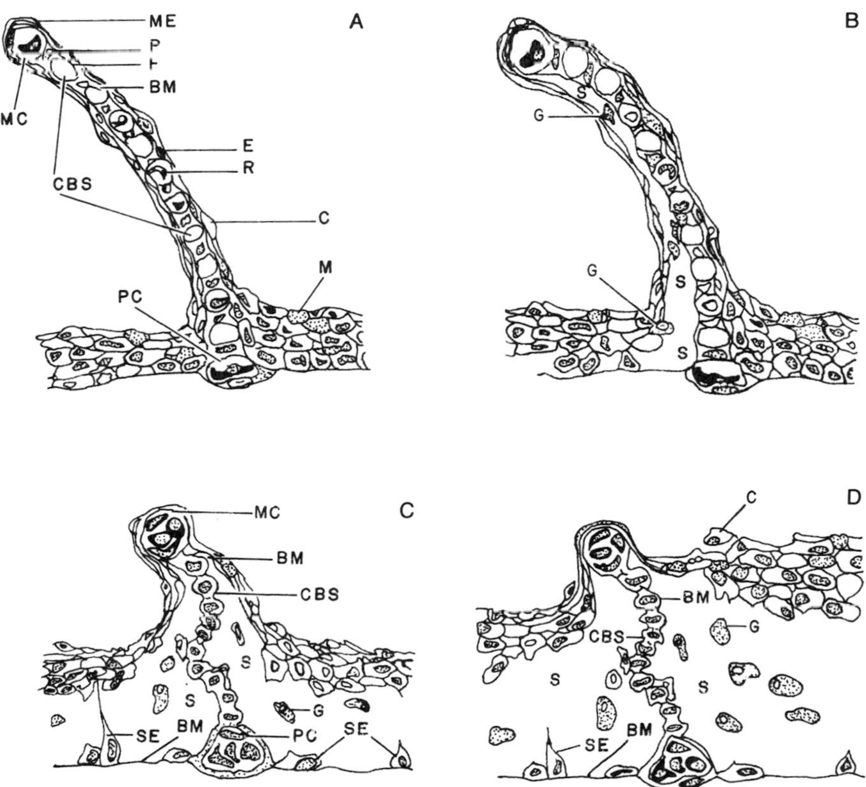

Figure 9-5. Diagrams of a secondary lamella (rainbow trout) in transverse section showing four stages of zinc-induced damage. (From Skidmore and Tovell, 1972.)
Figure 9-5A. Control tissue before exposure to zinc.
Figure 9-5B. After 60% of estimated survival time. Separation of lamellar epithelium and shortening of lamella apparent.
Figure 9-5C. After 90% of estimated survival time. Occlusion of central blood spaces and enlargement of epithelial cells apparent.
Figure 9-5D. At death. Occlusion of central blood spaces completed, subepithelium space enlarged.
Abbreviations: BM: basement membrane, C: chloride cell, CBS: central blood space, E: epithelial cell, F: pillar cell flange, G: granulocyte, M: mucous cell, MC: marginal channel, ME: marginal endothelial cell, P: pillar cell, PC: proximal channel, R: red blood cell, S: subepithelial space, SE: stretched epithelial cell.

mium. The non-stoichiometric accumulation of heavy metals by the carrier protein metallothionein could lead to zinc deficiencies in humans. Symptoms of deficiency include delayed healing, suppression of enzymatic activity, and immune response.

References

Adams, T.G., G.J. Atchison, and R.J. Vetter. 1980. The impact of an industrially contaminated lake on heavy metal levels in its effluent stream. *Hydrobiologia* **69**:187–193.

Anderson, R.V., and J.E. Brower. 1978. Patterns of trace metal accumulation in crayfish populations. *Bulletin of Environmental Contamination and Toxicology* **20**:120–127.

Ayling, G.M. 1974. Uptake of cadmium, zinc, copper, lead and chromium in the Pacific oyster, *Crassostrea gigas,* grown in the Tamar river, Tasmania. *Water Research* **8**:729–738.

Badsha, K.S., and M. Sainsbury. 1977. Uptake of zinc, lead and cadmium by young whiting in the Severn estuary. *Marine Pollution Bulletin* **8**:164–166.

Badsha, K.S., and M. Sainsbury. 1978. Aspects of the biology and heavy metal accumulation of *Ciliata mustela. Journal of Fish Biology* **12**:213–220.

Banat, K., U. Förstner, and G. Muller. 1974. Experimental mobilization of metals from aquatic sediments by nitrilotriacetic acid. *Chemical Geology* **14**:199–207.

Baudouin, M.F., and P. Scoppa. 1974. Acute toxicity of various metals to freshwater zooplankton. *Bulletin of Environmental Contamination and Toxicology* **12**:745–751.

Benes, P., and E. Steinnes. 1974. *In situ* dialysis for the determination of the state of trace elements in natural waters. *Water Research* **8**:947–953.

Bothner, M.H., P.J. Aruscavage, W.M. Ferrebee, and P.A. Baedecker. 1980. Trace metal concentrations in sediment cores from the Continental Shelf off the south-eastern United States. *Estuarine and Coastal Marine Science* **10**:523–541.

Brković-Popović, I., and M. Popović. 1977. Effects of heavy metals on survival and respiration rate of tubificid worms: Part I—Effects on survival. *Environmental Pollution* **13**:65–72.

Brooks, R.R., and D. Rumsey. 1974. Heavy metals in some New Zealand commercial sea fishes. *New Zealand Journal of Marine and Freshwater Research* **8**:155–166.

Brown, B.E. 1977. Effects of mine drainage on the River Hayle, Cornwall. A: Factors affecting concentrations of copper, zinc and iron in water, sediments and dominant invertebrate fauna. *Hydrobiologia* **52**:221–233.

Brown, J.R., and L.Y. Chow. 1977. Heavy metal concentrations in Ontario fish. *Bulletin of Environmental Contamination and Toxicology* **17**:190–195.

Brugmann, L. 1981. Heavy metals in the Baltic Sea. *Marine Pollution Bulletin* **12**:214–218.

Buikema, A.L., Jr., C.L. See, and J. Cairns, Jr. 1977. Rotifer sensitivity to combinations of inorganic water pollutants. Virginia Water Resources Research Center Bulletin 92, 42 pp.

Butterworth, J., P. Lester, and G. Nickless. 1972. Distribution of heavy metals in the Severn estuary. *Marine Pollution Bulletin* **3**:72–74.

Cammarota, V.A., Jr. 1980. Production and uses of zinc. *In*: J.O. Nriagu (Ed.), *Zinc in the environment, Part I, Ecological cycling.* Wiley, New York, pp. 1–38.

Canadian Minerals Yearbooks. 1970–1979. Publishing Center, Department of Supplies and Services, Ottawa, Ontario.

Chapman, G.A. 1978. Toxicities of cadmium, copper, and zinc to four juvenile stages of chinook salmon and steelhead. *Transactions of the American Fisheries Society* **107**:841–847.

Eide, I., S. Myklestad, and S. Melsom. 1980. Long-term uptake and release of heavy metals by *Ascophyllum nodosum* (L.) Le Jol. (Phaeophyceae) *in situ. Environmental Pollution (Series A)* **23**:19–28.

Elgmork, K., A. Hagen, and A. Langeland. 1973. Polluted snow in southern Norway during the winters 1968–1971. *Environmental Pollution* **4**:41–52.

Elzerman, A.W., D.E. Armstrong, and A.W. Andren. 1979. Particulate zinc, cadmium, lead, and copper in the surface microlayer of southern Lake Michigan. *Environmental Science and Technology* **13**:720–725.

Emerson, R.R., D.F. Soule, M. Oguri, K.Y. Chen, and J. Lu. 1976. Heavy metal concentrations in marine organisms and sediments collected near an industrial waste outfall. *In: Proceedings of International Conference on Environmental Sensing and Assessment,* Volume **I**, Las Vegas, Nevada, September 14–19, 1975. pp. 1–5.

Farmer, G.J., D. Ashfield, and H.S. Samant. 1979. Effects of zinc on juvenile Atlantic salmon *Salmo salar:* acute toxicity, food intake, growth and bioaccumulation. *Environmental Pollution* **19**:103–117.

Flos, R., A. Caritat, and J. Balasch. 1979. Zinc content in organs of dogfish (*Scyliorhinus canicula* L.) subject to sublethal experimental aquatic zinc pollution. *Comparative Biochemistry and Physiology* **64C**:77–81.

Franzin, W.G., and G.A. McFarlane. 1980. An analysis of the aquatic macrophyte, *Myriophyllum exalbescens,* as an indicator of metal contamination of aquatic ecosystems near a base metal smelter. *Bulletin of Environmental Contamination and Toxicology* **24**:597–605.

Giesy, J.P., J.W. Bowling, and H.J. Kania. 1980. Cadmium and zinc accumulation and elimination by freshwater crayfish. *Archives of Environmental Contamination and Toxicology* **9**:683–697.

Grieve, D., and K. Fletcher. 1977. Interactions between zinc and suspended sediments in the Fraser River estuary, British Columbia. *Estuarine and Coastal Marine Science* **5**:415–419.

Hahne, H.C.H., and W. Kroontje. 1973. Significance of pH and chloride concentration in behavior of heavy metal pollutants: mercury(II), cadmium(II), zinc(II), and lead(II). *Journal of Environmental Quality* **2**:444–448.

Harding, J.P.C., and B.A. Whitton. 1977. Environmental factors reducing the toxicity of zinc to *Stigeoclonium tenue. British Phycological Journal* **12**:17–21.

Harding, J.P.C., and B.A. Whitton. 1978. Accumulation of heavy metals by *Lemanea* in European rivers affected by mining. *British Phycological Journal (Abstract)* **13**:200–201.

Harding, J.P.C., and B.A. Whitton. 1981. Accumulation of zinc, cadmium, and lead by field populations of *Lamanea. Water Research* **15**:301–319.

Hart, B.T., and S.H.R. Davies. 1981. Trace metal speciation in the freshwater and estuarine regions of the Yarra River, Victoria. *Estuarine, Coastal and Shelf Science* **12**:353–374.

Hodson, P.V., and J.B. Sprague. 1975. Temperature-induced changes in acute toxicity of zinc to Atlantic salmon (*Salmo salar*). *Journal of the Fisheries Research Board of Canada* **32**:1–10.

Horowitz, A., and B.J. Presley. 1977. Trace metal concentrations and partitioning in zooplankton, neuston, and benthos from the south Texas outer continental shelf. *Archives of Environmental Contamination and Toxicology* **5**:241–255.

Ireland, M.P., and R.J. Wootton. 1977. Distribution of lead, zinc, copper and manganese in the marine gastropods, *Thais lapillus* and *Littorina littorea*, around the coast of Wales. *Environmental Pollution* **12**:27–41.

Landy, M.P., and D.A. Peel. 1981. Short-term fluctuations in heavy metal concentrations in Antarctic snow. *Nature* **291**:144–146.

Legittimo, P.C., G. Piccardi, and F.Pantani. 1980. Cu, Pb, and Zn determination in rainwater by differential pulse anodic stripping voltammetry. *Water, Air, and Soil Pollution* **4**:435–441.

Mackay, N.J., M.N. Kazacos, R.J. Williams, and M.I. Leedow. 1975. Selenium and heavy metals in black marlin. *Marine Pollution Bulletin* **6**:57–60.

Manly, R., and W.O. George. 1977. The occurrence of some heavy metals in populations of the freshwater mussel *Anodonta anatina* (L.) from the River Thames. *Environmental Pollution* **14**:139–154.

Mantoura, R.F.C., A. Dickson, and J.P. Riley. 1978. The complexation of metals with humic materials in natural waters. *Estuarine and Coastal Marine Science* **6**:387–408.

Mathis, B.J., and T.F. Cummings. 1973. Selected metals in sediments, water, and biota in the Illinois River. *Journal Water Pollution Control Federation* **45**:1573–1583.

Maxfield, D., J.M. Rodriguez, M. Buettner, J. Davis, L. Forbes, R. Kovacs, W. Russel, L. Schultz, R. Smith, J. Stanton, and C.M. Wai. 1974. Heavy metal content in the sediments of the southern part of the Coeur d'Alene lake. *Environmental Pollution* **6**:263–266.

Melhuus, A., K.L. Seip, H.M. Seip and S. Myklestad. 1978. A preliminary study of the use of benthic algae as biological indicators of heavy metal pollution in Sørfjorden, Norway. *Environmental Pollution* **15**:101–107.

Mills, W.L. 1976. Water quality bioassay using selected protozoa, II. The effects of zinc on population growth of *Euglena gracilis*. *Journal of Environmental Science and Health* **All**:567–572.

Montgomery, J.R., and R.J. Santiago. 1978. Zinc and copper in 'particulate' forms and 'soluble' complexes with inorganic or organic ligands in the Guanajibo River and coastal zone, Puerto Rico. *Estuarine and Coastal Marine Science* **6**:111–116.

Moore, J.W. 1979. Diversity and indicator species as measures of water pollution in a subarctic lake. *Hydrobiologia* **66**:73–80.

Myklestad, S., I. Eide, and S. Melsom. 1978. Exchange of heavy metals in *Ascophyllum nodosum* (L.) Le Jol. *in situ* by means of transplanting experiments. *Environmental Pollution* **16**:277–284.

Nriagu, J.O. 1979. Global inventory of natural and anthropogenic emissions of trace metals to the atmosphere. *Nature* **279**:409–411.

Nriagu, J.O., and R.D. Coker. 1980. Trace metals in humic and fulvic acids from Lake Ontario sediments. *Environmental Science and Technology* **4**:443–446.

Nriagu, J.O., H.K.T. Wong, and R.D. Coker. 1981. Particulate and dissolved trace metals in Lake Ontario. *Water Research* **15**:91–96.

Pagenkopf, G.K., and D. Cameron. 1979. Deposition of trace metals in stream sediments. *Water, Air, and Soil Pollution* **11**:429–435.

Peyton, T., A. McIntosh, V. Anderson, and K. Yost. 1976. Aerial input of heavy metals into an aquatic ecosystem. *Water, Air, and Soil Pollution* **5**:443–451.

Renfro, W.C. 1973. Transfer of ^{65}Zn from sediments by marine polychaete worms. *Marine Biology* **21**:305–316.

Roth, I., and H. Hornung. 1977. Heavy metal concentrations in water, sediments, and fish from Mediterranean coastal area, Israel. *Environmental Science and Technology* **11**:265–269.

Shiber, J.G. 1980. Metal concentrations in marine sediments from Lebanon. *Water, Air, and Soil Pollution* **13**:35–43.

Sillén, L.G., and A.E. Martell. 1971. *Stability constants of metal-ion complexes, Supplement No. 1.* Special Publication No. 25, The Chemical Society, London, 865 pp.

Sinley, J.R., J.P. Goettl, Jr. and P.H. Davies. 1974. The effects of zinc on rainbow trout (*Salmo gairdneri*) in hard and soft water. *Bulletin of Environmental Contamination and Toxicology* **12**:193–201.

Skei, J.M., and N.B. Price, S.E. Calvert, and H. Holtendahl. 1972. The distribution of heavy metals in sediments of Sørfjord, West Norway. *Water, Air and Soil Pollution* **1**:452–461.

Skidmore, J.F., and P.W.A. Tovell. 1972. Toxic effects of zinc sulphate on the gills of rainbow trout. *Water Research* **6**:217–230.

Skipnes, O., T. Roald, and A. Haug. 1975. Uptake of zinc and strontium by brown algae. *Physiologia Plantarum* **34**:314–320.

Smith, M.J., and A.G. Heath. 1979. Acute toxicity of copper, chromate, zinc, and cyanide to freshwater fish: effect of different temperatures. *Bulletin of Environmental Contamination and Toxicology* **22**:113–119.

Steinberg, C. 1980. Species of dissolved metals derived from oligotrophic hard water. *Water Research* **14**:1239–1250.

Tabata, K. 1969. Studies on the toxicity of heavy metals to aquatic animals and the factors to decrease the toxicity—II. The antagonistic action of hardness components in water on the toxicity of heavy metal ions. *Bulletin of the Tokai Regional Fisheries Research Laboratory* **58**:215–232.

Tessier, A., P.G.C. Campbell, and M. Bisson. 1980. Trace metal speciation in the Yamaska and St. François rivers (Quebec). *Canadian Journal of Earth Sciences* **17**:90–105.

Trollope, D.R., and B.Evans. 1976. Concentrations of copper, iron, lead, nickel and zinc in freshwater algal blooms. *Environmental Pollution* **11**:109–116.

United States Minerals Yearbooks. 1901–1979. Bureau of Mines, US Department of the Interior, Washington, D.C.

Van Loon, J.C., and R.J. Beamish. 1977. Heavy-metal contamination by atmospheric fallout of several Flin Flon area lakes and the relation to fish populations. *Journal of the Fisheries Research Board of Canada* **34**:899–906.

Vinikour, W.S., R.M. Goldstein, and R.V. Anderson. 1980. Bioconcentration patterns of zinc, copper, cadmium and lead in selected fish species from the Fox River, Illinois. *Bulletin of Environmental Contamination and Toxicology* **24**:727–734.

Wilson, A.L. 1976. Concentrations of trace metals in river waters, a review. Technical Report No. 16, Water Research Centre, Medmenham Laboratory and Stevenage Laboratory, U.K.

Young, D.R., T.K. Jan, and T.C. Heesen. 1978. Cycling of trace metal and chlorinated hydrocarbon wastes in the southern California Bight. *In:* M.L. Wiley (Ed.), *Estuarine interactions.* Academic Press, New York, pp. 481–496.

Young, D.R., T.K. Jan, G.P. Hershelman. 1980. Cycling of zinc in the nearshore marine environment. *In*: J.O. Nriagu (Ed.), *Zinc in the environment.* Wiley, New York, pp. 297–335.

Zitko, V., and W.G. Carson. 1977. Seasonal and developmental variation in the lethality of zinc to juvenile Atlantic salmon (*Salmo salar*). *Journal of the Fisheries Research Board of Canada* **34:**139–141.

10
Impact of Heavy Metals in Natural Waters

The discharge of heavy metal wastes into receiving waters may result in numerous physical, chemical, and biological responses. These can be separated into two broad categories: (i) effects of the environment on the metal, and (ii) effects of the metal on the environment. The first category emphasizes that conditions in receiving waters may lead to a change in the speciation and toxicity of metals. Such conditions include differential input of anthropogenic and geochemical material, quality of industrial effluents, and concentration of chelators and suspended solids. Biological responses under the second category are often equally diverse. Depending on environmental conditions, there may be a change in density, diversity, community structure, and species composition of populations. The nature and extent of change depends largely on the concentration of heavy metal species in the water and sediment. Hence, physicochemical processes within effluents and natural waters have a major, albeit indirect, effect on biological responses.

The purpose of this chapter is to describe the effect which heavy metal discharges have on physical, chemical, and biological processes within natural waters. There is an emphasis on point source discharges, reflecting their relative importance in environmental studies. In addition, environmental managers and monitoring agencies usually have a high level of involvement with the control of point source discharges.

Physico-Chemical Impact

The aquatic environment is characterized by (i) longitudinal variations in suspended solids, colloidal particles, and natural and synthetic ligands and

(ii) vertical variations in redox conditions, degree of mixing, and densities of living organisms. The fate of metals in natural waters is heavily dependent upon these variables. Transformations such as methylation and reduction to the metallic form constitute effects of the environment on metals. Similarly downward movement of metals to the bottom of natural water bodies results from scavenging by suspended solids and concomitant sedimentation. Organic ligands (natural/synthetic) and chlorides complex metals, reducing the sorption process and increasing residence time in water. In essence, speciation of metals is determined by the environment, and changes in speciation are responses to alterations.

Distribution and transport of metals in aquatic environments are primarily controlled by the sediment and water column, respectively. Mobilization of metals, or lack of it, from bed sediments depends upon the physical texture and chemical nature of sediments, which in turn determine the amount and strength of metal binding. The physico-chemical composition of the water column determines the mode(s) of transport of metals, such as particulate, colloidal, dissolved ionic, and dissolved complexed forms. Geological and anthropogenic inputs make up the character of the active surface sediments and the carrier compartment, the water column. Thus the natural waters of the world are bound to vary in their effect on metal distribution and transport. Six typical systems, studied over a long term, are discussed below.

Ottawa River

The Ottawa River, with an overall length of 1113 km and a mean annual flow rate of 2100 m^3 sec^{-1} (smaller than the Rhine but larger than the Arkansas), originates in the drainage area of Lake Timiskaming (Canada) and finally merges with the St. Lawrence River at Montreal (Figure 10-1). The river drains an area of some 143,400 km^2 in Ontario and Quebec, once glaciated and now containing extensive deposits of fine clay. The river is used extensively for water supply, recreational activities, municipalities, and industries (mainly pulp and paper). Historically, the river has received substantial input of organic materials from industrial and logging operations. The wood fragments have become an integral part of the Ottawa River sediments.

A 4.8 km (3 mile) reach, immediately downstream of the City of Ottawa, was chosen for a five-year multi-disciplinary study on the distribution and transport of mercury (Figure 10-1). The section offered varied environments for studying the natural clearance of a once mercury-laden environment. The river was categorized into six physical and biological compartments. They were (i) water, including all material that would pass through a 0.45 μm filter, (ii) bed sediments, mainly organic materials and microorganisms, (iii) suspended material, including biological and non-biological material filterable from water through a 0.45 μm filter, but not included in

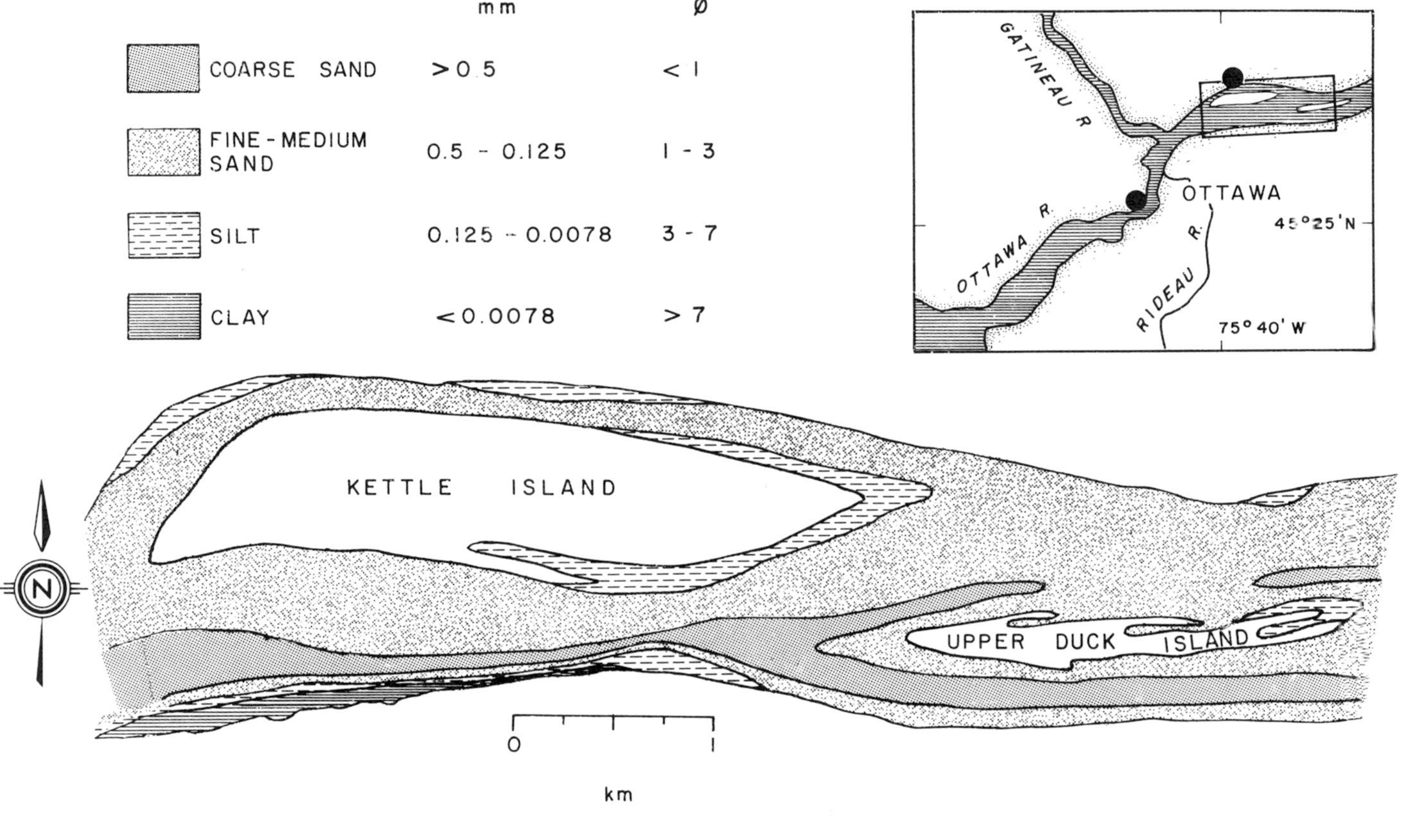

Figure 10-1. Map of study area and grain size distribution of bed sediment in the Ottawa River. Insert shows location of pulp/paper mills. (From Ramamoorthy and Massalski, 1979.)

Table 10-1. Mercury distribution and transport in the Ottawa River (1973).

	Total mass, summer (kg)	Total Hg conc.[a] (ppb)	Organic fraction (ppb)	Total Hg (g)		Annual flow through system (kg)	
				Inorg.	Org.	Mass	Mass of Hg
Water	23×10^9	~0.03	See text	350	—	6.6×10^{13}	1.3×10^3
Bed sediment[b]	0.33×10^9	41	0.01	13,530	135	8.3×10^7	3.4
Suspended solids	222×10^3	440	~0.3	98	29	6.4×10^8	2.90×10^2
Benthic invertebrates	11.1×10^3	220	~0.3	2.4	0.7	—	—
Higher plants	65.4×10^3	100	0.20	6.5	1.3	—	—
Fish[c]	6.0×10^3	180	0.85	1.1	0.9	—	—

Source: Ottawa River Project Group (1979).

[a] Hg concentrations quoted on wet-weight basis for water, invertebrates, plants, and fish; dry-weight basis for sediments and suspended material. [b]Sediments regarded as 4 cm deep for calculations. The amount disturbed on an annual basis is larger than this. [c]Fish Hg concentration quoted for yellow perch. Other species were significantly higher.

other categories, (iv) macroinvertebrates, mainly benthic species and including those associated with plants, (v) higher plants, including root, submerged and emerged parts, and (vi) fish. An extensive program was carried out between 1972–1977, to determine the annual average mercury concentrations in each compartment (Ottawa River Project Group, 1979). The results for a typical year are summarized in Table 10-1.

The bulk of mercury distribution and transport was associated with bed sediments, suspended solids, and the water column. Since an insignificant fraction of mercury was associated with the biota, transport through living material was nondetectable. However, this small fraction of methylmercury was of considerable sociological and political importance because of its toxicity. Hence, it was important to quantify and understand the production and movement of methylmercury in the aquatic environment. By contrast, total mercury and sediment types provided an estimate of the longevity of the problem. The concentration of methylmercury was markedly elevated in higher organisms, but the concentration of total mercury in different components of the system (excluding water) was approximately similar.

Although most of the mercury was associated with bed sediment, sediment transport was slow (Table 10-1). Clearance was due to desorption and subsequent transport by the flowing bulk water. Mercury levels in sediments decreased almost exponentially (Table 10-2). The inferred half-life in most areas was 0.95–1.05 years, reaching 2.6 years below a pulp mill. Measurements of sediment/water partition coefficients were adequate for the evaluation of total mercury burden and its fate in the river.

The Ottawa River study showed that the water column containing low molecular weight organic compounds (Ramamoorthy and Kushner,

Table 10-2. Decrease of mercury concentrations in sediment and water of the Ottawa River.

Region[a]	Area ($10^3 m^2$)	Total Hg (ppb, dry wt) 1972	1973	1974	1975	1976	Inferred half-life[b] (years)
1	550	123	20	29	25	5	0.95
2	696	380	312	222	130	170	2.6
3	3522	58	27	14	8	10	0.98
4	130	403	62	107	—	—	1.05
5	140	420	129	98	58	50	0.95
Water	—	0.038	0.013	~0.01	~0.02	—	—

Source: Ottawa River Project Group (1979).

[a] Regions: (1) North Channel above C.I.P. Mill (sand, woodchips); (2) North Channel below C.I.P. Mill (sand, woodchips, silt, sludge); (3) main channel (coarse sand); (4) Kettle Island Bay and shores of Upper Duck Island (medium to fine sand); (5) Ontario shore (clay). [b]Half-life calculated on basis of first 3 years. Many 1975 and 1976 values were at levels regarded as background.

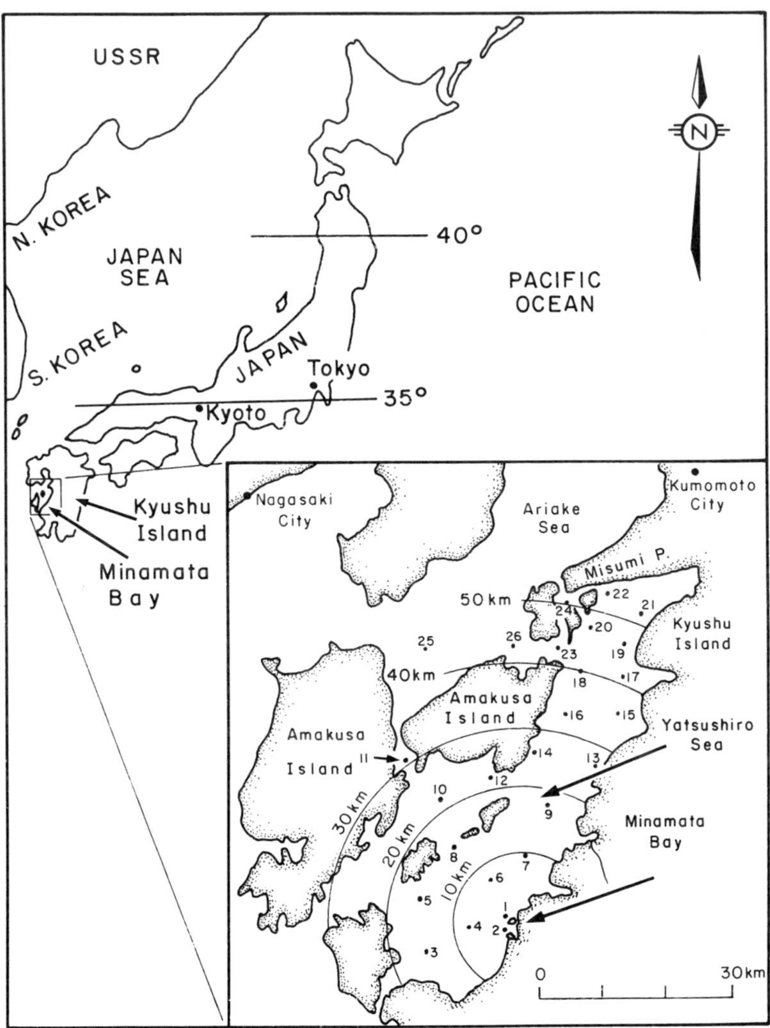

Figure 10-2. Yatsushiro Sea and Minamata Bay showing location of sampling stations. (From Kudo *et al.,* 1980.)

1975a,b) was much more important in the transport of mercury than suspended solids or bed sediment transport itself. It was predicted that biologically available mercury will decrease with a half-life of 1 or 2 years and the mercury in long-lived biota will follow with a delay of the order of their own lifetime.

Minamata Bay

Studies similar to the Ottawa River Project were made on the movement of mercury from Minamata Bay into the Yatsushiro Sea (Kudo *et al.* 1980).

Table 10-3. Increase of total mercury in the Yatsushiro Sea between 1975 and 1978.

Distance from source (km)	Area (km^2)	Amounts of sediments in top 4 cm ($\times 10^3$ metric tons)	Hg conc in 1975** (g. metric $tons^{-1}$)	Amounts of Hg in 1975 (metric tons)	Hg increase 1975–78 (metric tons)
Minamata Bay	3	1,500*	25–774	150***	
3–5	8	520	2.500	1.30	2.25
5–10	64	4,160	0.675	2.81	4.36
10–20	203	13,200	0.340	4.50	5.93
20–30	213	13,800	0.230	3.18	3.28
30–40	61	3,960	0.185	0.73	0.57
40–50	48	3,120	0.170	0.53	0.28
50–	36	2,340	0.160	0.38	0.16
Total	633	41,100	—	13.43	16.83

Source: Kudo *et al.* (1980).
* Contaminated sediments (over 25 mg kg^{-1}) estimated by Kumamoto Prefecture. **Average or estimated by Dispersion Curve. ***Estimation by Kumamoto Prefecture.

Minamata Bay is small (3 km^2 area), located at the southwestern coast on Kyushu Island of Japan (Figure 10-2). The severe tidal effect in the bay exchanges about 20% of the total volume of water (or 6 million tons) twice a day. Yatsushiro Sea is a small inland water surrounded by the Amakusa Islands (Figure 10-2).

Mercury concentrations in bed sediments of the Yatsushiro Sea outside Minamata Bay were determined at 26 sampling sites during 1975 and subsequent years (Figure 10-2). The study showed accelerated movement of mercury (about 16.8 metric tons) associated with Minamata Bay sediments during 1975–1978 (Table 10-3). Clearance rate, expressed as half-life, of mercury for different water systems in the US, Canada and Japan is given in Table 10-4.

Table 10-4. Clearance rate (half-life) of mercury for different water systems.

Locations	Half-life (years)	Water retention period (years)
Field observations		
Ottawa River (Canada)	1.5–2.7	0.0008
Lake Washington (US)	1.25	2.4
Lake Michigan (US)	22	0.12
Minamata Bay	18.2	0.007
Laboratory experiments		
Ottawa River sediments (Canada)	2.1–177	0.0003

Source: Kudo *et al.* (1980).

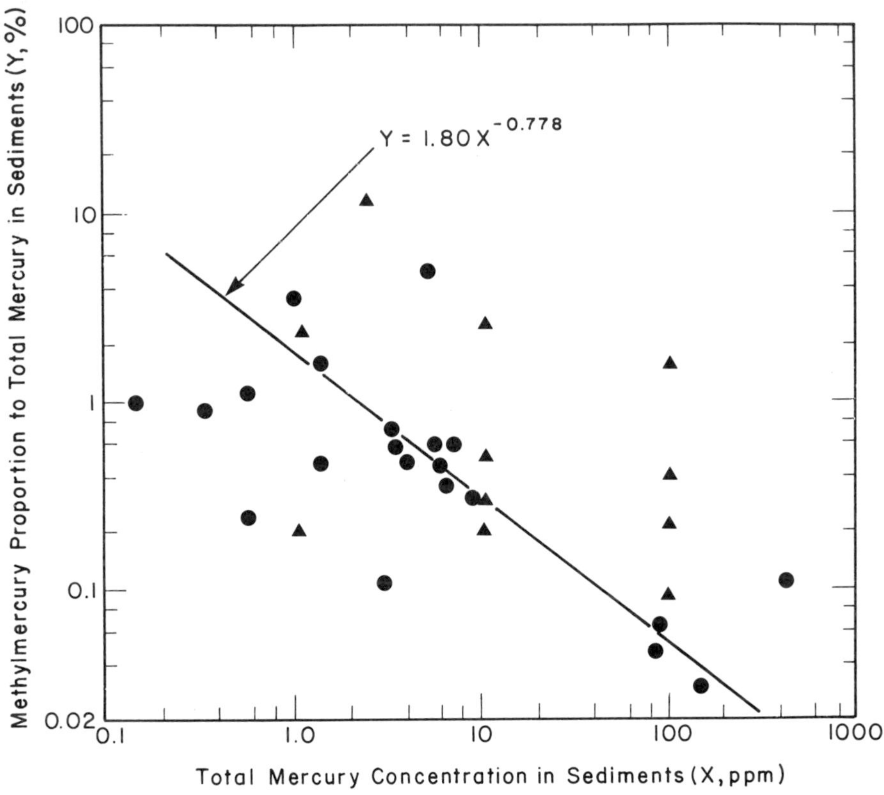

Figure 10-3. Mercury increase in bed sediments of the Yatsushiro Sea. (From Kudo *et al.* 1980.)

Information on the proportion of methylmercury to total mercury from 21 field surveys (closed circles) and 11 equilibrium studies (closed triangles) were fitted into a straight line to estimate the mercury increase in the Yatsushiro Sea (Figure 10-3). Using the equation obtained, the amounts of methylmercury in bed sediments for the period 1975–1978 were calculated (Table 10-5). While methylmercury in sediments of the sea increased by 5.8% per year, total mercury increased at the rate of 42% per year. Accelerated movement of mercury out of Minamata Bay during 1975–1978 (17 metric tons) compared to the period 1960–1975 (9 metric tons) was possibly due to increased commercial navigation in and out of the bay.

Taking the volume of water exchanged by tidal action, the half-life of a pollutant in Minamata Bay would be 1.6 days, assuming thorough mixing. But field data clearly showed that sediment contamination was still high even after 20 years. This indicated that water movement itself, though considerable, did not account for the clearance of mercury from the bay. This differs from other environments such as the Ottawa River, where large

Table 10-5. Amounts (metric tons) of methylmercury in the Yatsushiro Sea

Distance from source (km)	Amount of Hg in 1975		Amount of Hg in 1978	
	Total	Methyl	Total	Methyl
3–5	1.30	0.011	3.55	0.015
5–10	2.81	0.069	7.17	0.085
10–20	4.50	0.187	10.43	0.226
20–30	3.18	0.179	6.46	0.207
30–40	0.73	0.049	1.30	0.056
40–50	0.53	0.038	0.81	0.042
50–	0.38	0.029	0.54	0.030
Total	13.43	0.562	30.26	0.661

Source: Kudo *et al.* (1980).

volumes of water usually generate a sediment transport (Ottawa River Project, 1977). The half-life of mercury in Minamata Bay was calculated to be 18.2 years. Desorption of mercury from sediments was slow, which accounts for hot spots of sediment contamination. It was predicted that half of the sediment–mercury load (estimated as 150 metric tons in 1975) will move out of the bay by the year 1996.

Ruhr River

The Ruhr River, a tributary to the Rhine River, is a typical example of western European rivers. The average flow is 77 $m^3\ sec^{-1}$ and the catchment area covers 4500 km^2 (Figure 10-4). The river basin is structured by numerous valleys with smaller tributary creeks and rivers. Flow changes may vary within a ratio of 1 : 1000. The river provides a population of 6 million people and a variety of industries with drinking and service water. Industries include wire mills, pickling plants, rolling mills, and about 300 metal plating plants. Industrial wastewater is mixed with domestic wastewater, treated and discharged into the river. Variations in industrial pre-treatment cause fluctuations in speciation of metals and hence the quality of the treated wastewater discharged into the river.

Imhoff *et al.* (1980) evaluated the origin, concentration, behaviour, and fate of heavy metals in the Ruhr between 1970 and 1978. Average concentrations of seven heavy metals at four sampling stations are given in Table 10-6. All concentrations peaked at Wetter, mainly due to the impoundment of lakes Baldeney and Kettwig. The middle and lower Ruhr carried a higher heavy metal load than the upper section.

Concentrations of metals at each sampling site were fitted to cumulative frequency curves to describe the quality of the river with varying flows. The

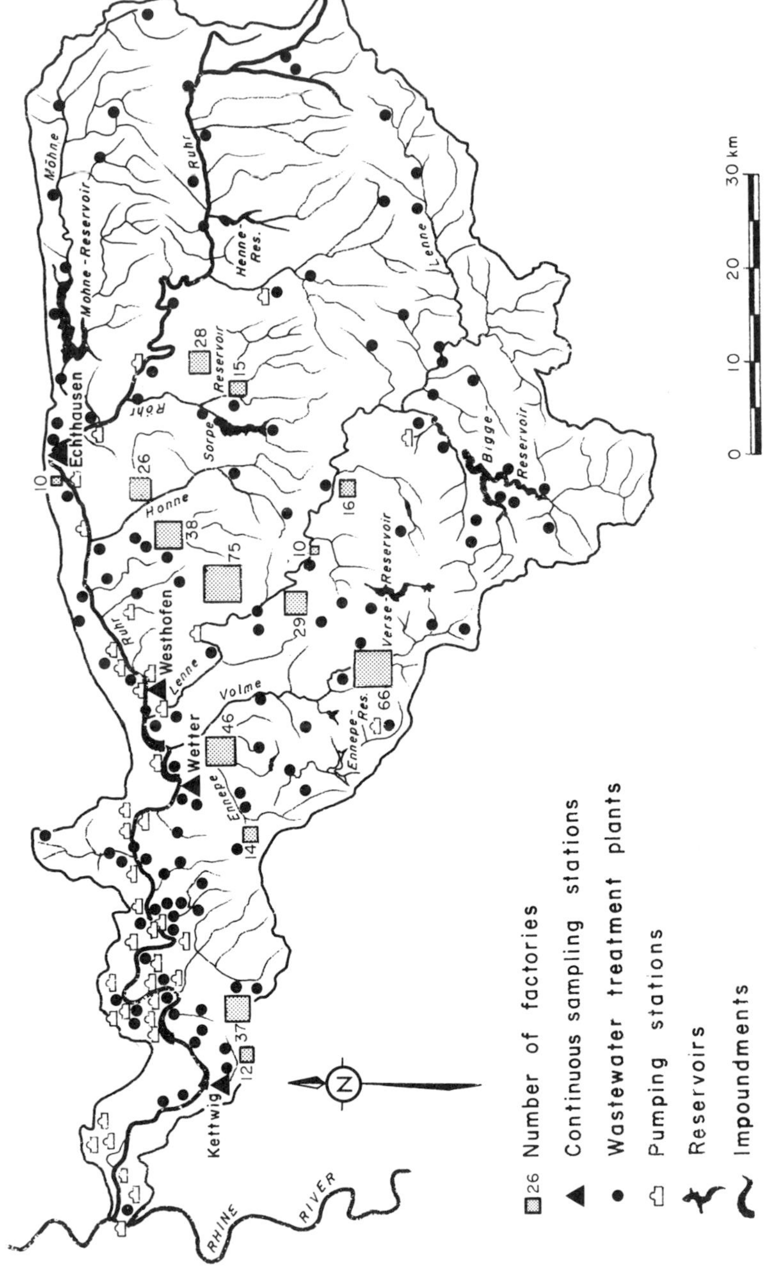

Figure 10-4. Ruhr catchment area with location of metal-working industries and sampling stations. (From Imhoff *et al.*, 1980.)

Table 10-6. Average concentrations ($\mu g\ L^{-1}$) of heavy metals in Ruhr water.

	Sampling station			
	Echthausen	Westhofen	Wetter	Kettwig
km from river mouth	127	95	82	22
	1972–74	1972–74	1972–78	1972–78
Zinc	69	102	226	136
Nickel	16	47	62	49
Copper	9	24	44	26
Chromium	6	16	20	10
Lead	21	17	17	10
Cadmium	1.4	2.1	2.3	1.4
Mercury	0.04	0.06	0.07	0.05

Source: Imhoff *et al.* (1980). Refer to Figure 10-4 for sampling station location.

curves were compared to the established upper imperative value (A 3I) of the European standards for potable waters. The fact that metal concentrations obtained from the 95% level of the cumulative frequency curves fell below the A 3I values indicates that the water was suitable for drinking purposes. Approximately 55% of the heavy metal load originated from wastewater and the rest was of geochemical origin. About 31% of the total load was retained by sediment and subsoil.

Metal concentrations were used to calculate a quotient (q):

$$q = \frac{C_{99\%}}{C_{1\%}}$$

where $C_{99\%}$ = concentrations not exceeded by 99% of all samples,
and $C_{1\%}$ = concentrations not exceeded by 1% of the samples.

The q values were divided into five classes according to metal concentration and change of the extent of variation on the way downstream. The change of gradient of the median value in the cumulative frequency curves was classified into four types (Table 10-7). For example, nickel belonged to class $q = 3-10$ and type I. That means (i) nickel concentration varied in a small range, even with changing flows and (ii) the dispersion of nickel data decreased with higher concentrations. This could be due to a great number and variety of nickel sources which provided equalization. Copper belonged to the same class as nickel; however, the extent of variation was independent of sampling station and average concentration. Lead and zinc belonged to the same class ($q = 3-10$) but to type III. Increasing ranges of variation with increasing concentrations were observed for lead and zinc. This indicated that few plants occasionally discharged vast quantities of these metals. Cadmium and chromium showed considerable variations and irregularities in regard to sampling station and average concentration. The greatest range

Table 10-7. Classification of the cumulative frequency curves of the analyzed elements.

Type	>100	30–100	10–30	3–10	<3
I				Ni	
				Co	Co
				Li	
					(Sr)
					K
				Na	
				B	B
II				Cu	
			Hg	Mn	
				(As)	
					Sr
					K
III				Pb	
				Zn	
				Fe	
				(Hg)	
		Be	Be		
IV			Cd	Cd	
	Cr		Cr	Cr	
				As	
			Se	Se	

Where

Type I: the gradient of the cumulative frequency curves increases with increasing average concentration

Type II: the cumulative frequency curves are almost parallel

Type III: the gradient of the cumulative frequency curves declines with increasing average concentration

Type IV: the gradient of the cumulative frequency curves varies irregularly with the average concentration

Source: Imhoff *et al.* (1980).

of variation with $q > 100$ was observed for chromium. Chromium was mainly discharged by metal-working industries.

When a heavy metal is discharged in varying quantities from a point source, its concentration diminishes and equalizes downstream of the site. This was reflected by decreasing q values. The concentration decline was due to sorption and sedimentation processes and also to dilution by uncontaminated ground and surface waters. The importance of the particulate fraction in metal behaviour was shown by the histogram of the ratio of dissolved to total concentrations of metals (Figure 10-5). Lead and iron were substan-

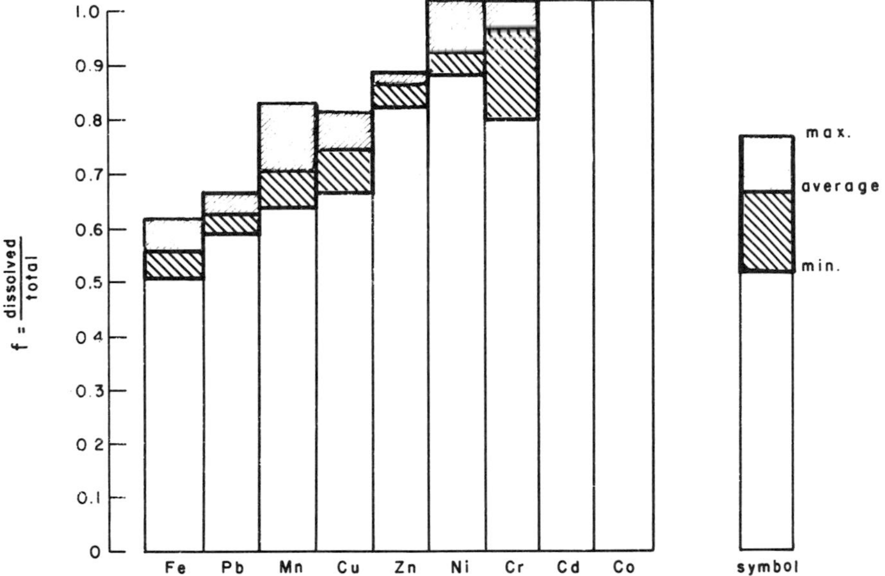

Figure 10-5. Relationship between dissolved and total heavy metal concentrations in Ruhr water. (From Imhoff *et al.*, 1980.)

tially bound to the particulates (average 38% and 45%, respectively). Cadmium and cobalt were exclusively in the dissolved forms. The ratio of dissolved to total metal followed the order:

$$Pb < Cu < Zn < Ni < Cr < Cd$$

On discharge, lead concentration in water declined most rapidly whereas cadmium declined slowly, with other metals ranging in the order indicated above. A heavy metal balance for the river system showed 45% input from geochemical origins and non-point sources, and 55% from municipal and industrial discharges. About 40% of the total heavy metal load of the Ruhr River was deposited in the sediments of the four impounding lakes in the middle and lower Ruhr. The population equivalent of heavy metals (metals (mg) capita^{-1} day^{-1}) was calculated from wastewaters in areas without metal-working industries and with correction for background levels in potable water. The highest rate was for zinc, next copper, and the lowest for chromium.

Rhine River

The Rhine extends through 10 different European countries and has an annual average discharge of 2200 m^3 sec^{-1}. Discharge is strikingly uniform throughout the year due to input from snow melting in the Alps in summer and rains from Basel in winter. This explains the extensive use of the Rhine

Table 10-8. Metal enrichment in Rhine sediment (mg kg^{-1}).

Element	Earth's crust (average)	River-clay, Rhine 1788	River-clay, Rhine 1975
Arsenic	5	13	54
Cadmium	0.2	0.3	30
Chromium	200	90	820
Copper	70	25	325
Lead	16	30	400
Mercury	0.5	0.2	10
Nickel	80	40	80
Zinc	130	100	1900

Sources: Van Driel (1979); Van der Veen and Huizenga (1980).

as a reliable source for drinking and industrial use. Enrichment of metals in sediments (Table 10-8) reveals significant industrial contamination.

Due to partitioning to the sediment phase, metal concentrations in the Rhine water are relatively low (average values: As 0.007, Cd 0.0015, Cr 0.023, Cu 0.012, Hg 0.0003, and Zn 0.1 mg L^{-1}), though maximum values exceed the limiting values set by IAWR (Van der Veen and Huizenga, 1980). In the Netherlands delta area (Rhine, Meuse, Scheldt, and Ems Rivers) metals were predominantly bound to the suspended matter. Most fine-grained material in the delta originated from the Rhine River, transporting 3.5 million tons in 73 km^3 of water. Suspended solids ($<16\ \mu m$) were primarily responsible for the transport of metals (Table 10-9), as noted for other rivers (Gibbs, 1977).

Mobilization of metals from suspended solids occurred in the Rhine estuary, as well as other estuarine environments. Organic matter and chloride, under vigorous tidal action, desorbed the metals from the suspended

Table 10-9. Annual transport of heavy metals (metric tons) in the Rhine.

Metal	Water	Fraction $< 16\ \mu m$	Water/$<16\ \mu m$ fraction ratio
Arsenic	375	500	1:1.3
Cadmium	125	105	1:0.8
Chromium	1250	2820	1:2.3
Copper	765	1355	1:1.8
Lead	695	1830	1:2.6
Mercury	42	53	1:1.3
Nickel	765	235	1:0.3
Zinc	11380	6705	1:0.6

Source: de Groot and Allersma (1975).

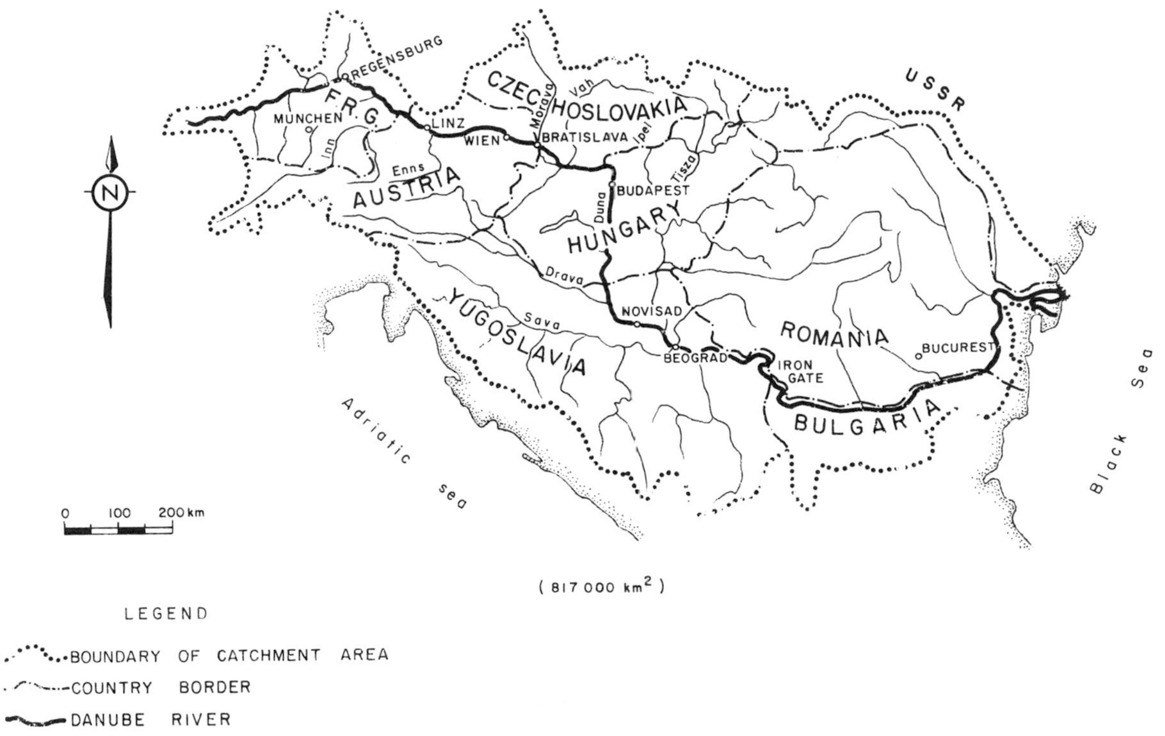

Figure 10-6. Catchment area of the Danube River. (From Benedek and László, 1980.)

solids. In the Rhine estuary, almost 100% of cadmium and mercury were mobilized, followed by copper, zinc, lead, chromium, and arsenic. Nickel mobilized at the 50% level, and elements such as lanthanum and manganese were not mobilized. Variations in mobilization among estuaries could be due to the varying character of fulvic acids, the mobilizing component of the organic matter. These acids vary from extremely aliphatic to the phenolic aromatic types.

Construction of dikes and dams in the mouthing areas of the Rhine stems tidal action. This increases the settling of suspended solids, minimizing mobilization of metals. Effect of estuarine environments on metal speciation has been discussed in earlier chapters.

Danube River

The Danube originates in the Federal Republic of Germany and intersects eight eastern and western European countries (Figure 10-6). Danube water is used for drinking and a variety of industrial purposes. Since increased use is anticipated with the opening of the Rhine–Maine–Danube navigational canal system in the 1980's, comprehensive monitoring and impact assessment programs are under development. At present the quality of water is satisfactory for multiple use over the entire length of the river. This is due to sorption, precipitation, and coprecipitation mechanisms removing metals from the dissolved phase (Table 10-10). In some sections of the river, there is a correlation between heavy metal concentrations and iron and manganese (Benedek and László, 1980). The recent increase in dissolved mercury concentrations is probably due to the appearance of complexing substances of biological and/or chemical origin (Table 10-11). Improved water treatment systems are planned to minimize the pollution load from industrial and domestic wastewaters.

Mississippi River

Presley *et al.* (1980) reported Pb and Cd pollution in sediments from an extensive area of the Mississippi delta. The Mississippi River carries about 60% of the total dissolved solids and 66% of the total suspended solids

Table 10-10. Concentrations* (mg kg^{-1}) of metals in Danube sediments of Szob-Budapest section.

Concentration	Hg	Cd	Pb	Zn	Mn	Fe
Minimum	0.2	7	100	400	1200	2500
Maximum	2.7	15	430	3000	3000	17000

Source: *Concentrations estimated from the figures in reference, Benedek and László (1980).

Table 10-11. Concentration ($\mu g\ L^{-1}$) of mercury and cadmium in Danube River at Szob.

Period	Mean concentration		Mean flow discharge
	Hg	Cd	($m^3\ sec^{-1}$)
1977	0.32	0.58	2350
1978	0.80	0.92	2110
1979	0.88	0.69	2330

Source: Benedek and László (1980).

transported to the oceans from the continental United States. Sediment Pb concentrations have increased by 70% over the past several decades from background levels (20 mg kg^{-1}). This is due to industrial discharges and increased use of leaded gasoline since 1940. This high suspended load (100–500 mg L^{-1}) and pH (7.5–8.1) provide a viable outlet for carrying 90% of the river metal load. The rest is carried out by the bed sediment, thus keeping the dissolved trace metal concentrations low and non-problematic (0–2 $\mu g\ L^{-1}$).

Conclusions

The modes of association and the resultant fate of metals is determined by the aquatic environment: the low molecular weight organic compounds in the water column (Ottawa River), bed sediment movement (Minamata Bay), suspended solids (Rhine River), and sedimentation (Ruhr River, Danube River). The equalization of heavy metal input from point-source discharges is due to the longitudinal dispersion and the sorption with subsequent sedimentation. The residence time of heavy metals in the water column will depend upon the extent of partitioning by the complexing processes. Thus surface reactions and complexation with organic and inorganic ligands will determine the physico-chemical forms and control the behaviour of heavy metals in rivers. Physico-chemical variations in sediments partition the metals in fractions of varying bioavailability.

Biological Impact

Aquatic Plants

Effect of heavy metal contamination on aquatic plants is highly variable. Although characteristic responses such as decreased diversity and density of populations generally occur in highly contaminated areas (Cairns *et al.*, 1972) there is much more inconsistent effect in moderately or lightly polluted areas. In addition, population response to heavy metals is significantly influenced by variations in natural environmental parameters, such as light and temperature (Rushforth *et al.*, 1981). Hence biological monitoring programs based on community criteria are subject to considerable intrinsic error. This implies that routine impact assessment and management of heavy metal discharges should not be solely based on density and diversity estimates.

Part of the variability associated with community oriented criteria is due to the ability of plants to adapt to heavy metals. As outlined in the previous chapters, such ability has been recorded for a number of species and metals. Hence, the presence of plants in potentially polluted waters may simply reflect adaptation to harsh environmental conditions. This could in turn lead to misinterpretation of the extent of ambient pollution. Another factor which may lead to inconsistency in community data is the ability of algae to form morphologically-indistinct races. Each of these races may have unique growth requirements and thus respond differently to environmental pollutants. Since traditional taxonomic methods are based on morphologic characters, microscopic analysis of species is not a consistently reliable monitoring tool.

The discharge of heavy metal wastes may also produce changes in physical conditions in receiving waters. These include alterations in particle size and organic content of the substrate and pH of the water. Aquatic plants respond to such perturbations by a decrease in density, species composition, and diversity. Hence, difficulty can be encountered in delineating effects of heavy metal contamination from those physical effects which are indirectly induced.

Whitton *et al.* (1981) described plant ecology in a seepage stream contaminated with wastes from abandoned mine tailings. Dissolved zinc levels varied by a factor of 2 over 24 h and decreased from an average of 21 mg L^{-1} at the source to 8 mg L^{-1} in an area 15 m downstream. The cyanobacterial alga *Plectonema gracillimum* and moss *Dicranella* sp. were dominant in the upper reaches, forming a laminated structure 3 cm thick. Further downstream the structure developed a white crust containing hydrozincite. Hence, the association became endolithic and survived at very high zinc levels (up to 370,000 mg Zn kg^{-1}). It was assumed that the growth and metabolic activity of the micro-organisms resulted in the precipitation of such large amounts of zinc. In addition the diurnal change in dissolved zinc levels was probably related to the shift in pH brought about by algal photosynthesis and release of CO_2.

Effects of contaminated bottom sediments on the growth cycles and diversity of benthic algal communities were determined in a shallow lake in northern Canada (Moore, 1981). Total metal concentrations in sediments were high near the waste source (metal mine) and fell rapidly moving away from this area (Table 10-12). Although algal densities in the zone of maximum contamination were generally low, the population did expand to near

Table 10-12. Total concentrations (mg kg^{-1} dry weight) of heavy metals in the sediments of Thompson Lake (Canada).

	Distance from waste source			
	0–1.0 km		1.1–3.0 km	
Metal	Average concentration	Range	Average concentration	Range
Mercury	0.44	0.05–1.32	0.10	0.04–0.23
Copper	85	25–175	32	30–50
Lead	30	2–110	<2	—
Zinc	115	65–255	87	40–110
Nickel	45	30–85	30	25–50

Source: Moore (1981).

control zone levels during the period of maximum growth (Figure 10-7). The dominant species were similar in all areas of the lake (Figure 10-7); however, the total number of species (63) near the mine was lower than elsewhere in the lake (111–132). Based on these data, it was concluded that (i) demonstrable impacts on the communities varied seasonally, (ii) no species could be designated as indicators of heavy metal contamination, and

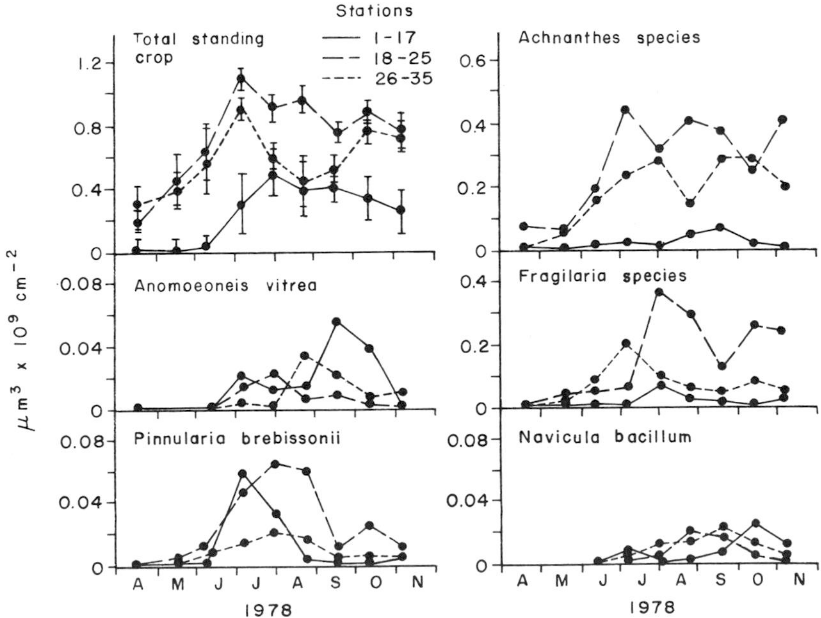

Figure 10-7. Seasonal changes in the densities of algae in Thompson Lake. Vertical bar represents 95% confidence limits. (From Moore, 1981.)

(iii) diversity indices were not reliable indicators of mildly contaminated sediments.

Hancock (1973) investigated algal ecology in a South African stream contaminated with mine wastes. The disposal techniques resulted in variable pH, turbidity, cyanide, and metal levels in receiving waters and thus provided an opportunity to determine the effects of interacting parameters on algal ecology. In the most heavily impacted area, pH ranged from 2.5 – 3.0 and conductivity from 260 to 3600 mmhos; in addition, the substrate surface was highly unstable due to the input of sand, resulting in the elimination of benthic species. Natural recovery commenced about 16 km below this area; however, during periods of high water, the zone of recovery was extended further downstream. The rate of rehabilitation increased on basic lava beds, which contributed to improvement in pH and dissolved solid levels. Indicators of recovery were the presence of the macrophyte *Potamogeton pectinatus,* the chlorophyte *Stigeoclonium lubricum,* and the diatom *Pinnularia* sp. It was also shown that high water periods increased turbidity, thereby decreasing algal densities and diversity. This seasonal factor had to be considered in the impact assessment.

Yan (1979) investigated the combined effects of heavy metals and low pH on the density and species composition of phytoplankton in several lakes contaminated with mine wastes. The brown alga *Peridinium inconspicuum* was dominant in all acidified lakes (pH $\leqslant$ 6.5), accounting for 30 – 55% of the total algal biomass. In more alkaline waters, it was replaced by a greater number of taxa, primarily flagellated chrysophytes. Overall, population biomass was better correlated with phosphorus concentration than pH (Figure 10-8). In addition, although copper and nickel concentrations in water were relatively high, there was only a weak correlation between biomass and heavy metal levels (Figure 10-8).

Invertebrates

Because invertebrates form such a diverse group, it is difficult to establish or predict a consistent response to heavy metals. Some of the difficulties involved in working with unicellular organisms include (i) the ability of some species to adapt to metals, and to form resistant spores, (ii) poor taxonomic descriptions which make identifications difficult, and (iii) imperfect knowledge of the natural life-history and ecology of most species. The latter point limits one's ability to distinguish natural variability from that induced by metals.

As indicated in earlier chapters, toxic effects of metals on multicellular invertebrates depend on a number of intrinsic factors such as age, size, sex, reproductive state, and nutritional status. These factors, combined with poor knowledge of natural life-history, again limit the usefulness of population-oriented data. In addition, aquatic insects show marked differences in abundance and diversity due to emergence of adults (Slobodchikoff and

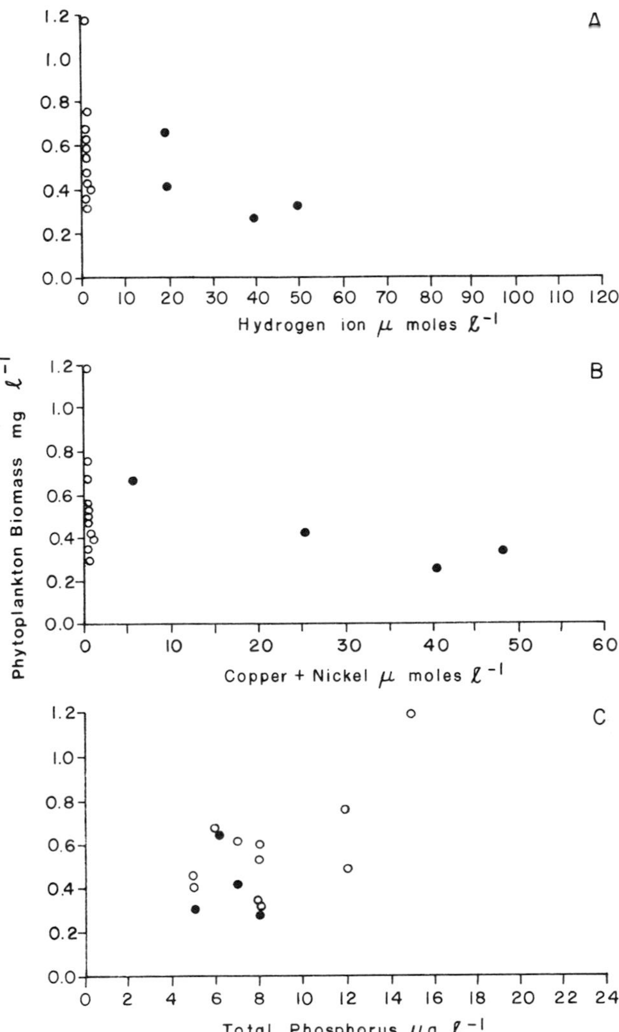

Figure 10-8. Biomass of phytoplankton in metal contaminated lakes in relation to (a) hydrogen ion activity, (b) sum of Cu and Ni levels, and (c) total phosphorus concentration. Lakes of pH <5.7 (●), lakes of pH >5.7 (○). (From Yan, 1979.)

Parrott, 1977; Friberg *et al.,* 1977; de March, 1976). Hence the use of this entire group in monitoring programs has to be carefully evaluated on a case by case basis. The ability of many species to live within the substrate at a depth which cannot be routinely monitored further limits the use of insects and other small species in impact assessment. Finally, it is difficult to distinguish the effects of contaminated water from those of sediments (Moore *et al.,* 1979). Heavy metals may readily desorb from sediments into interstitial water, thus influencing invertebrate survival; however, the effects of desorption on overlying water quality would probably not be measured during routine monitoring. In addition, since heavy metals persist in sediments over long periods, effect on the biota may simply reflect cumulative, poor waste disposal techniques at an earlier time.

Despite these difficulties, there have been numerous attempts to monitor and assess the effects of heavy metal discharge on aquatic environments. For example, Winner *et al.* (1980) found that macroinvertebrate community structure in a stream (Elam's Run) exhibited a predictable, grade response to metal pollution. Near the effluent source (metal-plating industry), copper, chromium, and zinc levels in water ranged up to 2.5, 0.32 and 0.65 mg L^{-1}, but fell rapidly within 1.0 km of this area (Table 10-13). Total insect density averaged 1300 m^{-2}, increasing to >2000 m^{-2} in downstream zones. The percentage of larval chironomids decreased much more rapidly (86–48%), whereas the abundance of trichopterans increased from 2 to 45%. Hence, it was suggested that the relative proportion of certain chironomid species in samples as an index of heavy metal pollution.

Effects of contaminated sediments on the density, diversity, and species composition of benthic invertebrates were determined in part of Great Slave Lake (Moore, 1979). Metal concentrations in sediments near the waste source were high, and decreased rapidly within 1 km of this area (Figure 10-9). Although benthic invertebrate densities were inversely related with metal levels, there was no apparent effect on diversity or species composition. These data possibly implied that population estimates provided a good assessment of the impact of the waste disposal method. However, since there was also a progressive increase in the organic content of the substrate, increasing food supply influenced the response of deposit feeders.

Although it is apparent that many species and community types cannot

Table 10-13. Maximum concentrations of heavy metals in water of Elam's Run (USA), and density and percentage abundance of insects.

	Station (Distance in km, above or below, waste discharge site)				
Metal (mg L^{-1})	1 (−0.07)	2 (+0.2)	3 (+0.8)	4 (+1.0)	5 (+1.2)
Copper	2.53	1.47	1.55	0.83	0.85
Chromium	0.32	0.20	0.23	0.05	0.02
Zinc	0.65	0.36	0.20	0.07	0.11
Density (insects m^{-2})	1300	1500	1300	2500	2100
% Chironomids	86	77	66	60	48
% Trichopterans	2	2	14	20	45

Source: Winner *et al.*, 1980.

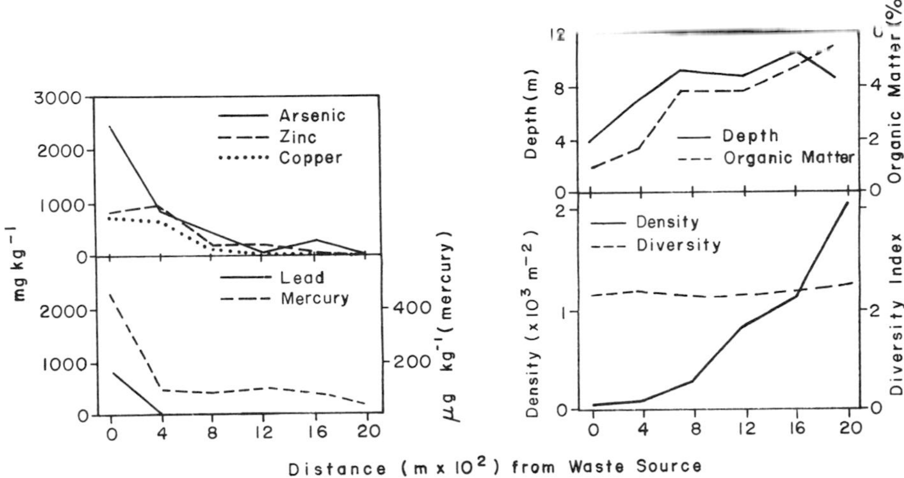

Figure 10-9. Concentrations of metals and organic material in sediments, water depth and density and diversity of benthic invertebrates near the waste source of a metal mine (Great Slave Lake). (From Moore, 1979.)

be used in monitoring, several investigators have consistently delineated zones of impacts using bivalve molluscs in both marine and freshwaters. Such species, including *Mytilus edulis* and *Anodonta* sp., generally feed from the overlying water and thus are not directly influenced by contaminated sediments. There is no tendency to burrow deep within the substrate or to exhibit marked cycles in abundance, such as found in insects. Most are readily distinguishable at the species level. While these qualitites make molluscs potentially good biomonitors, there is some difficulty in convincing resource managers of their value. For example, one could question the wisdom of modifying major waste treatment system based on impact data using molluscs and other invertebrates. This point is given further consideration in the following chapter.

Fish

Long-term, detailed investigations are generally required to obtain reliable data on the density, productivity, and diversity of fish populations. Hence, such criteria are not a part of most monitoring and impact studies, and are restricted to the assessment of the effects of major industrial projects on aquatic environments. As indicated in previous chapters, however, reduced growth and fecundity are often taken as a measure of the response of fish to metals. Although such data are relatively easy to obtain, it may be difficult to distinguish the influence of reduced food supply from the toxic effects of heavy metals in impacted waters. A possible scenario would be the elimination of invertebrates from contaminated sediments despite innocuous metal levels in the water. This situation might indicate to the unwary that metal pollution caused the reduction in growth and fecundity. Another complicating factor is the increase in acidification of water which often accompanies waste discharge. Once again fecundity and growth can be affected, either directly through pH stress or indirectly through reduction in food supply. The same can be said about the effects on fish of siltation and destruction of other components of the habitat.

The interaction of various polluting factors with fish populations is illustrated by a study of the river Ebbw Fawr, Wales (Turnpenny and Williams, 1981). This river was rendered devoid of fish by industrial pollution more than 100 years ago. Total zinc, lead, copper, and suspended solid levels in water ranged up to 6.0, 4.6, 0.5 and 2500 mg L^{-1}, respectively. Dissolved oxygen, pH, and conductivity also probably reduced fish survival (Figure 10-10). In 1971, a waste treatment system was installed at a major polluting source (steel plant), thus permitting fish to re-enter the river. Brown trout were the first to appear, followed by eels, stoneloach, and three-spine stickleback. Maximum population densities of these species (0.13, ≤0.01, 0.08 and ≤0.01 fish m^{-2}, respectively) were about 1/10 of

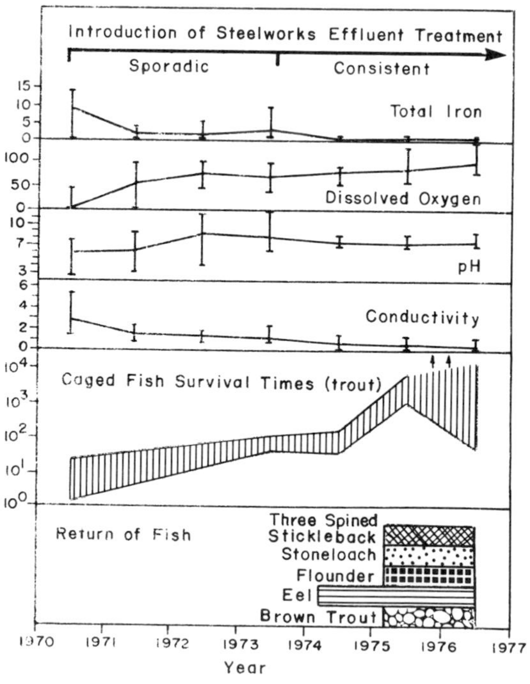

Figure 10-10. Changes in water quality, ranges in survival times (hours) of caged fish, and return of fish to the Ebbw Fawr. Total iron expressed as 10^3 mg L^{-1}, dissolved oxygen as percentage saturation, and conductivity as $\times 10^3 \mu S\ cm^{-1}$. (From Turnpenny and Williams, 1981.)

those recorded in control rivers. However, the growth and condition factor of brown trout were comparable or greater than those recorded in unpolluted British rivers. It was concluded that: (i) the main source of recolonization was tributary rivers, (ii) small species were incapable of rapid recolonization due to their restricted movement, and (iii) sublethal levels of toxic materials probably caused some species to avoid recolonization of the river. In additon, since the natural substrate was damaged by silt deposition, feeding and reproduction would be limited. Hence, the population could not return to normal levels.

Although considerable attention has been recently focused on acid-rain, significant levels of heavy metals are also carried in precipitation. Hagen and Langeland (1973) found that zinc and lead concentrations in snow on Norwegian lakes averaged 0.08 and 0.05 mg L^{-1}, respectively. This resulted in an elevation of zinc and lead levels in lake water to 0.04 and 0.02 mg L^{-1}, respectively. The metals were probably chronically toxic to fish and their food; however, the toxic effects of low pH and oxygen levels could not be readily distinguished from those of metals.

Multiple Level Impacts

There have been several studies into the effects of heavy metals on a combination of plant and animal species in both marine and freshwaters. Because the data generated by such investigations are relatively diverse, some of the shortcomings outlined in the three previous sections have been overcome. In general, these studies consist of a description of residue levels in water, sediments and biota, and population characteristics of one or more species. Some examples are listed below.

Lake Päijänne, Finland. This lake is located in an industrial area of south central Finland. It is the second largest in Finland, with a length, width, and maximum depth of 135 km, 20 km and 56 m, respectively (Figure 10-11). Pulp mill wastes and urban sewage are released into the northern and central parts of the lake. Särkkä *et al.* (1978a) showed that total mercury residues in sediments at stations 1, 2 and 3 (Figure 10-11) were similar (0.20–0.62 mg kg^{-1}) but decreased 2–4 fold in the southern part of the lake. Highest concentrations were always found in deep depositional basins, regardless of station. The lake plankton, consisting mainly of crustaceans, rotifers, and diatoms, showed relatively little contamination (0.02–0.72 mg kg^{-1} dry weight) (Särkkä *et al.* 1978b). Cyclopoid copepods were dominant in the polluted northern part of the lake, whereas harpacticoids were most common in clean water areas (Särkkä, 1975). The diversity of the plankton was always greater in this latter area.

Low mercury residues (0.01–0.14 mg kg^{-1} dry weight) were found in aquatic plants (Särkkä *et al.* 1978c) and consequently both herbivorous and planktivorous fish were only mildly contaminated with mercury (Table 10-14). Concentrations were high in all species of fish-eating birds, reaching

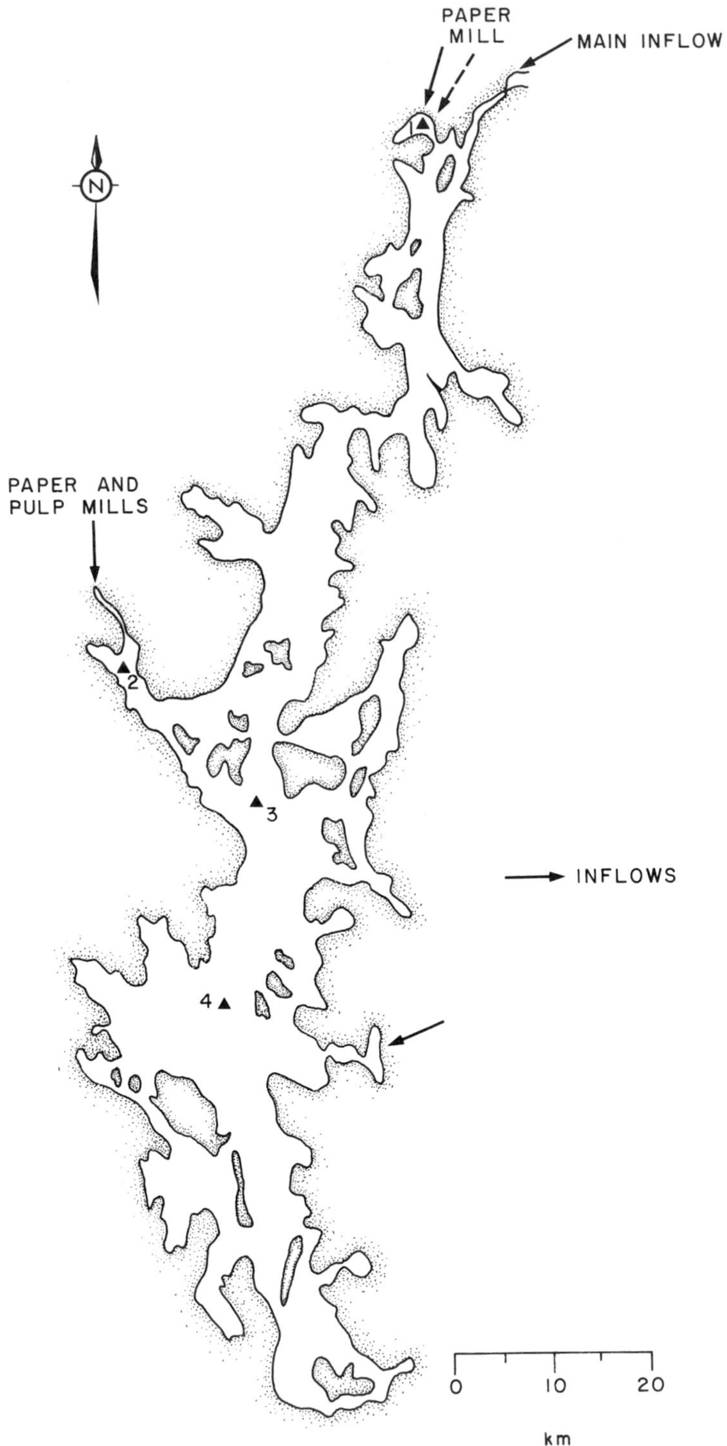

Figure 10-11. Lake Päijänne showing sampling stations (From Särkkä *et al.*, 1978b.)

14.6 mg kg^{-1} wet weight in muscle and 82.3 mg kg^{-1} in livers (Särkkä *et al.* 1978d). It was concluded that only about 25% of the discharged mercury could be found in the sediments, implying that most of the mercury was bound in the biota (Särkkä *et al.*, 1978a). Although high residues were found at various trophic levels, Särkkä and co-workers did not construct a Hg mass balance for the system to confirm this point.

Fjords, Norway. Several of the fjords along the west coast of Norway receive waste discharge from smelters, pulp and paper mills, chemical plants, and municipalities. Because the fjords have a large volume of deep sea water with only a small outlet to the open ocean, metals concentrate in sediment and biota. Skei (1978) reported that total mercury concentrations in sediments of Gunnekleivfjorden ranged up to 350 mg kg^{-1} dry weight. Although such levels were obtained in the immediate vicinity of a chlor-alkali plant, collections made 2 km down from the effluent discharge yielded concentrations of 90–110 mg kg^{-1}. In Sörfjord, lead, zinc, copper, and cadmium reached maximum concentrations of 30,500, 118,000, 12,000 and 850 mg kg^{-1}, respectively, in sediments (Skei *et al.*, 1972).

Metal residues in the brown alga *Ascophyllum nodosum* in Hardangerfjord were always greatest near a metal mine at the head of the fjord, whereas at the outlet, they were near background. Minimum–maximum levels for lead, zinc, copper, cadmium, and mercury were <3–95, 240–3700, 3–160, 0.7–16 and 0.05–20 mg kg^{-1}, respectively. A similar rate of decline in levels was recorded for zooplankton, mussels *Mytilus edulis,* and cod (Stenner and Nickless, 1974; Skei *et al.* 1976). Based on these studies, it was concluded that transport of mercury in the fjords by absorption on suspended sediment was probably negligible, reflecting the low turbidity of the water. Contaminants were dispersed by (i) turbulent diffusion and (ii) sorption by plankton,

Table 10-14. Concentration (mg kg^{-1} wet weight) of total mercury and methylmercury in fish from Lake Päijänne.

	Total mercury			Methylmercury		
	Average	Range	Sample size	Average	Range	Sample size
Whitefish	0.42	0.20–0.78	21			
Vendance	0.42	0.18–0.98	100			
Smelt	0.63	0.15–2.94	76			
Pike	1.07	0.23–3.96	315	1.07	0.32–2.45	64
Bream	0.34	0.05–0.81	261	0.22	0.06–0.44	19
Crucian carp	0.27	0.06–0.64	6			
Ide	0.35	0.16–0.59	8			
Roach	0.50	0.11–1.13	297	0.51	0.30–1.18	27
Burbot	1.51	0.14–4.34	121			
Ruffe	0.59	0.18–1.57	36			
Pikeperch	1.09	0.10–2.98	27			
Perch	0.63	0.10–4.68	506	0.75	0.17–3.62	47

Source: Hattula *et al.* (1978).

which was subsequently dispersed by a brackish surface flow. The ingestion of highly contaminated algae and autochthonous detritus was an important link in the transfer of mercury to animals.

References

Benedek, P. and F. László. 1980. A large international river: the Danube. *Progress in Water Technology* **13:**61–76.

Cairns, J., Jr., G.R. Lanza, and B.C. Parker. 1972. Pollution related structural and functional changes in aquatic communities with emphasis on freshwater algae and protozoa. *Proceedings of the Academy of Natural Sciences of Philadelphia* **124:**79–127.

de Groot, A.J., and E. Allersma. 1975. Field observations on the transport of heavy metals in sediments. *In*: P.A. Krenkel (Ed.), *Heavy metals in the aquatic environment.* Pergamon Press, Oxford, England, pp. 85–95.

de March, B.G.E. 1976. Spatial and temporal patterns in macrobenthic stream diversity. *Journal of the Fisheries Research Board of Canada* **33:**1261–1270.

Friberg, F., L.M. Nilsson, C. Otto, P. Sjöström, B.W. Svensson, B.J. Svensson and S. Ulfstrand. 1977. Diversity and environments of benthic invertebrate communities in south Swedish streams. *Archiv fuer Hydrobiologie* **81:**129–154.

Gibbs, R.J. 1977. Transport phases of transition metals in the Amazon and Yukon rivers. *Geological Society of America Bulletin* **88:**829–843.

Hagen, A., and A. Langeland. 1973. Polluted snow in southern Norway and the effect of the meltwater on freshwater and aquatic organisms. *Environmental Pollution* **5:**45–57.

Hancock, F.D. 1973. Algal ecology of a stream polluted through gold mining on the Witwatersrand. *Hydrobiologia* **43:**189–229.

Hattula, M.L., J. Särkkä, J. Janatuinen, J. Paasivirta, and A. Roos. 1978. Total mercury and methyl mercury contents in fish from Lake Päijänne. *Environmental Pollution* **17:**19–29.

Imhoff, K.R., P. Koppe, and F. Dietz. 1980. Heavy metals in the Ruhr River and their budget in the catchment area. *Progress in Water Technology* **12:**735–749.

Kudo, A., S. Miyahara, and D.R. Miller. 1980. Movement of mercury from Minamata Bay into Yatsushiro Sea. *Progress in Water Technology* **12:**509–524.

Moore, J.W. 1979. Diversity and indicator species as measures of water pollution in a subarctic lake. *Hydrobiologia* **66:**73–80.

Moore, J.W. 1981. Epipelic algal communities in a eutrophic northern lake contaminated with mine wastes. *Water Research* **15:**97–105.

Moore, J.W., V.A. Beaubien, and D.J. Sutherland. 1979. Comparative effects of sediment and water contamination on benthic invertebrates in four lakes. *Bulletin of Environmental Contamination and Toxicology* **23:**840–847.

Ottawa River Project Group. 1977. Distribution and transport of pollutants in flowing water ecosystems. Final Report. University of Ottawa–National Research Council of Canada, Ottawa.

Ottawa River Project Group. 1979. Mercury in the Ottawa River. *Environmental Research* **19:**231–243.

Presley, B.J., J.H. Trefry, and R.F. Shokes. 1980. Heavy metal inputs to Mississippi delta sediments. *Water, Air, and Soil Pollution* **13:**481–494.

Ramamoorthy, S., and D.J. Kushner. 1975a. Heavy metal binding sites in river water. *Nature* **256**:399–401.

Ramamoorthy, S., and D.J. Kushner. 1975b. Heavy metal binding components of river water. *Journal of the Fisheries Research Board of Canada* **32**:1755–1766.

Ramamoorthy, S., and A. Massalski. 1979. Analysis of structure localized mercury in Ottawa river sediments by scanning electron microscopy/energy dispersive x-ray microanalysis technique. *Environmental Geology* **2**:351–357.

Rushforth, S.R., J.D. Brotherson, N. Fungladda, and W.E. Evenson. 1981. The effects of dissolved heavy metals on attached diatoms in the Uintah Basin of Utah, USA. *Hydrobiologia* **83**:313–323.

Särkkä, J. 1975. Effects of the pollution on the profundal meiofauna of Lake Päijänne, Finland. *Aqua Fennica* **1975**:3–11.

Särkkä, J., M.L. Hattula, J. Janatuinen, and J. Paasivirta. 1978a. Mercury in sediments of Lake Päijänne, Finland. *Bulletin of Environmental Contamination and Toxicology* **20**:332–339.

Särkkä, J., M.L. Hattula, J. Janatuinen, and J. Paasivirta. 1978b. Mercury and chlorinated hydrocarbons in plankton of Lake Päijänne, Finland. *Environmental Pollution* **16**:41–49.

Särkkä, J., M.S. Hattula, J. Janatuinen, and J. Paasivirta. 1978c. Chlorinated hydrocarbons and mercury in aquatic vascular plants of Lake Päijänne, Finland. *Bulletin of Environmental Contamination and Toxicology* **23**:361–368.

Särkkä, J., M.L. Hattula, J. Janatuinen, J. Paasivirta, and R. Palokangas. 1978d. Chlorinated hydrocarbons and mercury in birds of Lake Päijänne, Finland 1972–74. *Pesticides Monitoring Journal* **12**:26–35.

Skei, J.M. 1978. Serious mercury contamination of sediments in a Norwegian semienclosed bay. *Marine Pollution Bulletin* **9**:191–193.

Skei, J.M., N.B. Price, S.E. Calvert, and H. Holtendahl. 1972. The distribution of heavy metals in sediments of Sörfjord, West Norway. *Water, Air, and Soil Pollution* 1:452–461.

Skei, J.M., M. Saunders, and N.B. Price. 1976. Mercury in plankton from a polluted Norwegian fjord. *Marine Pollution Bulletin* **7**:34–36.

Slobodchikoff, C.N., and J.E. Parrott. 1977. Seasonal diversity in aquatic insect communities in an all-year stream system. *Hydrobiologia* **52**:143–151.

Stenner, R.D., and G. Nickless. 1974. Distribution of some heavy metals in organisms in Hardangerfjord and Skjerstadfjord, Norway. *Water, Air, and Soil Pollution* **3**:279–291.

Turnpenny, A.W.H., and R. Williams. 1981. Factors affecting the recovery of fish populations in an industrial river. *Environmental Pollution* (*Series A*) **26**:39–58.

Van der Veen, C., and J. Huizenga. 1980. Combating river pollution taking the Rhine as an example. *Progress in Water Technology* **12**:1035–1059.

Van Driel, W. 1979. Zware metalen in zuiveringsslib en in rivierslib. *Landbouwkundig tijdschrift* **91**:177–182.

Whitton, B.A., N.L. Gale, and B.G. Wixson. 1981. Chemistry and plant ecology of zinc-rich wastes dominated by blue-green algae. *Hydrobiologia* **83**:331–341.

Winner, R.W., M.W. Boesel, and M.P. Farrell. 1980. Insect community structure as an index of heavy-metal pollution in lotic ecosystems. *Canadian Journal of Fisheries and Aquatic Sciences* **37**:647–655.

Yan, N.D. 1979. Phytoplankton community of an acidified, heavy metal-contaminated lake near Sudbury, Ontario: 1973–1977. *Water, Air, and Soil Pollution* **11**:43–55.

11
Monitoring and Impact Assessment Approaches

The discharge of heavy metal wastes has many obvious impacts on aquatic systems. There may be an increase in residue levels in water, sediments, and biota, decreased productivity, and increase in exposure of humans to harmful substances. Some of us are less aware of the effect of the environment on the fate of metals. As outlined in earlier chapters, such changes may decrease the toxicity of wastes, or at least remove them from immediate contact with humans. Given this diversity of effects, it is reasonable to suggest that environmental problems must be viewed in a broad context. Such diversity implies that methods used to manage, monitor, and assess heavy metal pollution should be equally complex. Although scientists often try to ignore social, political, economic, legal, and administrative forces, the success or failure of monitoring and assessment programs may also hinge on these factors (Tinkham, 1974). It has been estimated that total pollution control measures were responsible for the closure of 75 plants in the USA between 1971 and 1975, affecting 13,600 workers (Edmunds, 1978). In addition, total expenditures for pollution control will be $69 billion by 1984, thereby increasing the inflation rate by 0.3–0.5% (Edmunds, 1978). The magnitude of these figures emphasizes that the recommendations arising from monitoring and impact assessment programs must be relevant, timely and cost-effective.

Aggravated environmental problems often reflect the misuse or misunderstanding of technology (Petak, 1980). This implies that improvement in environmental conditions will depend on the development of integrated plans which fully describe existing conditions and predict future consequences. Hence, a good scientific data base should help managers use technology more effectively. Political, economic, and social factors are largely outside the control of environmental managers. This creates a situation where decisions, made on incomplete information, may be overturned on non-scientific grounds. Thus management can involve a degree of risk from both political and scientific forces. As outlined by Håkanson (1980) and others, risk-assessment and cost–benefit appraisals should be part of monitoring and impact assessment proposals. Obviously, managers need to maintain a balance in their knowledge of scientific, political, economic, and social factors.

During the last decade, there has been an attempt to school environmental managers in the complexity of response of the environment to heavy metal discharges. Some of the recommendations for improvement in monitoring and impact assessment programs include development of:

1. Microcosms and mesocosms to simulate natural conditions (Cairns, 1979, 1981);
2. Carbon cycling analysis of algae, detritus, invertebrates, fish, water, and sediments (Doremus *et al.,* 1978);
3. Environmental effects sequence planning, combining environmental indices and estimates of input (Gevirtz and Rowe, 1977);
4. Water quality management indices, including water chemistry data, an index of the need for abatement and the need for management planning (Truett *et al.,* 1975);
5. Screening mechanisms in which effects on abilities of species to interact with the environment are identified (Hansen, 1981).

These proposals have one major common feature: a comprehensive, multidisciplinary approach to problem solving. Although differing in scope, there is a consistent need for input from chemists, technologists, ecologists, and toxicologists.

The purpose of the following sections is to review and evaluate some of the criteria which managers may use during the formulation of monitoring and assessment programs. There is an emphasis on biological and chemical criteria, which should be used in program development.

Chemical Criteria

Reliable analytical measurements of pollutants are essential to provide information to ensure the quality of the aquatic environment and health of the public. Currently there is so much variation in overall environmental measurements that data, in most cases, are not valid for inter-laboratory comparisons. Variations in the techniques of sampling and storage, sample holding time prior to analysis, and analytical methods contribute to the incompatibility of data. There is a great need for standardizing the preanalytical as well as analytical procedures to generate a meaningful stock of data. In September 1978, the American Chemical Society's committee on Environmental Improvement directed its sub-committee on Environmental Analytical Chemistry to develop a set of guidelines to improve the overall quality of environmental analytical measurements (ACS, 1980). Analytical information is used for a broad range of purposes such as regulatory, environmental fate, and risk assessments, and decision-making processes. Hence it may require differing ranges of analytical reliability, but certainly an adequate degree of accuracy and precision must be guaranteed.

The guidelines recognized the fact that analytical chemistry is progressing toward more precise and accurate methods of analysis resulting in variations in equipment, personal capabilities, and analytical methods. The guidelines include (1) analytical planning, (2) quality control and assurance, (3) sampling protocol, (4) measurement procedure, (5) data handling, and (6) laboratory certification.

Analytical Planning

Data generation in any monitoring program must be based on a sound plan which has evaluated (i) the relationship between information needed and the conclusions drawn, (ii) the required data quality, and (iii) the need to establish the relevant analytical specifications. This requires a good understanding of the nature and magnitude of environmental problems and a thorough knowledge of the analytical methodologies. The planning team should include the field crew, laboratory analysts, a statistician evaluating the data, scientists who will be interpreting the data, and the manager who will use the information in decision-making processes. The scope and general outline of the program and the approved protocols for the execution of every stage of the monitoring program should be specified in the plan.

Quality Control

Quality control is an internal program maintained by the analytical laboratory to document the quality of data. This is done through (i) maintenance of skilled personnel, validated methods, and laboratory facilities, (ii) provision of representative samples and control, (iii) periodic calibrations, (iv) use of quality control samples with documented histograms, (v) duplicate checks and split samples, and (vi) checks on recovery of spikes from filtered and unfiltered water samples, and other matrices.

Quality Assurance

Quality assurance is a check on the quality control system, performed occasionally by a person outside of the normal operation. This will provide an independent assessment of monitoring operations and data evaluation, through on-site system surveys, response to user complaints, proficiency audit, interlaboratory comparisons, and sample exchange programs.

Quality assurance should not be interpreted as a policing action to harass the analysts. It is a corrective program to identify and remedy problem areas in the analysis. The performance of an analytical laboratory cannot be assessed unless there is a continuing recorded history of quality control. Data of unknown quality are invalid for any end use and thus become a drain on the resources of a monitoring program. The end use will determine the degree of quality required in the data; enforcement action or decision on health-related studies will require high quality analysis. On the other hand, pollution trend monitoring studies can accept information of lesser quality. It is the prime responsibility of management to set the monitoring objectives and the intended use of the data to be collected. Problems arise when a set of analytical data is used for different objectives, ignoring the level of quality it was originally intended to meet. More often than not less data of good quality is better than high volume data of unknown quality (Hauser, 1979). Environmental managers should be aware of this fact and insist on quality assurance in monitoring programs. Otherwise there is a risk of mandating unnecessary and expensive compliance programs.

Interlaboratory Comparisons

Interlaboratory comparisons are an essential part of the quality assurance program. They should be made mandatory for all analytical methods used in the monitoring programs intended for government decision-making processes. Interlaboratory comparison studies for the analysis of priority metals in water should include (i) blank samples, (ii) one synthetic low standard of metal ion in deionized-distilled water, (iii) a pair of identical duplicates of synthetic high standard of metal ion in deionized-distilled water, (iv) unfiltered environmental water sample, unspiked and spiked, and (v) filtered environmental sample, unspiked and spiked. Samples should be preserved by approved chemical additives. Analytical problems associated with blank determinations, calibration at low and high range, accuracy and internal precision, recovery from synthetic and natural water samples, effect of matrices on the extraction and analytical procedures, detection limit, and effect of sample holding time on the analysis (if any) are identified from statistical analysis of the data. Youden's paired analysis technique (Youden and Steiner, 1975) using identical duplicates should be used to detect and distinguish systematic and random errors. Such analysis should prove useful in individual method assessment and upgrading of applied analytical techniques by the participating laboratories. Thus it is essential that all environmental analytical laboratories (commercial, provincial/state and federal government) participate in continuing interlaboratory comparisons to improve the quality and compatibility of reported data. Participating laboratories should be asked to follow routine analytical procedures currently used in their laboratories. There should be no special status to the interlaboratory samples nor should analysis be conducted under research conditions or with specially calibrated apparatus. The objective of the study is to give the participating analysts and their laboratory staff an opportunity to evaluate their own performance and also compare it to that of other laboratories. Participating laboratories should be identified only by an assigned number and the true identity of the laboratories should be kept strictly confidential by the director of the study. The laboratories can, if they want, release their identity to clients or potential clients.

Similar studies can be conducted to compare the methods of digestion, extraction, and analysis used on other matrices and the accuracy and precision associated with these procedures.

Sampling Protocol

The quality of analytical information relies heavily on the validity of the sampling procedures. A sampling program should include rationale for the (i) choice of sampling sites, (ii) frequency of sampling, (iii) sample size to allow for the heterogeneity of the environmental samples, (iv) sample handling, pre-treatment, and storage, and (v) sample holding time prior to analysis.

Many trace element analyses, especially in the field of oceanography and marine biology, are unreliable due to gross sample contamination (Maienthal and Becker, 1976). This could arise during the sampling stage or/and subsequent procedural contamination. The values being reported are progressively lower as techniques and methodologies improve. Positive errors are common, due to failure to provide proper blank corrections. Proper materials and techniques are currently available to provide viable long-term stored samples for trace element analysis in most matrices. It is important that these practices are extended to the field situation to maintain continuity of quality.

Proper choice of the container, method of cleaning and conditioning of the containers, sampler design and storage by pH control, and preservation through chemical addition, refrigeration, and freezing will preserve the integrity of the samples prior to analysis. However, it is advisable to minimize the sample holding time to maximize the reliability of the results.

Biological tissue and fluid sampling present greater difficulties. Use of stainless steel implements is frought with dangers of contamination in analysis of trace elements (Maienthal and Becker, 1976). Contamination by copper, manganese, chromium, nickel, iron, cobalt, silver, tin, antimony, and tantalum at levels of 1.7, 0.64, 11, 12, 20, 0.24, 0.012, 0.46, 0.069, and 1.2 mg kg^{-1}, respectively from needle biopsies has been reported (Versieck *et al.,* 1973). Similar observations were made by Speecke *et al.* (1975) on the use of Meneghini biopsy needles and, to lesser extent, surgical blades. Possibility of using laser beams on hard and soft tissues and platinum-rhodium alloy needles with Kel-F hubs has been considered. A quartz or glass knife for cutting fish tissue is preferable to a conventional stainless steel knife.

Measurement Procedures

In the last few years, analytical chemistry has changed its emphasis from measurements of properties to determination of extremely low concentrations of a specific chemical moiety or element. This transition from milligrams to picograms with concomitant increased interferences from the matrix has been a challenge to analysts. This has resulted in new methods which continue to be modified and changed. Hence it is essential to adopt uniform methods to avoid judgement errors in litigation. Unfortunately, many laboratories resist changing to new instrumentation and new techniques on the basis of cost and work load. Regulatory agencies such as EPA (US) and Environment Canada insisted on standardizing analytical methods. The criteria used in selecting the methods (Ballinger, 1979) are: (i) the methods should meet the needs of the monitoring agency in terms of specificity, precision, and accuracy, in the presence of interfering materials usually encountered in environmental samples, (ii) the method should utilize the equipment and skills normally available in modern environmental laboratories, (iii) use of specialized instrumentation or techniques (such as GC/MS) must be justified by analytical or/and monitoring needs for unambiguous identification and quantitation of specific pollutants at very low concentrations, and (iv) the selected method should be rapid enough for large volume analysis on a routine basis and be able to be adopted by the majority of the laboratories.

Thus the monitoring agency has the responsibility of striking a balance between the need for high quality measurements and the economic burden placed upon the reporting laboratory. Contributions from other agencies such as Standard Methods for the Examination of Water and Wastewater, the ASTM Annual Book of Standards, the US Geological Survey of Methods for Collection and Analysis of Water Samples for Dissolved Minerals and Gases, etc., are included in the survey for standardization of methods. The selected methods for use are published in manuals prepared by EMSL (Environmental Monitoring Systems Laboratory), Cincinnati, and in the Methods Manual for Chemical Analysis for Water and Wastes, Environment Canada.

Applications for alternate procedures for a specific monitoring program should be submitted to the agency for approval, with statistical evidence to demonstrate compatibility with the approved procedure. Monitoring agencies approve only the best procedures or acceptable alternatives, but do not endorse specific instruments for monitoring (Ballinger, 1979).

Data Handling

Electronic data handling eliminates transcription errors and miscalculations. However, the system should be periodically tested with known data of sufficient accuracy and precision.

The results should include:

1. Estimates of measurement variability through precision and accuracy;
2. Estimates of concentration variability through mean, number of measurements, and standard deviation;
3. Limit of detection (usually 3σ above the measured average blank);
4. Limit of quantitation (usually 10σ above the blank);
5. Repeat analysis of questionable low or high results;
6. Recovery efficiency of spiked samples;
7. Reference to the standard analytical method and modifications, if any;
8. Full description, if any, of new methodology used, with exhaustive results on the validity of the method;
9. Method of calculation; and
10. Tabulation of possible positive and negative interferences.

In addition, new data for each sample including all field and laboratory informations as to the time and date of collection, preservation, transfer, sample number, pre-analytical manipulations, and instrumental responses should be permanently filed for future reference.

In conclusion, meaningful analytical data on priority metals can emerge only if all the following three conditions are met. They are (i) use of sensitive, specific, and validated analytical methods, (ii) use of standardized sampling and measurement techniques, and (iii) use of on-going quality assurance programs to demonstrate and monitor the quality of data generated. Openness to inspection for intrinsic defects and merits is vital to an analytical laboratory involved in monitoring and impact assessment of the environment. No laboratory or individual should undertake to generate environmental analytical results unless they realize the seriousness of the impact and commit themselves to provide reliable results.

Laboratory Certification

The Safe Drinking Water Act of the US (PL 93-523) enacted in December, 1974, stipulates that "samples may be considered only if they have been analyzed by a laboratory approved by the State." For those states which have not accepted the primary enforcement authority of the Act, the EPA regional administrator must approve the laboratory (Ballinger, 1979). In contrast, the Clean Water Act of 1972 did not require certification of a laboratory, resulting in no formal certification program. However, a set of criteria has been developed for use in interim certification of laboratories analyzing public water supplies. The criteria include required facilities and equipment, use of validated procedures, implementation of quality control programs, proper data handling and reporting, and annual performance on a set of evaluation samples.

It is highly desirable that the participating analytical laboratories be certified on the set of criteria mentioned above and renewed bi-annually through performance samples. Also it is recommended that split-samples be periodically exchanged between laboratories, to monitor the compatibility of the reported data. This is not intended to eliminate interlaboratory Quality Control Studies but rather to determine whether the laboratories are following through the recommendations of the studies.

Metal Speciation

The significance of heavy metal speciation in aquatic transport and bioavailability has been recognized in the last few years. The following recommendations should be given serious consideration in future projects, for meaningful monitoring surveys and laboratory toxicity studies to generate valid results that can be extrapolated to the natural environment.

(1) Improvement of analytical techniques and methodology to detect low levels in complex matrices including tissues, sediments, and body fluids.

(2) In addition to total metal analysis the samples should be analyzed for the toxic metal species such as redox species, chelated, particulate, and methylated species, for meaningful interpretation of aquatic toxicity data.

For example, the toxicity of arsenic and chromium depends greatly on the chemical form and valence: arsine and trivalent inorganic arsenic and hexavalent chromium are the most toxic species. Chelated forms of heavy metals such as copper, cadmium, mercury, etc. are less toxic than the unbound ions. Metals such as lead, copper, and nickel show a strong affinity to active surfaces such as particulates which might vary in toxicity depending on the route of ingestion. Expression of results as a fraction of dissolved to total metal concentration will result in better correlation to toxicity data. Most environmentally significant lead compounds are sparingly soluble or insoluble and hence the total metal concentrations of drastically acid-digested samples are not valid. Metals like mercury, arsenic, and to some extent lead undergo environmental methylation forming highly toxic methyl derivatives. Though their concentration with respect to the total metal may be less than 0.1%, their analysis is extremely important in assessing health effects on humans.

(3) For valid extrapolation, it is important that environmentally significant compounds of the heavy metals are chosen for the laboratory toxicity studies. In many studies, for the sake of experimental convenience and manipulation, ionic and soluble compounds of heavy metals which are never encountered in the natural environment are employed for the toxicity assessment of the metal. Examples include the use of nitrates and acetates of mercury, copper, and lead. The difference in anions can change the character of the metal. For example, nitrates of mercury and copper are highly ionic whereas the chlorides of the metals (commonly found in aquatic environment) are covalent. There is a marked difference in lipid solubility between ionic and covalent compounds. In addition, mercuric acetate photochemically produces greater amount of methylmercury than found in nature. This will introduce enormous bias into the laboratory toxicity experiments.

(4) It is important to match the environmental conditions in laboratory

experiments such as the inclusion of particulate phase, organic matter, etc. They determine the speciation of metals and their bioavailability.

(5) In laboratory experiments on heavy metal toxicity, the ratio of free to total metal concentrations should be determined to distinguish between the toxicity of the free metal ion versus the bound metal form. This bound fraction could be bound to physical surfaces or chemical moieties of the medium.

(6) Identification and quantitation of organic chelates and their direct and indirect effects on toxicity data should be included in field and laboratory studies.

(7) Physico-chemical characterization of sediments for their grain size, surface area, organic content, and cation exchange capacity and their subsequent effect on the sorption-desorption process is essential in evaluating the sinking capacity in overload situations and the longevity of the contamination.

(8) The bioavailability fraction(s) of sediment bound metals should be determined through fractionation in order to implement viable decontamination or reclamation program(s).

(9) As outlined in earlier chapters, many metals undergo a change in speciation as freshwater flows to the sea. Although cadmium undergoes one of the greatest shifts, other examples include zinc, copper, and lead. Large scale desorption of metals from suspended solids may also occur when rivers pass into relatively clear lakes, thereby increasing the availability and toxicity of metals to the biota. It is therefore desirable that the development of monitoring programs in freshwater consider the concomitant, albeit delayed, effects in coastal marine waters and lakes. While this approach may be too difficult to carry out in many rivers, relatively comprehensive programs have been developed for some systems. For example, chemical and biological monitoring occur throughout the River Rhine, despite the fact that it flows through ten European countries (Massing, 1980).

Biological Criteria

Traditional biomonitoring methods have often involved the use of community-oriented criteria such as diversity indices, species composition, productivity, and density of algae, invertebrates and, to a lesser degree, fish (Cairns, 1981). Although reliable data have been obtained in some studies, there are numerous technical/scientific inconsistencies involved with the use of such criteria, as discussed in earlier chapters. Hence, there are grounds to suggest that community-oriented data be used only in specific/isolated circumstances, in which investigators are fully aware of the ecology and predicted response of biota to metals. Otherwise, managers are put in a position of trying to distinguish natural biotic variability from adaptive responses and inhibition of productivity. In addition, it is difficult to see how managers could make major decisions based on community-oriented data for algae and invertebrates. For example, could we expect a major improvement in waste treatment methods used in a metal mine if the only demonstrable effects were decreased density, productivity, and diversity of insects and diatoms. The answer is obviously no, implying that the quality of biological information should be upgraded to effect a more positive response. Biological monitoring has also often involved residue analysis in plants and invertebrates. Some of the best examples include the use of *Mytilus edulis* and *Ascophyllum nodosum* in coastal marine waters. Although reproducible results can be obtained, it is once again improbable that resource managers would make major decisions based on such data.

Another factor which has to be considered in the development of impact assessment programs is the immediate source of pollutant to the biota. For example, elevated mortality of fish eggs near an industry could reflect poor water quality. However, desorption of metals from the substrate and concomitant resorption by the eggs would also lead to mortality. A similar situation would occur for a host of benthic invertebrate and algal species near the same plant. Since sediments accumulate metals over an extended period, toxic responses may reflect poor waste disposal techniques of earlier times. Hence, low productivity could not be considered indicative of poor performance of present waste treatment methods.

Comparable difficulties can be encountered if the chemical components of monitoring programs are emphasized at the expense of ecological considerations. For example, the transformation of aquatic larval insects to airborne adults generally leads to a mass reduction in invertebrate densities at various times of the year. A non-ecologist could easily construe this as mortality due to metal pollution, rather than natural biotic variability. The same situation may occur as fish move into and out of specific monitoring sites.

Dead fish often attract considerable attention from the public and media, which in turn influences politicians and eventually waste disposal methods. Living, but contaminated fish, have a lower profile, whereas those suffering from a population decline due to innocuous factors (e.g., reduced spawning sites) receive little or no attention. Hence monitoring programs could generate data useful to managers if there was an emphasis on mortality or decreased commercial value of priority species due to chemical residues in tissues. In fact, considerable effort has been spent in many countries to reintroduce or improve the quality of fisheries in rivers and lakes. These include examples from the Ebbw Fawr, Wales (Turnpenny and Williams, 1981) and the River Thames, UK (Black and Morrison, 1979). Similarly, routine environmental monitoring in some states includes analysis of fish tissues (Wallin and Schaeffer, 1979).

There have been a number of recent attempts to use chronic/sublethal criteria as monitors of heavy metal pollution. Examples include altered opercular activity, respiration rate, and behaviour in fish (Branson *et al.*,

1981; Gruber *et al.*, 1981; van der Schalie *et al.*, 1979). Another potential tool is the measurement of microbial bioluminescence using commercially available instruments. As outlined in earlier chapters, such methods are subject to a great deal of biotic variability (e.g., age, sex, size, and reproductive state). Similarly respiration and opercular activity may either increase or decrease, depending on the rate and sequence of exposure. These difficulties, combined with the fact that managers would once again have trouble in effectively using such data, suggest that chronic/sublethal criteria cannot be used on a widespread basis.

Traditional methods of biological monitoring in aquatic systems do not directly consider effects on *Homo sapiens* or other mammalian/avian species. This has to be considered a major shortcoming, in view of the high toxicity, mutagenicity, and carcinogenicity of many priority metals. Traditional methods often have little concern for the difference in effects which specific metals may have on poikilotherms and homeotherms. For example, nickel is not particularly toxic to aquatic biota but is a potential carcinogen to *H. sapiens.* A comparable situation exists for many organic chemicals (Branson, 1980). Similarly, there may be little consideration of toxic mechanisms or differences in the site of uptake. This deficiency would be particularly relevant to a metal like cadmium, which accumulates in liver and kidney but not in muscle.

It is perhaps obvious to suggest that individuals involved in monitoring of heavy metals should place less emphasis on effects on algae and invertebrates, and more on fish and *H. sapiens.* There is no reason why standard biological criteria cannot include testing for mutagenicity using bacterial assays. This approach has already been partially adopted by the Environmental Protection Agency, USA (Stara and Krivak, 1980). Data can be generated rapidly, at minimal cost, and be of immediate relevance to *H. sapiens.* Other tests, employing cell lines and various microorganisms are also available. Similarly, it is important for managers to carefully assess potential toxic effects to vertebrates of individual metals. In some jurisdictions, effluent criteria are developed in such a way as to consider only total levels (all metals combined). This approach may completely obscure toxic mechanisms and may disregard vulnerable components of the population.

References

American Chemical Society Committee on Environmental Improvement. 1980. Guidelines for data acquisition and data quality evaluation in environmental chemistry. *Analytical Chemistry* **52:**2242–2249.

Ballinger, D.G. 1979. Quality assurance update. Part II. *Environmental Science and Technology* **13:**1362–1366.

Black, P., and A. Morrison. 1979. Perspectives from three years experience of regional water services in Thames water authority. *Water Resources Bulletin* **15:**1578–1588.

Branson, D.R. 1980. Prioritization of chemicals according to the degree of hazard in the aquatic environment. *Environmental Health Perspectives* **34**:133–138.

Branson, D.R., D.N. Armentrout, W.M. Parker, C.V. Hall, and L.I. Bone. 1981. Effluent monitoring step by step. *Environmental Science and Technology* **15**:513–518.

Cairns, J., Jr. 1979. Hazard evaluation with microcosms. *International Journal of Environmental Studies* **13**:95–99.

Cairns, J., Jr. 1981. Biological monitoring, Part VI—Future needs. *Water Research* **15**:941–952.

Doremus, C., D.C. McNaught, P. Cross, T. Fuist, E. Stanley, and B. Youngberg. 1978. An ecological approach to environmental impact assessment. *Environmental Management* **2**:245–248.

Edmunds, S. 1978. Trade-offs in assessing environmental impacts. *Environmental Management* **2**:391–401.

Gevirtz, J.L., and P.G. Rowe. 1977. Natural environmental impact assessment: A rational approach. *Environmental Management* **2**:213–226.

Gruber, D., J. Cairns, Jr., and A.C. Hendricks. 1981. Computerized biological monitoring for demonstrating wastewater discharge. *Journal Water Pollution Control Federation* **53**:505–511.

Håkanson, L. 1980. An ecological risk index for aquatic pollution control. A sedimentological approach. *Water Research* **14**:975–1001.

Hansen, S.R. 1981. Screening for toxic effects on interspecies interactions: A mechanistic or an empirical approach? *Archives of Environmental Contamination and Toxicology* **10**:597–603.

Hauser, T.R. 1979. Quality assurance update. Part I. *Environmental Science and Technology* **13**:1356–1361.

Maienthal, E.J., and D.A. Becker. 1976. A survey of current literature on sampling, sample handling, and long term storage for environmental materials. National Bureau of Standards Technical Note 929, US Department of Commerce, Washington, D.C., 34 pp.

Massing, H. 1980. The River Rhine—Transnational river basin management developing programme to meet new challenges. *Progress in Water Technology* **13**: 77–91.

Petak, W.J. 1980. Environmental planning and management: The need for an integrative perspective. *Environmental Management* **4**:287–295.

Speecke, A., J. Hoste, and J. Versieck. 1975. Sampling of biological materials. *In*: P.D. Lafleur (Ed.), *Proceedings of the 7th IMR Symposium on Accuracy in Trace Analysis: Sampling, Sample Handling, and Analysis.* National Bureau of Standards Special Publication No. 422, Washington, D.C.

Stara, J.F., and J. Krivak. 1980. The US program to meet water quality standards. *Progress in Water Technology* **13**:267–275.

Tinkham, L.A. 1974. The public's role in decision-making for federal water resources development. *Water Resources Bulletin* **10**:691–696.

Truett, J.B., A.C. Johnson, W.D. Rowe, K.D. Feigner, and L.J. Manning. 1975. Development of water quality management indices. *Water Resources Bulletin* **11**:436–448.

Turnpenny, A.W.H., and R. Williams. 1981. Factors affecting the recovery of fish populations in an industrial river. *Environmental Pollution* (*Series A*) **26**:39–58.

van der Schalie, W.H., K.L. Dickson, G.F. Westlake, and J. Cairns, Jr. 1979. Fish bioassay monitoring of waste effluents. *Environmental Management* **3**:217–235.

Versieck, J., A. Speecke, J. Hoste, and F. Barbier. 1973. Trace contamination in biopsies of the liver. *Clinical Chemistry* **19**:472–475.

Wallin, T.R., and D.J. Schaeffer. 1979. Illinois redesigns its ambient water quality monitoring network. *Environmental Management* **3**:313–319.

Youden, W.J., and E.H. Steiner. 1975. *Statistical Manual of the Association of Official Analytical Chemists.* AOAC, Washington, D.C., 88 pp.

12
Politics and the Environmental Manager

Principles

Politics is a very powerful force which the manager must recognize and work with. The word "politics" means "of guiding or influencing government policy" which is relevant to the government of the state (federal, provincial or municipal) and society, including industry, and the interaction between any or all of these entities. Furthermore, the span of environmental management extends from the concept stage, through planning, development, operation and finally to reclamation and rehabilitation. The latter phases could be from one year to a hundred years or more later, especially in the case of a mining operation where the final point is the closing of the mine, its reclamation and the rehabilitation of the people.

The progressive environmental manager's cognizance of these matters is a means of advancing mankind's quality of life, resource development, and economic conditions, and of protecting or improving the environment. This does not mean altering current trends in resource development. It entails developing and practising effective methods whereby a balance can be struck between resource management, environmental protection, and the quality of life. This role or objective will be achieved through the effective planning of efficient policies, programs, and services. In essence, the role of environmental management is to emphasize prevention rather than treatment because such a principle is logical, practical, and economical. With

---

This chapter was prepared by E.E. Ballantyne.

environmental matters, this means coordination, comprehensive input, and long-term planning to ensure a good quality of life in the years to come. It simply and firmly documents that there can be concurrent strip mining for coal, oil sands, or gravel and adequate environmental protection along with providing a good quality of life for man. The environmental manager has an unalterable obligation to manage in such a way that there are funds provided to reclaim water and land to a biological production capability as good or better than when the mining commenced. In some cases, a crop of energy (e.g., coal or oil sands) could be developed followed by reclamation and agricultural development.

Society in total, including environmental managers, scientists, educational institutions, industry, and governments will avoid many problems if it is recognized that in general terms diverse components of the environment have direct or indirect impact on humans. This end point must not be forgotten. Any man-made additive, including metals, pesticides and other organics, must be analyzed and assessed as to its impact. This does not simply involve acute effects but should involve the longer term or chronic symptoms of intoxication of the living tissues of man, animals, and plants. Water has to be recognized as food for man, animals, and plants. Too often in the past, improved economics of production has been the main consideration in adding biocides to the food chain system. Again a balance has to be struck between production and conservation.

History records too many negative effects of man's inability to effectively manage the environment. However man can improve on nature, especially with the increased production of food, e.g., with irrigation, fertilization, better culture methods, improvements in plant breeding, and the recognition that agricultural and other food production is a business and a science. Also, one must acknowledge that all pollution is not man-made. Nature has its own share, such as excessive mercury in some lakes and rivers.

Because the topic of environmental management is of such an enormous magnitude and complexity, the foregoing may err on the side of simplicity but the basic principle still remains that the environmental manager is the key factor in preventing deleterious hazards to man, animals, and plants. His efforts with policies, programs, and services must be based on scientifically proven facts. At this point society has to be understanding, often charitable and supportive, because all scientific facts are not known. This produced the transition from the Naive Era to the Adjustment Era whereby extensive development, expansion, and funding of new research institutions took place at a cost of many millions of dollars. At some institutions there is an encouraging and effective interdisciplinary team approach to environmental research and services. Biologists, agrologists, chemists, medical doctors, veterinarians, pathologists, foresters, soil and water experts, librarians, and many more work together as needed on a given research project. Such a team can cover all aspects of man's life chain. One might say "what has this to do with politics?" As politics is also the art and science of government the above mentioned interdisciplinary research process covers the needs of several governmental departments in discharging their responsibilities to the public by providing more scientific data as the base for better government policies, programs, and services. Such departments deal with food production, environment, forestry, fish and other aquatic life, wildlife, human and animal health, soil, water, energy, and air and water quality.

Studies and research such as mentioned above are not only a result of the emergence and concentration of environmental interest in the 1970's. The subject and the politics go back into the history of mankind. Early civilizations had their social customs and governments influenced and shaped by environmental protection and control aspects. These include the location of cities, towns and villages, irrigation of arid areas to produce food and fibre, drainage, sewage systems, diversion of water, erosion control and others—mainly for the survival of man. Society today still has the same objective but it has become more intensive, complex and comprehensive due to population expansion whereby the air, water, and land has to be shared by billions of people instead of only millions. New industrial technology has brought drastic advancements coupled with more pollution problems. Thus there has to be clean-up measures for past mistakes (often due to lack of appropriate knowledge) as well as the development of new knowledge and its rigid application through society's demand via government laws and standards.

The environmental manager is aware that a high percentage of the public is much better informed now on pollution matters as they affect occupational health and leisure time activities. Often the latter concerns water resources for drinking, recreational sports, and fishing. All society is not willing to overlook pollution in favour of job emphasis. Society wants the facts on which pollution prevention and control can be developed and maintained. There is often no hesitation in expressing opinions, pro or con, on projects affecting the quality of life, now or in the future. This is good provided it is just not an emotional reaction. Some choose to ignore facts in telling an environmental manager what to do. However mankind, over time, has a check and balance process which brings out reasoned mature support for management systems. Finally, society is willing to have its governments spend money on research, administration, and enforcement.

Methods

There are many methods by which the environmental manager can obtain the views of the public, or conversely expose the public to his views, ideas, and information. These are usually relevant to a proposal to develop industries which have some impact on the environment or society.

(1) Elected members of federal, provincial/state, and municipal councils and/or their agencies relevant to the topic or proposed works are a major avenue to the politics of environmental management. Some governments

have established agencies to specifically conduct public hearings and make recommendations of action. Funds have been provided to enable society to prepare adequate briefs about its concerns, pro and con. Thus the public have an input to the decision-making process. The public also acts as a watch dog on the actions or proposed actions of governments. This is legitimate as these elected officials are working in a logical and practical manner in the public interest for balanced long-term development of the community. Public reaction on an issue has led to the defeat of a by-law or an individual in the next election, or necessitated plebiscites and hearings.

(2) In the early 1970's many governments throughout the world, due to the emergence of public concern about environmental protection, established departments of environment. Intergovernmental committees on such matters as air and water quality worked together and with industry, public associations, and universities to develop practical and minimum quality standards. These were based on protecting the health of man, animals, and plants, and improving the environment. International committees or commissions focused on common problems such as improving the quality of water in the Great Lakes and preventing and controlling acid rain. Within Canada there have been and are several federal–provincial committees working on common denominator subjects of current and long-term necessity. For example, the Mackenzie River Basin Pipeline Study Group spent several millions of dollars analyzing the potential impact of proposed oil and gas pipelines from the Mackenzie Delta to southern Canada via the Mackenzie River Basin. Public hearings were conducted in the area and across Canada. Pipeline proponents, governments (federal, territorial, and municipal), native peoples, and others made contributions, for and against all or parts of the proposal. This was important as the purpose of impact assessments is to assist the final decision makers in arriving at their decision. Although industry and the territorial and municipal governments supported the proposal, the National Energy Board decided not to proceed with development in the immediate future. It was suggested that the north was not yet ready for the social effects of such a major project. If the decision had been "yes", the submissions would have assisted in delineating the protective measures to be observed by the pipeline owners to eliminate concerns about the fauna, flora, peoples, water quality, soil erosion, etc., and to establish river crossings, contingency measures in case of pipeline breaks, the best route, etc. This would be the "balance process" between resource development, environmental protection, and the quality of life, or in other words the "politics" of guiding or influencing government policy.

(3) Politics have been helpful in analysing, recommending and deciding on acceptable solutions to significant environmental problems. Consider the location of pipeline corridors to connect to interprovincial, interstate and international pipelines and/or refineries and chemical plants. Consultants have been commissioned to study the matter and to propose the best or alternate routes with supporting data. Consideration is given to using the

poorest food-producing forestry land, the best river crossings, potential refinery or chemical plant locations, new or expanded towns, and water availability. Meetings are often held with provincial/state and municipal government officials, agricultural, forestry and wildlife associations, industry, individuals, especially on the route, and professional associations. Instead of numerous pipelines taking different routes, initial installations follow a common designated corridor which can handle additional pipelines for oil and gas, electric power lines, highways, and other transportation needs. Other advantages are (i) costs to government and industry can be easily calculated, and (ii) construction and staffing needs can fit the policy of the government for long-term plans of decentralization of industry and services. The environmental manager is also helped by the reduction in the amount of money spent on environmental impact statements. In this case the role of government, which is to concentrate on prevention of environmental problems and to act on long-term economic development, is fulfilled.

(4) Regardless of the type of proposal, the proponent should make presentations at a public meeting or meetings. Experience has shown that public opposition is often due to misunderstandings, based on lack of knowledge. Therefore clear visual aids and brief printed material (not highly technical) should be presented. Meetings should be held in the planning stage, thereby permitting the public to react, pro and con, in an intelligent and mature manner, based on facts and not on rumors or a mixture of the two. The proponent must show how the proposed works are in keeping with the legal environmental standards on air, water, and land quality. Thus the public can consider their general interests, such as aquatic life, food production, forestry, wildlife, and the health of man, animals, and plants. It is better that problems be solved directly with the proponent so that the ramifications of the proposal are well sorted out before the application is made to the decision-making body in government for a license to construct and operate the said works. However, if the public vs. proponent problems are not solved at the local level, the government licensing agency shall obtain comments via briefs and public hearings as input for the decision-making process. This can lead to amending the application to provide a balance between resource development, environmental protection, and the quality of life.

On occasion, resolution of these matters is difficult and emotional, especially if the displacement of people is involved. Such problems are particularly aggravated when the land has been in the family for generations, in spite of the fact that good food-producing soil can be saved for the reclamation stage. It is easy enough to be objective and say that a crop of energy can be obtained, followed by reclamation, and then the use of the land for the renewable resources (e.g., food production) can go on to infinity. However, it is not always that simple even with generous compensation which frequently is not the case.

(5) The government (federal, provincial/state, or municipal) is often the

environmental manager. The public (or society) makes its views known through delegations from individuals, associations, societies, groups, and other governments. It is usual to submit a brief and verbal explanation. Even governments have their differences on such matters as sewage and water systems, town planning, industrial plant locations, and park developments. Often the ramifications go far beyond the local government involved. Downstream users also have a keen interest in quality and quantity of water. In Western Canada, the provinces of Alberta, Saskatchewan and Manitoba, plus the government of Canada, work together through the Prairie Provinces Water Board to obtain the best knowledge possible by studies, research, and continuous monitoring in order that sound and logical decisions can be made on water use.

The possible water diversion from one watershed to another to meet projected society and industrial needs is a problem for the environmental manager. This is too large a topic to detail here. The answer is not simplistic on the basis that water is water only. Diversions can have positive and negative effects on climate, industrial growth, the location and development of towns and cities, aspects of aquatic life moving from one water system to another, and irrigation. Thus the environmental manager's work is not only of a current nature but involves the politics of long-term and complex decisions which may affect local, provincial, interprovincial, national, and international matters. In essence, water is life and must be considered seriously.

(6) The media, i.e., TV, radio, newspapers, and other forms, are a powerful force in the politics of influencing or guiding government and corporate policy. Most input is in the form of reporting and on occasion editorials. The impact is by repeatedly contacting a high percentage of the public. If facts are reported accurately, there shall be no complaint from any government or corporation. Conversely if the reporting is inaccurate or misleading, one observes that a correction is usually not given equal status in headlines, thereby damaging public understanding.

(7) The environmental manager, especially government, must be careful but fair in dealing with the "squeaky wheel" syndrome. Policy changes should not be made to placate someone unless there is beneficial effect on a majority of the people involved. Otherwise government by minority will rule, which is contrary to democratic principles. Nevertheless, an environmental manager can not always afford to ignore complaints until their validity is confirmed and reviewed. Often good ideas originate from such a process, especially in a period of rapidly changing technical knowledge.

(8) Often arguments have stated that the cost of reclamation is prohibitive for the consuming public to bear when such costs are added to the final retail product. Let us examine an example. Reclamation at an oil sands strip mine will commence approximately 10 years after mining commences. If there was a levy of $0.05 a barrel of synthetic crude oil for that period of time at a plant producing 100,000 barrels a day, it would amount to $5,000 a day,

or $1,500,000 for a 300 day year (without counting the accummulated interest). Such a trust fund would be a formidable nest-egg to undertake reclamation. Also there would continue to be $5,000 a day going into the fund until the plant closed in 20 to 30 years. Breaking it down another way would mean that, at $0.05 per gallon of unrefined oil, the consumer at the filling station would pay $0.00125 on a gallon of gasoline. That does not look like an exorbitant charge on the consumer to provide for returning oil sands area to a biological status as good or better than when mining commenced. The same type of levy can take care of coal, strip mines, and gravel pits to provide for good environmental management. Where rehabilitation of a whole town is potentially involved then the environmental manager may have to review the adequacy of the levy.

Summary

The environmental manager whether in government, a corporation or a landowner, has a responsibility to diligently work towards improving the balance of resource development, environmental protection, and the quality of life. The world shall be passed on from generation to generation as a good place in which to live whether 5 or 500 years from now. Special environmental efforts should be made concerning food production, living standards, and all renewable resources. This is the essence of survival and requires more research to enable the best input that society can make for maintaining and improving the environment. Politics are important in achieving this goal with the environmental manager as the key to a logical, practical, mature, and balanced effort in decision making and administration. This involves putting the emphasis on the prevention of hazards to the health of man, animals, and plants via water, air, and land. The funding for reclamation and rehabilitation is all part of this process in fairness to future generations. Good environmental politics and management will be the guarantee of success. Man cannot live alone as society becomes more dependent and interlocking with each of its components and the environment.

Appendix A
Summary of Current Production and Uses of Metals

Compound	Formula	World production (1971–80, × 10^3 metric tons)	Uses
White arsenic	As_2O_3	41.1*	alloying agent, medicinal, biocide, wood preservative
Metallic arsenic	As	—	alloying agent, batteries
Calcium arsenate	$CaHAsO_4$	—	insecticide
Sodium arsenite	$NaAsO_2$	—	herbicide
Lead arsenate	$PbHAsO_4$	—	insecticide
Chromated copper arsenate	reaction product of $H_3AsO_4 + Na_2HAsO_4 + Na_2Cr_2O_7 + CuSO_4$	—	wood preservative
Fluorchrome arsenate phenol	$CrF_2OH(C_6H_3)AsO(OH)_2$	—	wood preservative
Lewisite	$Cl{-}CH{=}CHAsCl_2$	—	warfare agent
Phenyl arseno benzene	$C_6H_5{-}As{=}O$	—	wood preservative
Metallic cadmium	Cd	150	electroplating, batteries, alloys
Cadmium sulfide	CdS	—	pigment, lubricant, photocells
Cadmium sulfate	$CdSO_4$	—	electrolyte, biocide, lubricant
Cadmium stearate	$(C_{18}H_{35}O_2)_2Cd$	—	plastic stablizer
Cadmium phosphors	$CdWO_4$,CdS	—	tubes (TV, X-ray, fluorescent)

* USA data withheld

Compound	Formula	World production (1971–80, × 10^3 metric tons	Uses
Cadmium selenide	$CdSe$	—	pigment, luminescent material
Cadmium nitrate	$Cd(NO_3)_2$	—	pigment, batteries, flashpowder
Chromite	$FeOCr_2O_3$	86200	ferroalloys, refractory
Sodium dichromate	$Na_2Cr_2O_7$	—	pigments, tanning agents, electroplating, oxidants, catalysts
Metallic copper	Cu	82500	electrical, alloys
Copper sulfate	$CuSO_4$	—	algicide
Copper arsenite	$CuHAsO_3$	—	insecticide, fungicide, wood preservative and as pigment
Copper chromate	$CuCrO_4$	—	fungicide, insecticide and in dyeing
Copper chloride	$CuCl_2$	—	industrial catalyst, in petroleum refining, feed additive
Cupric acetoarsenite	$Cu(C_2H_3O_2)_2$, $3Cu(AsO_2)_2$	—	insecticide, wood preservative and pigment
Cupric acetate	$Cu(CH_3COO)_2$	—	fungicide, catalyst in rubber aging and as pigment
Cupric basic carbonate	$CuCO_3$, $Cu(OH)_2$	—	seed treatment, paint and varnishes, additive in animal and poultry feeds
Cupric selenide	$CuSe$	—	semi conductors
Cupric stearate	$(C_{18}H_{35}O_2)_2Cu$	—	antifouling agent, rubber aging, as catalyst
Cupric sulfate	$CuSO_4$	—	algicide, herbicide, dyeing industry and a variety of industrial uses
Metallic lead	Pb	38000	storage batteries, metal products
Organometallic lead	Pb——R	—	biocides, preservatives
Tetraethyllead	$Pb(C_2H_5)_4$	—	antiknock agent in gasoline
Lead stearate	$(C_{18}H_{35}O_2)_2Pb$	—	synthetic polymers
Lead chromate	$PbCrO_4$	—	highway safety paints
Metallic mercury	Hg	7000	thermometers, barometers, manometers, electrical conductors, coolant
Mercuric sulfate	$HgSO_4$	—	catalyst
Mercuric chloride	$HgCl_2$	—	antiseptic
NaHgX	$NaHgX$	—	chloralkali process
Phenylmercury compounds	C_6H_5HgX	—	preservatives, fungicide in seed treatment
Organomercurials	Hg——R	—	catalysts, slimicides

Compound	Formula	World production (1971–80, × 10^3 metric tons	Uses
Metallic nickel	Ni	700000	alloys with copper, iron, steel electroplating, electroforming
Nickel dust	Ni	—	hydrogenation reactions
Nickel oxide	NiO	—	storage battery
Metallic zinc	Zn	63245	diecast products, alloys, galvanizing
Zinc dust	Zn	—	printing, dyeing, purifying fats, precipitating Ag and Au
Zinc oxide	ZnO	—	catalyst in vulcanization of rubber, paints
Zinc sulfide	ZnS	—	pigment
Zinc sulfate	$ZnSO_4$	—	rayon fibres
Zinc chloride	$ZnCl_2$	—	wood preservative

Appendix B
Physical and Chemical Terms Cited in This Book

Alkylation Transfer of an alkyl group (C_nH_{2n+1}) to a metal atom

Anthropogenic Any material originating from human activity

Apatite Crystalline calcium hydroxyphosphate

Arylation Transfer of an aryl group (C_6H_5) to a metal atom

Bonding Energy Constant (k) A constant for the strength of the bond between metal and sediment

Chelator A ligand with more than one binding site.

Colloids Particles of size range 0.1–0.45 μm.

Complex A simple ligand with one binding site

Concentration Factor (CF) Quotient relating concentrations of pollutant in two different phases. For example, biota/water

Conditional Stability Constant Stability constant valid for a given set of conditions, such as pH, and ionic strength. Cf., stability constant

Contamination Natural and/or man-induced adulteration of abiotic and/or biotic substrates.

Cumulative Formation Constant Overall stability constant. Example $\beta_2 = K_1 \times K_2$

Desorption Release of surface bound pollutants from solid matrices

Disproportionation Non-stoichiometric breakdown of a compound

Eh Redox potential

Electropositive Oxidation potential higher than hydrogen

Fractionation Separation of a compound(s) into different physical and chemical components

Fulvic Acid An acid–alkali soluble humic material, originating from the breakdown of lignin and tannins

Hard Acid Behaviour Metals prefering an oxygen-ligand environment, forming electro-valent bonds

Hexacoordinate Reaction Binding of a ligand through six coordinate bonds

Humic Acid An acid insoluble component of humic material, with molecular weight greater than fulvic acid

Ligand A molecule containing a donor atom capable of forming a bond with a metal

Ligand, Neutral A ligand without electrical charge

Ligand, Polydentate A ligand with multiple donor atoms

Metal, Dissolved Metal species which pass through a 0.45 μm membrane filter

Metal, Filterable Metal species retained by a 0.45μm membrane filter

Metal, Total Sum of dissolved and filterable metal

Metallothionein A low molecular weight ($\sim$ 1000) metal binding protein with 25–30% sulfhydryl group

Methylation Transfer (biotic and/or abiotic) of a methyl group (CH_3) to a metal atom

Non-dialysable Solute species retained by dialysis membranes

Oxyphilic Affinity to oxygen donor atoms

Partition Coefficient Ratio of the distribution of a metal species among different phases, such as sediment and water

Sesquioxide Oxides of the formula M_2O_3

Soft Acid Behaviour Metals preferring nitrogen, sulfur, or selenium donor atoms

Solubility Product A constant relating the degree of solubility of a compound in a given solvent

Sorption Reversible binding of a pollutant to a solid matrix

Sorption Maximum (b) The maximum concentration of a pollutant sorbed per unit mass

Stability Constants ($K_1 - K_n$) The formation of a complex or chelate

Sulfophilic Affinity to sulfur donor atoms

Uptake Non-reversible accumulation of a pollutant

Appendix C
Common and Scientific Names of Fish Cited in This Book

Arctic char *Salvelinus alpinus*
Arctic cod *Boreogadus saida*
Atlantic salmon *Salmo salar*

Black crappie *Pomoxis nigromaculatus*
Black bullhead *Ictalurus melas*
Black marlin *Makaira indica*
Bluegill *Lepomis machrochirus*
Bream *Abramis brama*
Brook trout *Salvelinus fontinalis*
Brown bullhead *Ictalurus nebulosus*
Brown trout *Salmo trutta*
Burbot *Lota lota*

Carp *Cyprinus carpio*
Channel catfish *Ictalurus punctatus*
Chinook salmon *Oncorhynchus tshawytscha*
Cod *Gadus morhua*
Coregonids *Coregonidae*
Crucian carp *Carassius carassius*

Dab *Pleuronectes limanda*
Deepwater cisco *Coregonus hoyi*
Dogfish *Scyliorhinus canicula*

Eel *Anguilla anguilla*
Emerald shiner *Notropis atherinoides*

Fathead minnow *Pimephales promelas*
Five-bearded rockling *Ciliata mustela*
Flatfish *Pleuronectes platessa*
Flounder *Pleuronectes flesus*

Goldfish *Carassius auratus*
Green sunfish *Lepomis cyanellus*
Guppy *Poecilia reticulata*

Haddock *Gadus aeglefinus*

Ide *Leuciscus idus*

Largemouth bass *Micropterus salmoides*
Lumpsucker *Cyclopterus lumpus*

Mosquito fish *Gambusia affinis*
Mummichog *Fundulus heteroclitus*
Muskellunge *Esox masquinongy*

Perch *Perca fluviatilis*
Pike *Esox lucius*
Pikeperch *Lucioperca sandra*
Plaice *Pleuronectes platessa*

Rainbow trout *Salmo gairdneri*
Roach *Rutilus rutilus*

Rock bass *Ambloplites rupestris*
Ruffe *Gymnocephalum cernua*

Sauger *Stizostedion canadense*
Shorthead redhorse *Moxostoma macrolepidotum*
Shorthorn sculpin *Myoxocephalus scorpius*
Slimy sculpin *Cottus cognatus*
Smelt *Osmerus eperlanus*
Sockeye salmon *Oncorhynchus nerka*
Spotted Wolffish *Anarhichas minor*
Sprat *Sprattus sprattus*
Steelhead *Salmo gairdneri*
Stoneloach *Noemacheilus barbatulus*

Three-spine stickleback *Gasterosteus aculeatus*

Vendance *Coregonus albula*

Walleye *Stizostedion vitreum*
White crappie *Pomoxis annularis*
Whitefish *Coregonus lavaretus*
White sucker *Catostomus commersoni*
Whiting *Merlangius merlangus*

Yellow perch *Perca flavescens*

Index

Springer Series on Environmental Management
Robert S. DeSanto, Series Editor

Natural Hazard Risk Assessment and Public Policy
Anticipating the Unexpected
by **William J. Petak** and **Arthur A. Atkisson**

This volume details the practical actions that public policy makers can take to lessen the adverse effects natural hazards have on people and property, guiding the reader step-by-step through all phases of natural disaster.

1982/489 pp./89 illus./cloth
ISBN 0-387-**90645**-2

Gradient Modeling
Resource and Fire Management
by **Stephen R. Kessell**

"[The] approach is both muscular enough to satisfy the applied scientist and yet elegant and deep enough to satisfy the aesthetics of the basic scientist . . . Kessell's approach . . . seems to overcome many of the frustrations inherent in land-systems classifications."

—*Ecology*

1979/432 pp./175 illus./27 tables/cloth
ISBN 0-387-**90379**-8

Disaster Planning
The Preservation of Life and Property
by **Harold D. Foster**

"This book draws on an impressively wide range of examples both of man-made and of natural disasters, organized around a framework designed to stimulate the awareness of planners to sources of potential catastrophe in their areas, and indicate what can be done in the preparation of detailed and reliable measures that will hopefully never need to be used."

—*Environment and Planning A*

1980/275 pp./48 illus./cloth
ISBN 0-387-**90498**-0

Air Pollution and Forests
Interactions between Air Contaminants and Forest Ecosystems
by **William H. Smith**

"A definitive book on the complex relationship between forest ecosystems and atomospheric deposition . . . long-needed . . . a thorough and objective review and analysis."

—*Journal of Forestry*

1981/379 pp./60 illus./cloth
ISBN 0-387-**90501**-4

Springer Series on Environmental Management
Robert S. DeSanto, Series Editor

Global Fisheries
Perspectives for the '80s
Edited by **B.J. Rothschild**

This timely, multidisciplinary overview offers guidance toward solving contemporary problems in fisheries. The past and present status of fisheries management as well as insights into the future are provided along with particular regard to the effects of the changing law of the sea.

1983/approx. 224 pp./11 illus./cloth
ISBN 0-387-**90772-6**

Heavy Metals in Natural Waters
Applied Monitoring and Impact Assessment
by **James W. Moore** and **S. Ramamoorthy**

This is a complete presentation on monitoring and impact assessment of chemical pollutants in natural waters, and provides a review of data, methods, and principles that are of potential use to environmental management and research experts.

1984/270 pp./48 illus./cloth
ISBN 0-387-**90885**-4

Landscape Ecology
Theory and Applications
by **Zev Naveh** and **Arthur S. Lieberman**
With a Foreword by Arnold M. Schultz, and an Epilogue by Frank E. Egler

This first English-language monograph on landscape ecology treats the subject as an interdisciplinary, global human ecosystem science, examining the relationships between human society and its living space.

1984/376 pp./78 illus./cloth
ISBN 0-387-**90849**-8